Diet and Nutrition: Metabolic Diseases

Diet and Nutrition: Metabolic Diseases

Guest Editor

Iñaki Elío

MDPI

Basel • Beijing • Wuhan • Barcelona • Belgrade • Novi Sad • Cluj • Manchester

Guest Editor
Iñaki Elío
Research Group on Foods,
Nutritional Biochemistry and
Health
Universidad Europea del
Atlántico
Santander
Spain

Editorial Office
MDPI AG
Grosspeteranlage 5
4052 Basel, Switzerland

This is a reprint of the Special Issue, published open access by the journal *Nutrients* (ISSN 2072-6643), freely accessible at: https://www.mdpi.com/journal/nutrients/special_issues/65S4G90O56.

For citation purposes, cite each article independently as indicated on the article page online and as indicated below:

Lastname, A.A.; Lastname, B.B. Article Title. *Journal Name* **Year**, *Volume Number*, Page Range.

ISBN 978-3-7258-5057-0 (Hbk)
ISBN 978-3-7258-5058-7 (PDF)
https://doi.org/10.3390/books978-3-7258-5058-7

Contents

About the Editor

Iñaki Elío

Iñaki Elío Pascual is Academic Director of the Degree in Human Nutrition and Dietetics at Universidad Europea del Atlántico and serves as the Editor-in-Chief of MLS Health and Nutrition Research. He holds a Doctorate in Health, Nutrition, and Food Projects; a Master's in Nutrition and Metabolism; and a Degree in Human Nutrition and Dietetics. Professor Elío has extensive experience as a registered dietitian–nutritionist, having worked in both clinical settings, such as the Hospital Universitari de Bellvitge, and in academia, where he has contributed to the development and supervision of clinical nutrition, digestive pathology, head and neck cancer patients, neurology and metabolic disease curricula.

His primary research interests focus on the complex relationships between diet, nutrition, and metabolic diseases, with special emphasis on prevention and management strategies for obesity, diabetes, and metabolic syndrome. Professor Elío has led multidisciplinary projects exploring the health benefits of the Mediterranean diet, the impact of ultra-processed foods, and the role of functional foods and nut consumption in elderly populations.

Currently, Professor Elío is involved in projects that utilize advanced methodologies such as multi-omics, computational models, eating disorders, and patient-derived organoids to better understand nutritional pathophysiology and develop more effective dietary interventions. He is also the leader of the E+DIETing_LAB project, which has been awarded for its innovative approach to nutrition education through digital simulation and international collaboration.

Professor Elío's work is driven by a passion for translating scientific discoveries into practical dietary recommendations and public health strategies. He has received recognition for his commitment to research, education, and interdisciplinary collaboration, and continues to inspire the next generation of nutrition professionals through his editorial and teaching roles.

Preface

It is our great pleasure to present this Reprint, a curated compilation of research and review articles that collectively advance our understanding of the complex interplay between diet, nutrition, and metabolic diseases. The subject of this collection is both timely and critical, as the incidence of metabolic disorders—including obesity, diabetes, metabolic syndrome, and related conditions—continues to rise globally, posing significant challenges to healthcare systems and public health initiatives.

The scope of this Reprint is intentionally broad, encompassing original studies and comprehensive reviews that address the multifaceted roles of dietary patterns, specific nutrients, and lifestyle interventions in the onset, progression, and management of metabolic diseases. This volume highlights not only the biological mechanisms underlying these conditions but also the practical applications of nutritional science in clinical and community settings. By featuring research that spans molecular insights, population-based studies, and real-world interventions, this Reprint seeks to bridge the gap between laboratory findings and practical solutions for improving metabolic health.

Our primary aim in assembling this collection is to provide a valuable resource for a diverse audience. Researchers will find up-to-date evidence and emerging themes that can inform future investigations. Clinicians are offered practical insights into how nutritional strategies can be integrated into patient care for metabolic disorders and public health professionals and policymakers will benefit from the evidence-based recommendations and preventive strategies discussed throughout the volume. Ultimately, we hope this Reprint will inspire new collaborations and innovations in the field of human nutrition and metabolic health.

The motivation for compiling this Reprint stems from the urgent need to address the growing burden of metabolic diseases worldwide. As Guest Editors, we are acutely aware of the challenges faced by both individuals and healthcare systems in managing these conditions. We believe that advancing scientific knowledge and promoting interdisciplinary dialog are essential steps toward developing effective, sustainable solutions.

We extend our deepest gratitude to all contributing authors for their dedication, expertise, and commitment to advancing the field. We also thank the reviewers, whose thoughtful evaluations and constructive feedback have ensured the high quality of this collection. In addition, we extend out appreciation to the Editorial team at Nutrients, whose unwavering support and professionalism have been instrumental in bringing this Reprint to fruition.

This Reprint is addressed to all those engaged in the study, prevention, and management of metabolic diseases—whether in research, clinical practice, or public health. We trust that the insights and evidence presented here will serve as a foundation for further progress and innovation in the field.

Iñaki Elío
Guest Editor

nutrients

MDPI

Editorial

Advancing Nutritional Science: Contemporary Perspectives on Diet's Role in Metabolic Health and Disease Prevention

Iñaki Elío [1,2,3]

1 Research Group on Foods, Nutritional Biochemistry and Health, Universidad Europea del Atlántico, 39011 Santander, Spain; inaki.elio@uneatlantico.es
2 Faculty of Health Sciences, Universidad Internacional Iberoamericana, Arecibo, PR 00613, USA
3 Department of Health, Nutrition and Sport, Universidad Internacional Iberoamericana, Campeche 24560, Mexico

1. Introduction

This Special Issue of Diet and Nutrition: Metabolic Diseases showcases cutting-edge research exploring the intersection between nutrition, dietary patterns, and public health. The contributions in this collection involve both fundamental and applied research, offering new insights into how nutrition can combat the growing global burden of non-communicable diseases [1]. The studies in this issue emphasize the critical role that diet plays in promoting metabolic health, preventing chronic diseases, and improving overall quality of life.

In recent years, nutrition has become a central focus in global health efforts, with a growing body of evidence demonstrating its impact on both individual and population-level outcomes [2,3]. This Special Issue encompasses several key themes, including the role of dietary interventions in managing metabolic disorders, the importance of nutrient timing and quality, and the broader implications of sustainable dietary practices.

2. An Overview of Published Articles

The study by Cubas-Basterrechea et al. [4] examined the association between vegetable intake and metabolic syndrome (MetS) in 264 older adults aged 65–79 years. Only 17% met the recommended daily intake of ≥2 servings (400 g) of vegetables. Adequate intake was linked to a 19% lower MetS prevalence (24.4% vs. 43.4%, $p < 0.05$), while inadequate consumption significantly increased MetS risk (OR: 2.21; $p = 0.035$). The authors emphasized the need to encourage vegetable consumption in older populations to reduce MetS prevalence.

Gwizdak et al. [5] investigated the dietary habits and nutritional knowledge of 297 women aged 18–45 with thyroid disorders, focusing on the impact of education level. Hypothyroidism and Hashimoto's disease were most common in younger women (18–25 years). Higher education correlated with better awareness of protein, carbohydrate roles, and nutrient–drug interactions ($p < 0.01$), as well as healthier cooking practices. Low water intake was linked to comorbidities like insulin resistance and cardiovascular disease ($p < 0.01$). The authors highlighted significant gaps in dietary knowledge and emphasized the need for tailored educational interventions to improve thyroid disorder management.

An study with mice Huang et al. [6] explored short-term zinc supplementation (30–90 ppm zinc sulfate in drinking water for one week) in C57/BL6J male mice. Despite no changes in body weight or food intake, zinc reduced visceral fat deposition and adipocyte size by **enhancing lipolysis** (↑ *Atgl*, *Hsl*, p-HSL, and PPARγ) and **suppressing lipogenesis** (↓ FASN protein). However, zinc increased serum insulin levels (*hyperinsulinemia*) and adipose inflammation markers (↑ *F4/80*, *Tnfa*). The authors propose short-term zinc supplementation as a potential strategy for fat reduction but caution its inflammatory side effects, suggesting future studies on intermittent dosing protocols.

The study by Sprenghini et al. [7] retrospectively examined adrenoleukodystrophy (X-ALD), a metabolic disorder caused by the dysfunctional peroxisomal beta-oxidation of very-long-chain fatty acids (VLCFAs) and the consumption of a Mediterranean diet low in VLCFAs in 36 patients and carriers at baseline and one year following the introduction of dietary restrictions The authors concluded that a diet low in VLCFAs is highly recommended as a supplement to treatment because it effectively lowers plasma VLCFA levels.

Another noteworthy contribution is the study by Liu et al. [8], which investigated the impact of dietary patterns on metabolic syndrome in young adults and how physical activity modulates this effect. A cross-sectional study was conducted at a health management center in Tianjin, China, from September 2022 to March 2023. Participants aged 18–35 years were recruited using convenience sampling. No significant associations were found for the Sugar–Processed and Alcohol–Meat patterns. Subgroup analysis revealed that the Legume–Nut pattern increased the risk of metabolic syndrome among those with irregular physical activity, whereas the Egg–Vegetable pattern decreased the risk. These findings highlight the significant influence of dietary patterns on the risk of metabolic syndrome in young adults and the modifying effect of regular physical activity, underscoring the need for targeted dietary and lifestyle interventions to prevent metabolic syndrome in this population.

Khoshkerdar et al. [9] examined how pregnancy causes significant changes in the mother's physiology and homeostatic regulation in rats. Pregnancy-related disruptions in maternal adaptations are linked to poor maternal diet; however, the impact of paternal diet on maternal health is less well established. This illustrates how patterns of maternal metabolism and gestation-related modifications to the mother's physiology can be impacted by the paternal diet at the time of mating.

The study by Rauch et al. [10] examined how administering a combination of ketone diester (KE) and medium-chain triglyceride (MCT) oil, known as KEMCT, to WAG/Rij rats eliminated the rise in blood glucose levels brought on by isoflurane anaesthesia and extended the recovery period. These findings imply that EKSs may be useful in reducing the adverse metabolic effects of isoflurane, including hyperglycaemia, in both sexes when administered as adjuvant therapy.

A laboratory study by Torrens et al. [11] showed that consuming carbohydrates prior to exercise significantly reduces the symptoms of McArdle disease (IWMD) and increases exercise tolerance in the early stages of exercise. Information regarding the type and quantity of carbohydrates consumed before exercise was gathered using an online survey method. Merely 17.5% of participants said that eating carbohydrates prior to exercise reduced or alleviated their symptoms of MD.

In the sixth article, Xu et al. [12] investigated how to improve the postprandial glycaemic response (PPGR) and postprandial insulin response (PPIR) after consuming biscuits made with 30% wheat flour and autoclaved BSG (ABSG) or fermented BSG (FBSG) in people with metabolic syndrome (MetS). The impact on the subjective appetite response, breath hydrogen (H2) and methane (CH4) concentrations, and postprandial lipid panel was also investigated. In comparison to the control group, the insulin level decreased 180 min after consuming FBSG ($p = 0.051$) and ABSG ($p = 0.010$). However, subjective appetite response, breath H2 and CH4 content, and postprandial lipid panel did not differ. In summary, eating biscuits with BSG can reduce the PPGR, and adding fermented BSG provides an additional benefit in controlling PPGR.

The study by Bugi et al. [13] used the "Food Neophobia Scale" (FNS) to measure neophobia in 34 individuals with phenylketonuria, who had previously received a diagnosis, and a control group that ranged in age from 7 months to 40 years. The statistics showed that 70.57% of the control group and 61.76% of PKU patients were neophobic. The current age of PKU patients, the time between birth and PKU diagnosis, and the educational attainment of their parents were all linked to food neophobia.

The study by Li et al. [14] investigated various stages of metabolic dysfunction-associated steatosic liver disease (MASLD) by prolonging the incubation period of human precision-cut liver slices (PCLSs). To develop MASLD, healthy human PCLSs were cultivated for up to 96 h in a medium supplemented with high levels of sugar, insulin, and fatty acids. Hepatic steatosis, which is defined by the accumulation of intracellular fat, was seen in PCLSs. A time-dependent effect on lipid metabolism seemed to be involved in the development of hepatic steatosis, with fatty acid absorption and storage initially increasing and lipid oxidation and secretion later decreasing. The authors concluded that a strong ex vivo model for MASLD can be created by incubating human PCLSs for an extended period, making it easier to find and assess possible treatment options.

In the ninth article, Alonso-Allende et al. [15] reviewed the main effects of inulin on human metabolic health, with a particular focus on the mechanisms of action of this prebiotic. Inulin supplementation contributes to body weight and BMI control, reduces blood glucose levels, improves insulin sensitivity, and reduces inflammation markers, mainly through the selective favouring of short-chain fatty acid (SCFA)-producing species from the genera Bifidobacterium and Anaerostipes. This review indicated that consuming inulin improves the gut microbiota and creates compounds through fermentation, which have beneficial metabolic effects.

The potential of royal jelly (RJ), a naturally occurring bee product that is high in bioactive components, was examined in the article by El Seedi et al. [16] as an alternate approach to treating metabolic illnesses. RJ demonstrates antimicrobial, estrogen-like, anti-inflammatory, hypotensive, anticancer, and antioxidant properties, among others. RJ's function in regulating immunological responses, boosting anti-inflammatory cytokines, and inhibiting important inflammatory mediators has been highlighted in recent studies. Despite these encouraging results, more research is required to fully comprehend the processes underlying RJ's therapeutic effects.

In the last article, Gao et al. [17] provided a comprehensive review of the application and mechanisms of probiotic-mediated gut microbiota homeostasis in skin care and offered a theoretical basis for the application of probiotics in skin care. Although the preventive properties and activities of topical probiotics helped to preserve skin homeostasis, their drawbacks and restrictions resulted in inflammatory skin disorders that are challenging to fully treat with topical probiotics. The effectiveness and side effects of internal probiotic formulations for the treatment of wounds, psoriasis, acne, atopic dermatitis, and numerous other skin issues are being investigated in several clinical trials.

3. Several Key Themes Emerge from the Research Presented

1. Dietary Patterns and Metabolic Health: In this issue, a cross-sectional study examines how food patterns affect metabolic syndrome in young individuals while simultaneously examining how physical activity can modulate this condition [8]. The intake of $\geq$2 servings (400 g) of vegetables lower the risk of MetS in a 19% in older adults aged 65–79 years [4]. A cross-sectional study stablished that Hashimoto's disease were most common in younger women (18–25 years), higher education correlated with better awareness of protein, carbohydrate roles, and nutrient–drug interactions [5]. These studies highlight the importance of tailored dietary strategies to improve metabolic and thyroid health across age groups. For young adults, future research should focus on how physical activity can optimize the benefits of healthy dietary patterns, such as the Egg-Vegetable diet, in reducing metabolic syndrome risk. Additionally, efforts to enhance vegetable intake in older adults and nutritional education in younger women with thyroid disorders are crucial for improving overall health outcomes.

2. Innovative Dietary Approaches: The studies in this issue explore novel nutritional interventions, such as the use of brewers' spent grain in biscuits to improve glycaemic response in metabolic syndrome patients [12]. These findings highlight the potential of repurposing food industry by-products to yield health benefits, which is a growing trend in sustainable nutrition.

3. Personalized Nutrition: The research on carbohydrate ingestion for individuals with McArdle disease underscores the importance of tailored nutritional strategies for rare metabolic disorders [11]. This research emphasizes the need for condition-specific dietary recommendations, aligning with broader trends in personalized nutrition.

4. Gut–Metabolic Health Connection: Multiple studies explore the intricate relationship between gut health and metabolic function. The review on probiotics and skin health via the gut–skin axis demonstrates the significant impacts of gut microbiota on overall health [17], supporting growing evidence of the gut microbiome's role in metabolic health.

5. Nutritional Management of Rare Disorders: Articles addressing conditions such as phenylketonuria [13] and adrenoleukodystrophy [7] provide valuable insights into the dietary management of rare metabolic diseases, highlighting the crucial role of nutrition in these complex conditions.

6. Metabolic Adaptations and Diet: Research on maternal adaptations to pregnancy and the impact of paternal diet offers intriguing perspectives on how nutrition influences metabolic health across generations [9].

7. Novel Research Models: The development of human precision-cut liver slices as a model for metabolic dysfunction-associated steatotic liver disease showcases an innovative approach to studying metabolic conditions and screening potential interventions [14].

8. Nutraceutical Potential: Reviews on compounds like inulin [15] and royal jelly [16] explore the therapeutic potential of natural products in addressing metabolic disorders, opening avenues for future research and potential interventions. The short-term zinc supplementation might reduce fat deposition in mice [6].

9. Metabolic Effects of Anesthesia and Ketone Supplements: This research highlights the potential of ketone supplements in mitigating the metabolic side effects of anesthesia, contributing to our understanding of metabolic adaptations in medical interventions [10].

4. Conclusions

This Special Issue not only enhances our understanding of the complex connections between nutrition, food, and metabolic disorders but also explores areas for future study and therapeutic application. The diverse topics covered in this issue illustrate the complexity of metabolic health and the importance of nutrition in preserving and restoring it.

We extend our sincere gratitude to the editorial team for their dedicated efforts in realizing this Special Issue, the peer reviewers for their insightful feedback, and all the authors for their invaluable research contributions.

Conflicts of Interest: The authors declare no conflicts of interest.

References

1. Saklayen, M.G. The Global Epidemic of the Metabolic Syndrome. *Curr. Hypertens. Rep.* **2018**, *20*, 12. [CrossRef] [PubMed]
2. Xiao, Y.-L.; Gong, Y.; Qi, Y.-J.; Shao, Z.-M.; Jiang, Y.-Z. Effects of Dietary Intervention on Human Diseases: Molecular Mechanisms and Therapeutic Potential. *Signal Transduct. Target. Ther.* **2024**, *9*, 59. [CrossRef] [PubMed]
3. Guo, Y.; Huang, Z.; Sang, D.; Gao, Q.; Li, Q. The Role of Nutrition in the Prevention and Intervention of Type 2 Diabetes. *Front. Bioeng. Biotechnol.* **2020**, *8*, 575442. [CrossRef] [PubMed]

4. Cubas-Basterrechea, G.; Elío, I.; González Antón, C.; Muñoz Cacho, P. Daily Intake of Two or More Servings of Vegetables Is Associated with a Lower Prevalence of Metabolic Syndrome in Older People. *Nutrients* **2024**, *16*, 4101. [CrossRef] [PubMed]
5. Gwizdak, P.; Szlacheta, P.; Łaskawiec-Żuławińska, D.; Grajek, M.; Krupa-Kotara, K.; Markowski, J.; Korzonek-Szlacheta, I. Dietary Habits, Nutritional Knowledge, and Their Impact on Thyroid Health in Women: A Cross-Sectional Study. *Nutrients* **2024**, *16*, 3862. [CrossRef] [PubMed]
6. Huang, X.; Jiang, D.; Zhu, Y.; Fang, Z.; Feng, B. Short-Term Zinc Supplementation Stimulates Visceral Adipose Catabolism and Inflammation in Mice. *Nutrients* **2024**, *16*, 3719. [CrossRef]
7. Spreghini, M.R.; Gianni, N.; Todisco, T.; Rizzo, C.; Cappa, M.; Manco, M. Nutritional Counseling and Mediterranean Diet in Adrenoleukodystrophy: A Real-Life Experience. *Nutrients* **2024**, *16*, 3341. [CrossRef] [PubMed]
8. Liu, J.; Lu, W.; Lv, Q.; Wang, Y.; Xu, X.; He, Y.; Chang, H.; Zhao, Y.; Zhang, X.; Zang, X.; et al. Impact of Dietary Patterns on Metabolic Syndrome in Young Adults: A Cross-Sectional Study. *Nutrients* **2024**, *16*, 2890. [CrossRef]
9. Khoshkerdar, A.; Eid, N.; Batra, V.; Baker, N.; Holmes, N.; Henson, S.; Sang, F.; Wright, V.; McLaren, J.; Shakesheff, K.; et al. Sub-Optimal Paternal Diet at the Time of Mating Disrupts Maternal Adaptations to Pregnancy in the Late Gestation Mouse. *Nutrients* **2024**, *16*, 1879. [CrossRef] [PubMed]
10. Rauch, E.; Ari, C.; D'Agostino, D.P.; Kovács, Z. Exogenous Ketone Supplement Administration Abrogated Isoflurane-Anesthesia-Induced Increase in Blood Glucose Level in Female WAG/Rij Rats. *Nutrients* **2024**, *16*, 1477. [CrossRef] [PubMed]
11. Torrens, S.L.; Parr, E.B.; McNulty, C.; Ross, L.; MacLaughlin, H.; Robergs, R.A. Carbohydrate Ingestion before Exercise for Individuals with McArdle Disease: Survey Evidence of Implementation and Perception in Real-World Settings. *Nutrients* **2024**, *16*, 1423. [CrossRef] [PubMed]
12. Xu, Y.; Leong, Z.N.; Zhang, W.; Jin, X.; Kong, J.W.; Chan, G.C.T.; Kim, J.E. Impact of Brewers' Spent Grain-Containing Biscuit on Postprandial Glycaemic Response in Individuals with Metabolic Syndrome: A Crossover Randomised Controlled Trial. *Nutrients* **2024**, *16*, 909. [CrossRef] [PubMed]
13. Bugi, M.-A.; Jugănaru, I.; Isac, R.; Simina, I.-E.; Munteanu, A.-I.; Mang, N.; Brad, G.-F.; Nicoară, D.-M.; Cîrnatu, D.; Mărginean, O. Factors Impacting the Reduction in Neophobia Prevalence in Phenylketonuria Patients. *Nutrients* **2024**, *16*, 768. [CrossRef] [PubMed]
14. Li, M.; Larsen, F.T.; van den Heuvel, M.C.; Gier, K.; Gorter, A.R.; Oosterhuis, D.; Bijzet, J.; de Meijer, V.E.; Ravnskjaer, K.; Nagelkerke, A.; et al. Metabolic Dysfunction-Associated Steatotic Liver Disease in a Dish: Human Precision-Cut Liver Slices as a Platform for Drug Screening and Interventions. *Nutrients* **2024**, *16*, 626. [CrossRef] [PubMed]
15. Alonso-Allende, J.; Milagro, F.I.; Aranaz, P. Health Effects and Mechanisms of Inulin Action in Human Metabolism. *Nutrients* **2024**, *16*, 2935. [CrossRef] [PubMed]
16. El-Seedi, H.R.; Salama, S.; El-Wahed, A.A.A.; Guo, Z.; Di Minno, A.; Daglia, M.; Li, C.; Guan, X.; Buccato, D.G.; Khalifa, S.A.M.; et al. Exploring the Therapeutic Potential of Royal Jelly in Metabolic Disorders and Gastrointestinal Diseases. *Nutrients* **2024**, *16*, 393. [CrossRef] [PubMed]
17. Gao, T.; Wang, X.; Li, Y.; Ren, F. The Role of Probiotics in Skin Health and Related Gut–Skin Axis: A Review. *Nutrients* **2023**, *15*, 3123. [CrossRef] [PubMed]

nutrients

MDPI

Article

Daily Intake of Two or More Servings of Vegetables Is Associated with a Lower Prevalence of Metabolic Syndrome in Older People

Gloria Cubas-Basterrechea [1,*], Iñaki Elío [2,3,*], Carolina González Antón [4] and Pedro Muñoz Cacho [5]

1 Dietetic Section, Hospital Universitario "Marqués de Valdecilla", 39008 Santander, Spain
2 Research Group on Foods, Nutritional Biochemistry and Health, Universidad Europea del Atlántico, 39011 Santander, Spain
3 Department of Health, Nutrition and Sport, Iberoamerican International University, Campeche 24560, Mexico
4 NEXO Multidisciplinar Center, Department of Nutrition and Dietetic, 04004 Almeria, Spain; carolinaglan@gmail.com
5 Teaching Department of Primary Care Management, Cantabrian Health Service, Instituto de Investigación Marqués de Valdecilla (IDIVAL), 39011 Santander, Spain; pedro.munoz@scsalud.es
* Correspondence: gcbasterrechea@cofcantabria.org (G.C.-B.); inaki.elio@uneatlantico.es (I.E.); Tel.: +34-617-597-045 (G.C.-B.); +34-942-244-244 (I.E.)

Abstract: Objectives: We sought to examine the correlation between the recommended consumption of at least two servings (400 g) of vegetables per day and the prevalence of metabolic syndrome (MetS) in an elderly population. Methods: This observational, cross-sectional, and descriptive study was conducted with 264 non-institutionalized people aged 65 to 79 years old. We adhered to the recommended guidelines for vegetable intake from the MEDAS-14 questionnaire, which has been validated for elderly populations at high cardiovascular risk. Diagnoses of MetS were made based on the criteria set forth by the International Diabetes Federation (IDF). Results: Among 264 individuals, who had a mean age of 71.9 (SD: 4.2) and comprised 39% men, the prevalence of MetS was 40.2%. A total of 17% of the participants adhered to the recommended vegetable consumption. Consuming the recommended amount of vegetables was correlated with a 19% reduction in the prevalence of MetS, to 24.4% from 43.4% among those with low vegetable consumption ($p < 0.05$). A main finding was that inadequate vegetable consumption was significantly associated with a higher prevalence of MetS (OR: 2.21; 95% CI: 1.06–4.63; $p = 0.035$), considering potential influences by nutritional (consumption of fruit and nuts) and socio-demographic (sex, age, and level of education) covariates. Conclusions: A beneficial inverse correlation was identified between the recommended vegetable intake and the prevalence of MetS. In contrast, inadequate vegetable consumption was revealed as an independent variable associated with the prevalence of MetS. Considering the very low adherence to the recommended vegetable intake we observed, encouraging increased vegetable consumption among older individuals, who have a high prevalence of MetS, is advisable.

Keywords: vegetables; metabolic syndrome; aged; Spain

1. Introduction

The global population is aging, with the proportion of individuals over 65 years of age having almost doubled, from 4.97% in 1960 to 9.54% in 2022, in the space of six decades [1]. The greatest percentage of people over 65 years of age (28.7%) is found in Japan, whereas the most aged populations in Europe are found in Italy (23.61%), Portugal (23.15%), and Finland (22.96%). Spain ranks 23rd in the world, with 20.3% of its population being over 65 years of age [1]. In Santander, Cantabria, in northern Spain, the percentage of people over 65 years of age in 2024 is also very high at 25.42% [2].

This high percentage of older people is associated with a high prevalence of metabolic risk factors involved in the development of metabolic syndrome (MetS) [3], which results in a high prevalence of MetS within this group [3–5]. MetS, as defined by the International Diabetes Federation (IDF) [6], is a set of five metabolic and inflammatory disorders: abdominal obesity, elevated triglycerides and blood glucose, increased blood pressure, and decreased high-density lipoprotein cholesterol (HDL-c). All of these conditions increase the risk of developing type 2 diabetes mellitus (DM2) and cardiovascular disease (CVD) in adults. A prerequisite for the diagnosis of MetS is abdominal obesity and at least two of the other four risk factors mentioned [6]. The pathogenesis of MetS is associated with genetic and acquired factors that trigger oxidative stress, cellular dysfunction, and systematic inflammation [7].

Modifiable factors such as dietary habits are key to the prevention and progression of MetS. Vegetables are widely designated as "protective foods" within healthy dietary patterns, such as the Mediterranean Diet (MedDiet), which has been confirmed to be one of the main strategies for the prevention and treatment of MetS [8,9]. The importance of vegetables in this way can be attributed to their richness in vitamins (vitamin C), minerals (potassium), essential fatty acids, amino acids, and bioactive components, such as phytochemicals (polyphenols) and fiber [10]. The low caloric value of vegetables, together with their high fiber and polyphenol contents, aids in maintaining a healthy body weight and the reduction in abdominal obesity [10,11]. Phytochemicals in particular confer health benefits due to their anti-inflammatory and antioxidant effects [12]. In consuming the recommended amount of vegetables in terms of the number of servings consumed and the frequency of their consumption, these dietary components have been associated with the prevention and treatment of chronic inflammation-related and MetS-associated pathologies [13–15]. In this sense, polyphenols enhance insulin sensitivity, regulating blood sugar levels [14]; polyphenols also lower blood pressure thanks to their potassium content [16] and prevent dyslipidemia due to their high fiber composition [17].

Instead, low vegetable consumption has been associated with the prevalence of three or more risk factors for MetS (abdominal obesity, arterial hypertension (AHT), hyperglycemia, and low HDL-c levels) [18]; consequently, low vegetable consumption is associated with an increased risk of MetS and CVD.

Therefore, an inadequate vegetable intake could constitute a modifiable factor for the presence of MetS in the large group of the older population [19,20], for whom vegetable consumption tends to be low [21–24] and MetS prevalence tends to be high. Measures to increase vegetable consumption among the elderly are justified on this basis to enable healthy aging [25].

The aim of this study was to examine the relationship between adherence to the recommended intake of at least two servings (400 g) of vegetables per day and the prevalence of MetS in an elderly population.

2. Materials and Methods

2.1. Study Design

This study employed observational, cross-sectional, and descriptive methods.

2.2. Participants

The study population comprised non-institutionalized elderly people aged 65 to 79 years old and living in Santander (Cantabria). In January 2023, the population within this age bracket totaled 31,334 individuals [26]. The Granmo v.7.10 software for finite populations was utilized to compute the sample size [27]. At an alpha risk of 5% and a beta risk of 20% with bilateral contrast, 73 individuals who consumed the recommended amount of vegetables ($\geq$2 servings/day) and 146 individuals who consumed <2 servings of vegetables/day were assessed in order to identify a minimum difference of 20% between patients with and without MetS. As indicated by previous studies, one of these groups was expected to reach a proportion of 40% [20]. The ARCOSENE approach was employed.

The study population consisted of patient groups from four physicians affiliated with three Primary Health Care Centers (PCCs) in Santander (Cantabria) under the Cantabrian

Health Service (CHS). To acquire the sample, a three-layer sampling method was employed: First, the three PCCs in Santander with the highest populations of patients aged 65 and older were identified. Subsequently, a purposive medical quota was sampled from each of the three designed centers; specifically, the opportunity to participate was offered to the coordinator of each of these PCCs, and from the center with the greatest number of patients within this age group, a second medical quota recommended by the coordinator of the PCC was chosen. Finally, a random sample of individuals, stratified by sex and age (65–79 years old), was selected using systematic sampling.

Following random and systematic sampling, this study commenced with 556 individuals; however, 292 were unable to participate for various reasons. Of these, 106 were excluded by physicians due to the applicability of the exclusion criteria, primarily mental and cognitive decline (>4 errors in the Pfeiffer test [28]) and acute issues with maintaining an upright posture. Letters were sent to the home addresses of the individuals selected via the CHS, informing them about this study, followed by telephone contact. However, 68 individuals were unreachable, and 65 declined to participate in this study. Ultimately, 53 people were excluded due to their CHS health cards lacking biochemical blood and/or prescription data from the past twelve months, as these data were pertinent to the diagnostic variables required to ascertain the prevalence of MetS according to the IDF criteria [6].

Consequently, after the application of the selection criteria, a final sample of 264 participants remained (men: 39%; women: 61%).

2.3. Socio-Demographic Characteristics

The patients' socio-demographic characteristics were examined, including sex, age group, marital status, and level of education. Three age groups were established: 65–69 years old, 70–74 years old, and 75–79 years old. Marital status comprised four categories: married/partnered, separated, widowed, and single. Their educational levels were classified as university, secondary school, primary school, or incomplete primary school.

2.4. Diagnosis of Metabolic Syndrome

The IDF criteria were employed for the diagnosis of MetS, which stipulate that abdominal obesity is a prerequisite. Waist circumference was measured in the midpoint between the lower costal border and the top edge of the iliac crest in the standing position [29]. Given the challenges in taking this measurement in elderly individuals, particularly those who are overweight or obese, an anatomical reference point was established 2.5 cm above the umbilicus. This reference point facilitated measurement and was previously identified as the most accurate indicator of abdominal adipose tissue in older people [30]. Abdominal obesity in European individuals is defined as a waist circumference of ≥94 cm in men and ≥80 cm in women. Furthermore, a diagnosis of MetS requires two or more of the following parameters to be present: arterial hypertension (AHT) (130/85 mmHg, either treated with medication or diagnosed); fasting hyperglycemia (≥100 mg/dL or a prior diagnosis of DM2 or treatment); hypertriglyceridemia (≥150 mg/dL or under treatment); and low HDL-c (<40 mg/dL in men and <50 mg/dL in women or under treatment).

2.5. Instruments

2.5.1. Adherence to Recommended Vegetable Consumption

The MEDAS-14 questionnaire, which has been validated for individuals aged 55–80 years old at elevated cardiovascular risk, is a fast tool to assess adherence to the MedDiet [31]. The brief screener containing 14 questions has a positive correlation with the frequency of consumption of items deemed healthy, such as vegetables, while demonstrating an inverse relationship with the intake of unhealthy foods, identifying items for improvement individually. Adherence to the MedDiet is deemed "good" if a test score is ≥9 points and "low" if it is ≤8 points [32]. We used the vegetable-related item from the MEDAS-14 questionnaire, which recommends the consumption of two or more servings of vegetables daily, with at least one of these servings including salad or raw vegetables. One serving constituted 200 g here, whereas side dishes and half servings were considered half portions. This recommendation for vegetable consumption aligns with that suggested by the Spanish Society of Community Nutrition (SENC) for older people, which is 2–3 servings per day, with each serving comprising 150–250 g [33]. To ascertain the participants' consumption of vegetables and enhance their understanding of portion sizes, the dietitians–nutritionists (interviewers) utilized images showing portions of 200 g (one portion) and 100 g (a half portion) of various vegetables, except for potatoes. The study population was categorized into two groups: individuals who consumed vegetables in the amount recommended ($\geq$2 servings per day) and those who had an inadequate consumption of vegetables below the recommended level (<2 servings per day).

2.5.2. Assessment of Diagnostic Parameters for Metabolic Syndrome According to the IDF Criteria

A SECA® model 203 (SECA, Hamburg, Germany) ergonomic tape with millimetric precision was utilized to measure the participants' waist circumference, based on which diagnoses of abdominal obesity were made based on a comparison of the results with the reference values established for the European population by gender [6]. Hypertension was determined by measuring patients' blood pressure using the OMRON M3® Comfort automatic arm blood pressure monitor (Omron, Shimogyo-ku, Kyoto, Japan) in accordance with a strict protocol and by comparing the results with established diagnostic values [6]; additionally, the inclusion of one or more type of antihypertensive medication on a patient's CHS health card facilitated diagnosis. Diagnoses of hyperglycemia were made utilizing fasting blood glucose data (and comparing them to the diagnostic values) and/or mentions of medication being prescribed for DM2 in the CHS health cards. The results on triglyceride levels in the blood and how they compared with the reference values [6] and/or triglyceride medications being listed on a patient's CHS health card were used to determine whether an individual had hypertriglyceridemia. Low HDL-c levels were detected using HDL-c blood data from patients' CHS health cards and comparing them to the diagnostic value established by the IDF criterion [6].

2.6. Statistical Analysis

Quality variables were characterized by frequencies and percentages, whilst quantitative variables were represented by the arithmetic mean and standard deviation in the descriptive analysis. To establish differences between qualitative variables, the chi-square test was used. The odds ratio was employed to measure the association between the independent variables and the dependent variable MetS, while statistical significance was assessed using the Wald test.

We employed logistic regression analysis to determine the variables associated with MetS. The recommended intake frequencies in the MEDAS-14 questionnaire were used as reference categories, as indicated in Table S1. Additionally, socio-demographic variables such as sex (reference women), age (years), and education level (reference university) were included. Variables with $p < 0.25$ in univariate analysis were incorporated into the multivariate analysis, adhering to the method established by Hosmer and Lemeshow [34] and applied by other authors [35]. The main independent variable was vegetable consumption, and the control variables were the consumption of fruits and nuts, which had shown an association with MetS in a previous study conducted in this same population, using the MEDAS-14 questionnaire [36]. Also, essential socio-demographic variables such as sex, age, and education level were included. To select the final model, the automatic variable selection procedure was used with the backward method (model M1). To compare the predictive capacity of the different models, the area under the ROC curve was used. It is difficult to label a certain range of ROC curve area magnitudes as "poor" and "good" because it depends on the disorder and clinical application [37]. In this case, it may be appropriate to consider an AUC greater than 0.70–0.75 as desirable.

For the analyses of the above data, we used SPSS 25 (IBM Co. Released 2017. IBM SPSS Statistics for Windows, Version 25.0, Armonk, NY, USA: IBM Corp) and MedCalc® Software version 22.023 (MedCalc Software Ltd., Ostend, Belgium; https://www.medcalc.org; accessed on 29 October 2024).

3. Results

3.1. Socio-Demographic Characteristics

Table 1 shows the socio-demographic characteristics of the population investigated in terms of gender, age group, marital status, and level of education.

Table 1. Socio-demographic characteristics.

	Total (n = 264) N	%
Gender		
Men	103	39.0
Women	161	61.0
Age group		
65–69	88	33.3
70–74	98	37.1
75–79	78	29.6
Marital status		
Married/partnered	167	63.2
Separated	16	6.1
Widowed	48	18.2
Single	33	12.5
Level of education		
University	107	40.5
Secondary school	74	28.0
Primary school	76	28,8
Incomplete primary school	7	2.7

3.2. Adherence to Recommended Vegetable Consumption

In the older group studied (n = 264), the level of adherence to the recommended intake of two or more servings of vegetables per day was 17%. According to gender, the percentage of women who consumed the recommended amount of vegetables (21.1%) was significantly higher than the percentage of men who did (10.7%) ($p < 0.05$) (Table 2). There were no other significant differences according to the socio-demographic variables (age group, marital status, and level of education) or MetS diagnostic variables (abdominal obesity, MetS diagnostic parameters, and number of MetS diagnostic variables).

Table 2. Vegetable consumption according to gender.

	<2 Servings/Day (n = 219) n (%)	≥2 Servings/Day (n = 45) n (%)	p	Total (n = 264) n (%)
Men	92 (89.3)	11 (10.7)	0.028	103 (39.0)
Women	127 (78.9)	34 (21.1)		161 (61.0)

p: chi-square test.

3.3. Prevalence of Abdominal Obesity

The mean waist circumference by gender was compared with the diagnostic values of abdominal obesity by gender. The average waist circumference in men (n = 103) was 102.6 (±SD: 10.9), whereas in women (n = 161) it was 90.7 (±SD: 13.6) ($p < 0.001$). In both cases, the mean waist circumferences exceeded the diagnostic thresholds for abdominal obesity (≥94 cm in men and ≥ 80 cm in women). The prevalence of abdominal obesity observed was 78.8%, without a significant difference by gender at 77.7% in men and 79.5% in women (Table 3). There were no significant differences by age group.

Table 3. Diagnostic parameters related to metabolic syndrome.

	Total (n = 264)	%
Abdominal obesity	208	78.8
Diagnostic parameters for MetS		
Arterial hypertension	210	79.5
Hyperglycemia	84	31.8
Hypertriglyceridemia	59	22.3
Low HDL-c levels	52	19.7
Number of variables for a diagnosis of MetS		
None	38	14.4
1	102	38.6
2	73	27.6
3	40	15.2
4	11	4.2
Metabolic syndrome	106	40.2

3.4. *Prevalence of Metabolic Syndrome*

The prevalence of MetS was 40.2% (Table 3), with a significantly higher rate in men (n = 103) at 47.6% compared to women (n = 161) at 35.4%, ($p < 0.05$). There were no statistically significant differences in the prevalence of MetS by age group.

In terms of the prevalence of diagnostic variables related to MetS, AHT was the most prevalent at 79.5%. The second highest prevalence was found for hyperglycemia at 31.8%. The prevalence of the remaining diagnostic variables for MetS was considerably lower, with a prevalence of 23.4% recorded for hypertriglyceridemia and a prevalence of low HDL-c of 19.7% (Table 3).

Concerning the number of diagnostic variables for MetS excluding abdominal obesity, it was observed that for 14.4% of the participants, none of the other diagnostic variables applied, and one other variable applied to 38.6% of them. Subsequently, two diagnostic variables for MetS applied to 27.6% of the participants; three variables applied to 15.2% of them; and the percentage of individuals to whom four diagnostic parameters applied was the lowest at 4.2% (Table 3).

3.5. *Association Between Vegetable Consumption and the Prevalence of Metabolic Syndrome*

Consuming two or more servings of vegetables per day, in line with the recommendations, was associated with a prevalence of MetS of 24.4%, whereas consuming fewer than two servings per day was associated with a 19% higher prevalence of MetS (43.4%), with a significant difference ($p < 0.05$) (Figure 1).

As shown in Table 4, in the univariate logistic regression analysis, of the 14 nutritional variables from the MEDAS-14 questionnaire, only three were found to be significant ($p < 0.25$): consumption of vegetables in the amount of <2 servings/day (OR: 2.37; 95%CI: 1.14–4.92; p = 0.021); consumption of fruit in the amount of <3 pieces/day (OR: 1.35; 95%CI: 0.82–2.23; p = 0.237); and consumption of nuts in the amount of <3 portions/week (OR: 2.04; 95% CI: 1.21–3.42; p = 0.007).

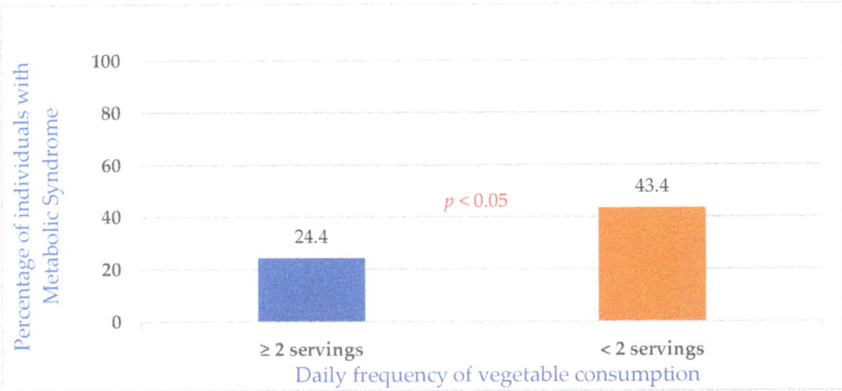

Figure 1. Association between the frequency of vegetable consumption and the prevalence of metabolic syndrome.

Table 4. Univariate logistic regression analysis of the potential independent variables with metabolic syndrome.

Variables	OR	95% CI	*p*
Use olive oil for cooking	2.27	0.37–13.83	0.373
Oil (≥4 tablespoons/day)	1.01	0.61–1.67	0.967
Vegetables (≥2 servings/day)	2.37	1.14–4.92	0.021
Fruit (≥3 pieces/day)	1.35	0.82–2.23	0.237
Red meat, hamburgers, sausage (<1 serving/day)	1.05	0.39–2.84	0.929
Butter, margarine, cream (<1 serving/day)	1.04	0.52–2.08	0.904
Carbonated/sweetened beverage (<1 serving/day)	1.38	0.67–2.85	0.380
Wine (≥7 glasses/week)	1.02	0.60–1.73	0.941
Legumes (≥3 servings/week)	0.80	0.48–1.35	0.411
Fish/seafood (≥3 portions/week)	1.24	0.76–2.03	0.393
Commercial pastries (<2 servings/week)	1.01	0.62–1.66	0.962
Nuts (≥3 portions/week)	2.04	1.21–3.42	0.007
Prefers chicken, turkey, or rabbit instead beef, pork, hamburgers or sausages	1.20	0.73–1.98	0.471
Sofrito with cooked vegetables, pasta, or rice (≥2 times/week)	0.76	0.42–1.38	0.368
Sex (reference women)	1.66	1.00–2.74	0.049
Age (years)	1.01	0.96–1.08	0.623
Level of education			
University (reference)	0.86	0.46–1.58	0.615
Secondary school	1.14	0.63–2.07	0.662
Primary school	1.12	0.24–5.24	0.889
Incomplete primary school			

p: Wald test; OR: odds ratio; and CI: confidence interval.

From the socio-demographic variables analyzed (sex, age, and level of education), only the sex (reference women) was significant: the men had a 66% higher prevalence of MetS than women (OR: 1.66; 95% CI: 1.00–2.74; $p = 0.049$).

For a multivariate regression logistic analysis, in the M0 model, the variables that had been significant in the univariate logistic regression analysis (vegetables, fruit, and nuts) and all the essential socio-demographic variables were included.

As shown in Table 5, when controlling for the previous variables, inadequate vegetable consumption was related to a higher presence of MetS on the edge of meaning (OR: 1.96; 95% CI: 0.98–4.17; $p = 0.082$). The consumption of nuts below the recommendation was associated with a higher prevalence of MetS (OR: 1.93; 95% CI: 1.13–3.28: $p = 0.016$).

Table 5. Multivariate regression logistic analysis of the M0 and M1 models.

Variables	M0 *			M1 **		
	OR	95% CI	*p*	OR	95% CI	*p*
Vegetables (≥2 servings/day)	1.96	0.98–4.17	0.082	2.21	1.06–4.63	0.035
Fruit (≥3 pieces/day)	1.20	0.71–2.02	0.490			
Nuts (≥3 servings/week)	1.93	1.13–3.28	0.016	1.95	1.15–3.29	0.013
Sex (reference women)	1.56	0.91–2.67	0.108			
Age (years)	0.98	0.92–1.04	0.442			
Level of education						
University (reference)						
Secondary school	0.86	0.45–1.64	0.654			
Primary school	1.17	0.62–2.22	0.629			
Incomplete primary school	1.43	0.28–7.14	0.667			

p: Wald test; OR: odds ratio; and CI: confidence interval. The area under the ROC curve (AUC): * M0: 0.641; ** M1: 0.61.

After applying the procedure of automatic variable selection using the backward method (model M1), only two variables were retained (low consumption of vegetables and nuts). Both variables were identified as independent contributors to MetS prevalence in this population: inadequate consumption of vegetables (OR: 2.21; 95%CI: 1.06–4.63; $p = 0.035$) and inadequate consumption of nuts (OR: 1.95; 95% CI: 1.15–3.29; $p = 0.013$).

The area under the ROC curve (AUC) values in the M0: 0.641 and M1: 0.61 were quite similar.

4. Discussion

4.1. Adherence to the Recommended Vegetable Consumption

The level of vegetable consumption in our study population was low, with merely 17% of the participants complying with the guideline on consuming at least two servings of vegetables per day. However, this low level of vegetable consumption aligns with previous findings on the dietary patterns of the population of Cantabria, as this region of Spain ranks third lowest in its vegetable consumption (at 48.10 kg/person/year) [38]. This diminished vegetable intake also aligns closely with the results of the PREDIMED-Plus study [39], which showed reduced vegetable consumption among the population of northern Spain.

The low consumption of vegetables in older people can be attributed to physiological factors such as loss of appetite, oral health problems, and tooth loss, as well as complications related to polypharmacy, mobility problems affecting shopping habits, and living alone [24].

The prevalence of adequate vegetable consumption was markedly low for both genders, although it was significantly higher in the women (21.1%) compared to the men (10.7%) at $p < 0.05$ (Table 2), as is shown in previous studies [21,23].

4.2. Prevalence of Metabolic Syndrome

The prevalence of MetS among the participants according to the IDF criteria was 40.2 [6], with similar data found for the population of Spanish people over 65 years of age enrolled in the ENRICA study (42.3%) [5]. Meanwhile, these data differ from those obtained in other parts of the world using different diagnostic criteria, such as Mexico (72.9%) [40], Brazil (66.1%) [41], Iran (51.7%) [42], and the United States (48.6%) [43].

MetS is not a disease but it is a chronic syndrome with a clustering of individual metabolic risk factors including abdominal obesity, hyperglycemia, hypertriglyceridemia, hypertension, and low high-density lipoprotein cholesterol levels that could increase the prevalence of DM2 and CVD [44]. For the diagnosis of MetS by the IDF criteria, abdominal obesity is indispensable as well as at least two of the four risk factors [6]. Therefore, the high prevalence of MetS obtained in this study is related to the fact that 78% of the people had abdominal obesity and 47% had two (27.6%), three (15.2%), or four (4.2%) diagnostic variables for MetS. The most prevalent diagnostic variables for MetS were AHT (79.5%) and hyperglycemia (31.8%) (Table 3). By sex, the prevalence of MetS in men was significantly higher (47.6%) than in women (35.4%) ($p < 0.05$). MetS is characterized by an increase in oxidative stress, which is associated with impaired inflammation, vascular disfunction, atherosclerosis, and a deregulation of the innate immune system [45]. It has been shown previously in several studies that AHT and hyperglycemia values allowed a better prediction of increased CVD risk than MetS itself in older people, due to the important role of insulin resistance in promoting chronic inflammation and atherosclerosis [46–49]. Also, a retrospective study in a large sample of the Korean population developed by Sung et al. showed a linear association between the number of diagnostic variables for MetS and risk of cardiovascular mortality, ranging from 1.99 for one variable to 2.98 for four–five variables [49].

4.3. Association Between Vegetable Consumption and the Prevalence of Metabolic Syndrome

An inverse association was found between vegetable consumption and the prevalence of MetS. This study revealed that the prevalence of MetS among individuals who consumed fewer than two servings of vegetables per day was 43.4% and was 19% higher than the prevalence of MetS among those who adhered to the recommended level (24.4%) ($p < 0.05\%$) (Figure 1). The most relevant finding was that inadequate consumption of vegetables acquires its own significance in being associated with a higher prevalence of MetS, which was equivalent to a 2.21 times higher risk of prevalence of MetS (OR: 2.21; 95%CI: 1.06–4.63; $p = 0.035$). This held even after controlling for the potential confounding influence of socio-demographic variables (sex, age, and level of education) and the dietary consumption of other relevant foods, such as fruit and nuts, analyzed by the MEDAS-14 questionnaire in a previous study in the same population [36], with low nut consumption identified as another independent variable for the presence of MetS (Table 5). Based on the area under the ROC curve (AUC) values, although it is not the objective of this study, the predictive capacity was not satisfactory in either of the two models (M0: 0.641 and M1: 0.61). Both models had low predictive capacity (<0.70) despite the fact that M1 had a smaller number of variables, which suggests that there are more relevant variables that have not been included, such as physical exercise, socioeconomic status, and total calorie intake. The recent study by Papaioannou et al. [20] revealed analogous findings in older people (65–70 years old), indicating that low vegetable consumption significantly increased the likelihood of MetS (OR: 1.47; 95% CI: 1.04–2.07). In addition, inadequate vegetable consumption takes on its own importance as the independent variable for the prevalence of MetS, independent of physical activity or a sedentary lifestyle in older people [20]. Both studies were conducted on older people and confirmed that inadequate vegetable consumption was significantly associated with a higher prevalence of MetS in this group of people. Even in those with comorbidities, the risk of MetS proved to be significantly lower in individuals with a high intake of white and red vegetables (OR: 0.77; 95% CI: 0.57–0.91) compared to those with a low intake of these types of vegetables [13].

Oxidative stress and inflammation are interconnected conditions that characterize the pathophysiology of MetS and diseases associated with it [45,50]. It is well known that the consumption of vegetables—which are characterized by a nutritional composition high in vitamins (mainly A, B, and C) and minerals (selenium and potassium), rich in fiber, and low in fat [51–53], along with bioactive phytochemical contents with antioxidant and anti-inflammatory properties (such as polyphenols)—influences the diagnostic parameters for MetS positively and thereby reduces MetS risk [15,54]. Polyphenols are secondary metabolites synthesized by plants that potentially influence metabolic processes [54], and the 8000 that have been identified thus far are divided into flavonoids (flavones, flavonols, isoflavones, flavanones, flavanols, and anthocyanins) and non-flavonoids (phenolic acids, stilbenes, and lignans) [54]. Meanwhile, vegetables are one of our main sources of vitamin C, the antioxidative and anti-inflammatory properties of which are associated with mechanisms of action that can reverse MetS [50].

The following paragraphs describe how adequate vegetable consumption affects the various metabolic risk factors of MetS.

The elevated prevalence of abdominal obesity identified in the population studied (78.8%) (Table 3) may facilitate adipose tissue dysfunction (due to hyperplasia and hypertrophy of the adipose tissue with the release of inflammatory mediators), insulin resistance, the development of metabolic comorbidities and a higher prevalence of MetS [55]. On the other hand, consuming vegetables is associated with a reduction in abdominal obesity due to their high fiber content, which increases intestinal volume and encourages slower food consumption, thereby increasing satiety and reducing caloric intake. This results in diminished absorption of metabolizable energy in the intestines and improves the intestinal microbiota, facilitating a slimmer phenotype [39]. The high antioxidant/anti-inflammatory content in vegetables may help to reduce the low-grade inflammation associated with visceral adiposity [56]. In addition, greater adherence to adequate vegetable consumption in older people had a confirmed association with a lower BMI and waist circumference after 7.1 years of follow-up in previous research [57]. In this sense, the high flavonoid content of vegetables, like that found, for example, in green leafy vegetables, broccoli, tomatoes, celery, and parsley, may contribute to weight maintenance [11].

Polyphenols are known for being anti-diabetic bioactive compounds [58] and boosting insulin sensitivity, which improves insulin resistance by reducing postprandial glycemia, modulating glucose transport, protecting pancreatic β-cells from damage, and affecting insulin signaling pathways [59]. The latter mechanism is particularly significant because chronic activation of pro-inflammatory pathways in target cells responsible for the action of insulin can bring about obesity, insulin resistance, and DM2 [60]. Moreover, the polyphenols in vegetables can improve frequent diabetic complications, such as vascular dysfunction and coronary diseases [61].

A recent meta-analysis confirmed that vegetable consumption was associated with a reduction in both systolic and diastolic blood pressure [62]. This beneficial effect on blood pressure has been linked to the high fiber content and bioactive components of vegetables; high flavonoid intake in particular has been connected to improved endothelial vasodilation [63] and reduced inflammation and oxidative stress [19,64].

Vegetables are abundant in components that prevent dyslipidemia, such as fiber, phytosterols, antioxidants, and polyphenols [65]. Given its viscosity, fiber content has been identified as a factor that can lower triglyceride levels [17] in slowing gastric emptying, disrupting fat emulsification and micelle formation in the gastrointestinal tract and thus reducing the availability of bile acids in circulation. In addition, soluble fiber can be fermented by the gut microbiota to produce metabolites such as short-chain fatty acids, which regulate genes related to lipid metabolism and the postprandial triglyceride response [17]. Our study examined tomato consumption, characterized by its richness in lycopene, a well-studied terpene-derived carotenoid, which was shown to have a positive association with HDL-cholesterol levels [66,67]. Due to its antioxidant/anti-inflammatory activity, lycopene was found to contribute to beneficial changes in the components of MetS [67,68].

The multi-morbidity that characterizes MetS leads to each of those diseases being aggravated (synergism) and concomitantly to increased mortality [69]. MetS and its associated complications, such as CVD, are also characterized by chronic low-grade inflammation [70]. Therefore, the consumption of vegetables at the recommended level may have an important role in ameliorating MetS-associated pathologies, although the exact mechanisms behind this are still unknown [19]. In this vein, a recent prospective study of 48,632 individuals followed for 14 years found that increased vegetable consumption was associated with a reduction between 11 and 12% in cardiovascular mortality risk [71].

An inverse relationship between vegetable consumption and inflammatory biomarkers (IL-6) has been identified in older people [19,25,65]. The rich antioxidant contents in vegetables are able to alleviate the excess production of free radicals caused by normal-aging-related oxidative stress [72], potentially promoting healthier aging and lowering age-related systemic inflammation [19,25].

Therefore, given that adequate vegetable consumption has been associated with a significant reduction in the risk of MetS [73], health strategies that encourage eating the recommended quantity of vegetables and an increased variety of types of vegetables must be implemented [74]; moreover, it is important to take the wide range of constituents into account when discussing the positive impact of vegetable consumption on MetS [75]. Also, how the method of preparation affects vegetables that are consumed (e.g., whether they are raw or cooked) should also be considered [22].

Given that this was an observational study, inherent limitations apply, and cause–effect relationships could not be determined as in prospective studies. For this reason, the participants' dietary habits were not analyzed temporally. The results of this study cannot be extrapolated to the general population of elderly people either, as geographical location influences dietary habits [23]; for example, only 17% of our sample adhered to the recommended vegetable intake, and this may not be the usual level of consumption in other places. Also, level of education is a socioeconomic factor that can moderate the relationship between vegetable intake and MetS [76], and in our sample, 40.5% had a high level of education (university level), a proportion higher than normal among older people. Therefore, it would be interesting to carry out studies in other geographical areas to uncover how vegetable consumption influences the development of MetS through longitudinal and cross-sectional research in a greater number of elderly people.

In this study, the potential influence of significant confounding variables associated with MetS, such as inadequate consumption of fruit and nuts, age, sex, and level of education, was taken into account to identify inadequate vegetable consumption as an independent factor for the presence of MetS. Therefore, another limitation arose in those important variables related to MetS, such as physical activity, socioeconomic status, total calorie intake, and risk habits (smoking, stress, and high alcohol consumption), which were not quantified in our study. Also, residual confounding from other dietary factors may have influenced the results.

5. Conclusions

The present study showed a beneficial association between an intake of at least two servings of vegetables per day and a lower prevalence of MetS. An important finding was that inadequate vegetable consumption was identified as an independent variable associated with the prevalence of MetS in older people, even when taking inadequate consumption of fruit and nuts and age, sex, and level of education into account. Promoting increased vegetable consumption through public health nutrition programs emerges as an appropriate strategy for enabling healthier aging in this population group with very low adherence to the recommended intake of vegetables and a high prevalence of MetS.

Supplementary Materials: The following supporting information can be downloaded at https://www.mdpi.com/article/10.3390/nu16234101/s1, Table S1: Types of consumption for the variables in the MEDAS-14 questionnaire.

Author Contributions: Conceptualization, G.C.-B., C.G.A. and P.M.C.; methodology, G.C.-B., I.E., C.G.A. and P.M.C.; software, G.C.-B., I.E., C.G.A. and P.M.C.; validation, G.C.-B. and P.M.C.; formal analysis, P.M.C.; investigation, G.C.-B., I.E., C.G.A. and P.M.C.; resources, G.C.-B., I.E., C.G.A. and P.M.C.; data curation, G.C.-B.; writing—original draft preparation, G.C.-B.; writing—review and editing, G.C.-B. and I.E.; visualization, G.C.-B., I.E., C.G.A. and P.M.C.; supervision, G.C.-B., I.E.,

C.G.A. and P.M.C.; project administration, G.C.-B. All authors have read and agreed to the published version of the manuscript.

Funding: This research received no external funding.

Institutional Review Board Statement: This study was conducted in line with the guidelines of the Declaration of Helsinki and approved by the Ethics Committee of the Valdecilla Institute for Health Research (IDIVAL) (2018-308, April 2019) of the Cantabrian Health Service (CHS; scientific–technical validation of the development of this study was obtained by the Primary Care Management (PCM) of the CHS).

Informed Consent Statement: Informed consent was obtained from all of the study participants and protection of their personal data and guarantee of their digital rights were ensured in accordance with Organic Law 3/2018 of 5 December.

Data Availability Statement: The data presented in this study are available upon request from the corresponding author. The data are not publicly available due to privacy protection.

Acknowledgments: An educational cooperation agreement was signed with the European Atlantic University Foundation, which established a collaboration for the development of the fieldwork (anthropometric measurements, interviews on habitual vegetable consumption, and blood pressure measurements) by the students of Human Nutrition and Dietetics. The authors would like to thank Beatriz Macías for her contribution to editing this article.

Conflicts of Interest: The authors declare no conflicts of interest.

References

1. World Bank. Población de 65 Años de Edad y Más, Total. Washington. 2022. Available online: https://datos.bancomundial.org/indicator/SP.POP.65UP.TO (accessed on 5 November 2024).
2. Ayuntamiento de Santander. Santander-Cifras [Internet Database]. 2024. Available online: https://santander.es/ciudad/santander/cifras (accessed on 5 November 2024).
3. Noubiap, J.J.; Nansseu, J.R.; Lontchi-Yimagou, E.; Nkeck, J.R.; Nyaga, U.F.; Ngouo, A.T.; Tounouga, D.N.; Tianyi, F.-L.; Foka, A.J.; Ndoadoumgue, A.L.; et al. Geographic distribution of metabolic syndrome and its components in the general adult population: A meta-analysis of global data from 28 million individuals. *Diabetes Res. Clin. Pract.* **2022**, *188*, 109924. [CrossRef] [PubMed]
4. Aguilar, M.; Bhuket, T.; Torres, S.; Liu, B.; Wong, R.J. Prevalence of the metabolic syndrome in the United States, 2003–2012. *JAMA* **2015**, *313*, 1973–1974. [CrossRef] [PubMed]
5. Guallar-Castillón, P.; Pérez, R.F.; López García, E.; León-Muñoz, L.M.; Aguilera, M.T.; Graciani, A.; Gutierrez-Fisac, J.L.; Banegas, J.R.; Rodriguez-Artalejo, F. Magnitude and management of metabolic syndrome in Spain 2008–2010: The ENRICA study. *Rev. Esp. Cardiol.* **2014**, *67*, 367–373. [CrossRef] [PubMed]
6. International Diabetes Federation. The IDF Consensus Worldwide Definition of Metabolic Syndrome [Monograph on Internet]. Brussels. 2005. Available online: https://pitt.edu/~super1/Metabolic/IDF1.pdf (accessed on 5 November 2024).
7. Fahed, G.; Aoun, L.; Bou Zerdan, M.; Allam, S.; Bou Zerdan, M.; Bouferraa, Y.; Assi, H.I. Metabolic syndrome: Updates on pathophysiology and management in 2021. *Int. J. Mol. Sci.* **2022**, *23*, 786. [CrossRef]
8. Babio, N.; Bulló, M.; Basora, J.; Martínez-González, M.A.; Fernández-Ballarts, F.; Márquez-Sandoval, F.; Molina, C.; Salas-Salvadó, J.; Nureta-PREDIMED Investigators. Adherence to the Mediterranean diet and risk of Metabolic Syndrome and its components. *Nutr. Metab. Cardiovasc. Dis.* **2009**, *19*, 563–570. [CrossRef]
9. Angelico, F.; Baratta, F.; Coronati, M.; Ferro, D.; Del Ben, M. Diet and metabolic syndrome: A narrative review. *Intern. Emerg. Med.* **2023**, *18*, 1007–1017. [CrossRef]
10. Kumar, D.; Kumar, S.; Shekhar, C. Nutritional components in green leafy vegetables: A review. *J. Pharmacogn. Phytochem.* **2020**, *9*, 2498–2502. Available online: https://phytojournal.com/archives?year=2020&vol=9&issue=5&ArticleId=12718 (accessed on 5 November 2024).
11. Bertia, M.L.; Rimm, E.B.; Mukamal, K.J.; Hu, F.B.; Willett, W.C.; Cassidy, A. Dietary flavonoid intake and weight maintenance: Three Prospective cohorts of 124 086 US Men and Women followed for up to 24 years. *BMJ* **2016**, *352*, i17. [CrossRef]
12. Zhang, H.; Tsao, R. Dietary polyphenols, oxidative stress and antioxidant and anti-inflammatory effects. *Curr. Opin. Food Sci.* **2016**, *8*, 33–42. [CrossRef]
13. Gai, Z.; Hu, S.; Gong, G.; Zhao, J. Recent advances in understanding dietary polyphenols protecting against hypertension. *Trends Food Sci. Technol.* **2023**, *138*, 685–696. [CrossRef]
14. Shahwan, M.; Alhumaydhi, F.; Ashraf, G.M.; Hasan, P.M.; Shamsi, A. Role of polyphenols in combating Type 2 Diabetes and insulin resistance. *Int. J. Biol. Macromol.* **2022**, *206*, 567–579. [CrossRef] [PubMed]
15. Nguyen, H.D.; Oh, H.; Kim, M.S. Higher intakes of fruits, vegetables, and multiple individual nutrients is associated with a lower risk of metabolic syndrome among adults with comorbidities. *Nutr. Res.* **2022**, *99*, 1–12. [CrossRef] [PubMed]

16. Muiesan, M.L.; Buso, G.; Rosei, C.A. Less sodium and more potassium to reduce cardiovascular risk. *Eur. Heart J. Suppl.* **2023**, *25* (Suppl. B), B108–B110. [CrossRef] [PubMed]
17. Lee, D.P.S.; Low, J.H.M.; Chen, J.R.; Zimmermen, D.; Actis-Goretta, L.; Kim, J.E. The influence of different foods and food ingredients on acute postprandial triglyceride response: A systematic literature review and meta-analysis of randomized controlled trials. *Adv. Nutr.* **2020**, *11*, 1529. [CrossRef]
18. Kityo, A.; Kaggwa, A. Fruit and vegetable intake, and Metabolic Syndrome components: The Uganda NCD risk factor survey. *Biol. Life Sci. Forum.* **2022**, *12*, 18. [CrossRef]
19. Papaioannou, K.G.; Kadi, F.; Nilsson, A. Consumption of vegetables is associated with systematic inflammation in older adults. *Nutrients* **2022**, *14*, 1765. [CrossRef]
20. Papaioannou, K.G.; Kadi, F.; Nilsson, A. Benefits of fruit and vegetable consumption on prevalence of metabolic syndrome are independent of physical activity behaviors in older adults. *Nutrients* **2022**, *14*, 263. [CrossRef]
21. Bolbinski, P.; Nascvimento-Souza, M.A.; Lima-Costa, M.F.; Viana Peixoto, S. Consumption of fruits and vegetables among older people: Findings from the ELSI-Brazil Study. *Cad. Saúde Pública.* **2023**, *39*, e001158122. [CrossRef]
22. Olaya, B.; Essau, C.A.; Moneta, M.V.; Lara, E.; Miret, M.; Martín-María, N.; Moreno-Agostino, D.; Ayuso-Mateos, J.L.; Abduljabbar, A.S.; Haro, J.M. Fruit and vegetable consumption and potential moderators associated with all-cause mortality in a representative sample of Spanish older adults. *Nutrients* **2019**, *11*, 1794. [CrossRef]
23. Clum, G.; Gustat, J.; O´Malley, K.; Begalieva, M.; Luckett, B.; Rice, J.; Johson, C. Factors influencing consumption of fruits and vegetables in older adults in New Orleans, Louisiana. *J. Nutr. Health Aging* **2016**, *20*, 678–684. [CrossRef]
24. Nicklett, E.J.; Kadell, A.R. Fruit and vegetable intake among older people: A scoping review. *Maturitas* **2013**, *75*, 305–312. [CrossRef] [PubMed]
25. Papaioannou, K.G.; Nilsson, A.; Kadi, F. Healthy diets rich in vegetables and systematic inflammation in older adults. *Biol. Life Sci. Forum.* **2022**, *12*, 13. [CrossRef]
26. Instituto Cántabro de Estadística (ICANE). Padrón Municipal de Habitantes de 2022: Por Grupos Quinquenales y Sexo: Municipio: Santander [Internet Database]. Available online: https://www.icane.es/munreport/dashboard (accessed on 5 November 2024).
27. GRANMO Sample Size and Power Calculator. Version 8.0. Available online: https://www.datarus.eu/ca/aplications/granmo/ (accessed on 5 November 2024).
28. Martínez de la Iglesia, J.; Dueñas Herrero, R.; Onís Vilches, M.C.; Aguado Taberné, C.; Albert Colomer, C.; Luque Luque, R. Adaptación y validación al castellano del cuestionario de Pfeiffer (SPMSQ) para detectar la existencia de deterioro cognitivo en personas mayores de 65 años. *Med. Clin.* **2001**, *117*, 129–134. [CrossRef] [PubMed]
29. World Health Organization (WHO). *Obesity: Preventing and Managing the Global Epidemic*; WHO Technical Report Series 894; WHO: Geneva, Switzerland, 2020. [CrossRef]
30. Guerra, R.S.; Amaral, T.F.; Marques, E.A.; Mota, J.; Restivo, M.T. Anatomical location for waist circumference measurement in older adults; a preliminary study. *Nutr Hosp.* **2012**, *27*, 1554–1561. [CrossRef]
31. Schröder, H.; Fitó, M.; Estruch, R.; Martínez-González, M.A.; Corella, D.; Salas-Salvadó, J.; Lamuela-Raventós, R.; Ros, E.; Salaverría, I.; Fiol, M.; et al. A short screener is valid for assessing mediterranean diet adherence among older Spanish men and women. *J. Nutr.* **2011**, *141*, 1140–1145. [CrossRef]
32. Sánchez-Tainta, A.; San Julián, B.; Martínez-González, M.A. *PREDIMED Date el Gusto de Comer Sano*; EUNSA (UNIVERSIDAD DE NAVARRA): Pamplona, Spain, 2015; ISBN 978-84-313-3076-7.
33. Aranceta, J. (Ed.) *Guía de la Alimentación Saludable Para Atención Primaria y Colectivos Ciudadanos*; SENC-Planeta: Madrid, Spain, 2018.
34. Hosmer, D.W.; Leweshow, S.; Sturdivant, R.X. Model-building strategies and methods for logistic regression. In *Applied Logistic Regression*, 3rd ed.; Whiley: Hoboken, NJ, USA, 2013; Chapter 4; pp. 89–151, ISBN 978-0470582473.
35. Hilbe, J.M. Chapter 5, Model development. In *Logistic Regression Models*; Chapman & Hall/CRC Press: Boca Raton, FL, USA, 2009; pp. 73–187, ISBN 978-1420075755.
36. Cubas-Basterrechea, G.; Elío, I.; Alonso, G.; Otero, L.; Gutiérrez-Bardeci, L.; Puente, J.; Muñoz-Cacho, P. Adherence to the Mediterranean Diet is inversely associated with the prevalence of Metabolic Syndrome in older people from the north of Spain. *Nutrients* **2022**, *14*, 4536. [CrossRef]
37. Zhou, X.; Obuchowski, N.A.; McClish, D.K. *Statistical Methods in Diagnostic Medicine*, 2nd ed.; John Wiley & Sons, Inc.: Hoboken, NJ, USA, 2011; p. 29, ISBN 9780470183144.
38. Ministerio de Agricultura, Pesca y Alimentación (MAPA). *Informe de Consumo Alimentario en España 2020*; Ministerio de Agricultura, Pesca y Alimentación (MAPA): Madrid, Spain, 2021. Available online: https://www.mapa.gob.es/es/alimentacion/temas/consumo-tendencias/informe-anual-consumo-2020-v2-nov2021-baja-res_tcm30-562704.pdf (accessed on 15 February 2024).
39. Cano-Ibañez, N.; Bueno-Cavanillas, A.; Martínez-González, M.A.; Corella, D.; Salas-Salvadó, J.; Zomeño, M.D.; García de la Hera, M.; Romaguera, D.; Martínez, J.A.; Barón-López, F.J.; et al. Dietary intake in population with metabolic syndrome: Is the prevalence of inadequate intake influenced by geographical area? Cross-sectional analysis from PREDIMED-Plus study. *Nutrients* **2018**, *10*, 1661. [CrossRef]
40. Ortíz-Rodríguez, M.A.; Yáñez-Velasco, L.; Carnevale, A.; Romero-Hidalgo, S.; Bernal, D.; Aguilar-Salinas, C.; Rojas, R.; Villa, A.; Tur, J.A. Prevalence of metabolic syndrome among elderly Mexicans. *Arch. Gerontol. Geriatr.* **2017**, *73*, 288–293. [CrossRef]

41. Oliveira, L.V.A.; Soares dos Santos, B.N.; Machado, I.E.; Carvalho Malta, D.; Velasques-Melendez, G.; Felisbino-Mendes, M.S. Prevalence of the Metabolic Syndrome and its components in the Brazilian adult population. *Ciência Saude Colect.* **2020**, *25*, 4269–4280. [CrossRef]
42. Farmanfarma, K.K.; Kaykhaei, M.A.; Adineh, H.A.; Mohammadi, M.; Dabiri, S.; Ansari-Moghaddam, A. Prevalence of metabolic syndrome in Iran: A meta-analysis of 69 studies. *Diabetes Metab. Syndr.* **2019**, *13*, 792–799. [CrossRef]
43. Hirode, G.; Wong, R.J. Trends in the prevalence of Metabolic Syndrome in the United States, 2011–2016. *JAMA* **2020**, *323*, 2526–2528. [CrossRef] [PubMed]
44. Wang, H.H.; Lee, D.K.; Portincasa, P.; Wang, D.Q.-H. Novel insights into the pathogenesis and management of the Metabolic Syndrome. *Pediatr. Gastroenterol. Hepatol. Nutr.* **2020**, *23*, 189. [CrossRef] [PubMed]
45. Montserrat-Mesquida, M.; Quetglas-Llabrés, M.; Capó, X.; Bouzas, C.; Mateos, D.; Pons, A.; Tur, J.A.; Sureda, A. Metabolic Syndrome is associated with oxidative stress and proinflammatory state. *Antioxidants.* **2020**, *9*, 236. [CrossRef]
46. Sergi, G.; Dianin, A.; Zanforlini, B.M.; Curreri, C.; Mazzochi, M.; Simons, L.A.; Manzato, E.; Trevisan, C. Gender differences in the impact of Metabolic Syndrome in older people: A systematic review and meta-analysis. *Nutr. Metab. Cardiovas. Dis.* **2020**, *30*, 1452–1464. [CrossRef]
47. Mozaffarian, D.; Kamineni, A.; Prineas, R.J.; Siscovick, D.S. Metabolic syndrome and mortality in older adults: The Cardiovascular Health Study. *Arch. Intern. Med.* **2008**, *168*, 969–978. [CrossRef]
48. Zanoli, L.; Di Pino, A.; Terranova, V.; Di Marca, S.; Pisano, M.; Di Quattro, R.; Ferrara, V.; Scicali, R.; Rabuazzo, A.M.; Fatuzzo, P.; et al. Inflammation and ventricular vascular coupling in hypertensive patients with metabolic syndrome. *Nutr. Metab. Cardiovasc. Dis.* **2018**, *28*, 122–129. [CrossRef]
49. Sung, K.C.; Rhee, E.J.; Ryu, S.; Kim, B.J.; Kim, B.S.; Lee, W.Y.; Oh, K.-W.; Kim, Y.B.; Chung, P.-W.; Kim, H.; et al. Increased cardiovascular mortality in subjects with metabolic syndrome is largely attributable to diabetes and hypertension in 159,971 Korean adults. *J. Clin. Endocrinol. Metab.* **2015**, *100*, 2606–2612. [CrossRef]
50. Wong, S.K.; Chin, K.Y.; Ima-Nirwana, S. Vitamine C: A review on its role in the management of metabolic syndrome. *Int. J. Med. Sci.* **2020**, *17*, 1625–1638. [CrossRef]
51. Hosseinpour-Niazi, S.; Mirmiran, P.; Mirzaei, S.; Azizi, F. Cereal, fruit and vegetable fiber intake and the risk of metabolic syndrome: A prospective study in the Tehran lipid and glucose study. *J. Hum. Nutr. Diet.* **2015**, *28*, 236–245. [CrossRef]
52. Kim, J.; Choi, Y.H. Physical activity, dietary vitamin C and metabolic syndrome in the Korean adults: The Korea National Health and Nutrition Examination Survey 2008 to 2012. *Public Health* **2016**, *135*, 30–37. [CrossRef]
53. Um, Y.J.; Oh, S.W.; Lee, C.M.; Kwon, H.T.; Joh, H.K.; Kim, Y.J.; Ahn, S.H. Dietary fat intake and the risk of metabolic syndrome in Korean adults. *Korean J. Fam. Med.* **2015**, *26*, 245–252. [CrossRef] [PubMed]
54. Ambrosselli, D.; Masciulli, F.; Romano, E.; Catanzaro, G.; Besharat, Z.M.; Massari, M.C.; Ferretti, E.; Migliaccio, S.; Izzo, L.; Ritieni, A.; et al. New advances in Metabolic Syndrome, from prevention to treatment: The role of diet and food. *Nutrients* **2023**, *15*, 640. [CrossRef] [PubMed]
55. Smith, U. Abdominal obesity: A marker of ectopic fat accumulation. *J. Clin. Investig.* **2015**, *125*, 1790–1792. [CrossRef] [PubMed]
56. Thomas, M.S.; Calle, M.; Fernandez, M.L. Healthy plant-based diets improve dyslipidemias, insulin resistance, and inflammation in metabolic syndrome, A narrative review. *Adv. Nutr.* **2023**, *14*, 44–54. [CrossRef]
57. Chen, Z.; Schoufour, J.D.; Rivadeneira, F.; Lamballais, S.; Ikram, M.A.; Franco, O.H.; Voortman, T. Plant-based diet and adiposity over time in a middle-aged and elderly population: The Rotterdam Study. *Epidemiology* **2019**, *30*, 303–310. [CrossRef]
58. Al-Ishaq, R.K.; Abotaleb, M.; Kubatka, P.; Kajo, K.; Büsselberg, D. Flavonoids and their anti-diabetic effects: Cellular Mechanisms and effects to improve blood sugar levels. *Biomolecules* **2019**, *9*, 430. [CrossRef]
59. Chen, L.; Pu, Y.; Xu, Y.; He, X.; Cao, J.; Ma, Y.; Jiang, W. Anti-diabetic and anti-obesity: Efficacy evaluation and exploitation of polyphenols in fruits and vegetables. *Food Res. Int.* **2022**, *157*, 111202. [CrossRef]
60. Tsalamandris, S.; Antonopoulos, A.S.; Oikonomou, E.; Papamikroulis, G.A.; Vogiatzi, G.; Papaioannou, S.; Deftereos, S.; Tousoulis, D. The role of inflammation in Diabetes: Current concepts and future perspectives. *Eur. Cardiol. Rev.* **2019**, *14*, 50. [CrossRef]
61. Sun, C.; Zhao, C.; Guven, E.C.; Paoli, P.; Simal-Gandara, J.; Ramkumar, K.M.; Wang, S.; Buleu, F.; Pah, A.; Turi, V.; et al. Dietary polyphenols as antidiabetic agents: Advances and opportunities. *Food Front.* **2020**, *1*, 18–44. [CrossRef]
62. Gibbs, J.; Gaskin, E.; Ji, C.; Miller, M.A.; Cappuccio, F.P. The effect of plant-based dietary patterns on blood pressure: A systematic review and meta-analysis of controlled intervention trials. *J. Hypertens.* **2021**, *39*, 23–47. [CrossRef]
63. Macready, A.L.; George, T.W.; Chong, M.F.; Alimbetov, D.S.; Jin, Y.; Vidal, A.; Spencer, J.P.E.; Kennedy, O.B.; Tuohy, K.M.; Minihane, A.M.; et al. Flavonoid-rich fruit and vegetables improve microvascular reactivity and inflammatory status in men at risk of cardiovascular disease—FLAVURS: A randomized controlled trial. *Am. J. Clin. Nutr.* **2014**, *99*, 479–489. [CrossRef] [PubMed]
64. Krzeminska, J.; Wronka, M.; Mlynarska, E.; Franczyk, B.; Rysz, J. Arterial hypertension—Oxidative stress and inflammation. *Antioxidants* **2022**, *11*, 172. [CrossRef] [PubMed]
65. Kim, H.; Lee, K.; Rebholz, C.M.; Kim, J. Plant-based diets and incident metabolic syndrome: Results from a South Korean prospective cohort study. *PLoS Med.* **2020**, *17*, e1003371. [CrossRef] [PubMed]
66. Takayagi, Y.; Obana, A.; Muto, S.; Asaoka, R.; Tanito, M.; Ermakov, I.V.; Bernstein, P.S.; Gellermannn, W. Relationships between skin carotenoid levels and metabolic syndrome. *Antioxidants* **2022**, *11*, 14. [CrossRef]

67. Mirahmadi, M.; Aghasizadeh, M.; Esmaily, H.; Baharara, H.; Mohammadi-bajgiran, M.; Assaran-Darban, R.; Ferns, G.A.; Hadizaheh, F.; Mobarhan, M.G. The effects of lycopene intake on blood pressure and lipid profile of patients with metabolic syndrome; a randomized clinical trial. *J. Food Nutr. Res.* **2023**, *11*, 602–607. [CrossRef]
68. Senkus, K.E.; Tan, L.; Crowe-White, K.M. Lycopene and metabolic syndrome: A systematic review of the literature. *Adv. Nutr.* **2019**, *10*, 19–29. [CrossRef]
69. Aggarwal, P.; Woolford, S.J.; Patel, H.P. Multi-morbidity and polypharmacy in older people: Challenges and opportunities for Clinical Practice. *Geriatrics* **2020**, *5*, 85. [CrossRef]
70. Silveira Rossi, J.L.; Barbalho, S.M.; Reverete de Araujo, R.; Bechara, M.D.; Sloan, K.P.; Sloan, L.A. Metabolic syndrome and cardiovascular diseases: Going beyond traditional risk factors. *Diabetes Metab. Res. Rev.* **2021**, *38*, e3502. [CrossRef]
71. Ghorbani, Z.; Noormohammadi, M.; Kazemi, A.; Poustchi, H.; Pourshams, A.; Martami, F.; Hashemian, M.; Malekzadeh, R.; Hekmatdoost, A. Higher intakes of fiber, total vegetables, and fruits may attenuate the risk of all-cause and cause-specific mortality: Findings from a large prospective cohort study. *Nutr. J.* **2023**, *22*, 60. [CrossRef]
72. Jiang, S.; Liu, H.; Li, C. Dietary regulation of oxidative stress in chronic metabolic diseases. *Foods* **2021**, *10*, 1854. [CrossRef]
73. Tian, Y.; Su, L.; Wang, J.; Duan, X.; Jiang, X. Fruit and vegetable consumption and risk of the metabolic syndrome: A meta-analysis. *Public Health Nutr.* **2017**, *21*, 756–765. [CrossRef] [PubMed]
74. López-González, L.; Becerra-Tomás, N.; Babio, N.; Martínez-González, M.Á.; Díaz-López, A.; Corella, D.; Goday, A.; Romaguera, D.; Vioque, J.; Alonso-Gómez, Á.M.; et al. Variety in fruits and vegetables, diet quality and lifestyle in an older adult mediterranean population. *Clin. Nutr.* **2021**, *40*, 1510–1518. [CrossRef] [PubMed]
75. Brandt, K. Commentary: Vitamin C and Metabolic Syndrome: A meta-analysis of observational studies. *Front Nutr.* **2021**, *8*, 810716. [CrossRef] [PubMed]
76. Mirmiran, P.; Bakhshi, B.; Hossaimpour, S.; Azizi, F. Does the association between patterns of fruit and vegetables and metabolic syndrome incidence vary according to lifestyle factors and socioeconomic status? *Endocr. Abstr.* **2020**, *70*, AEP441. [CrossRef]

nutrients

MDPI

Article

Dietary Habits, Nutritional Knowledge, and Their Impact on Thyroid Health in Women: A Cross-Sectional Study

Przemysław Gwizdak [1], Patryk Szlacheta [2], Daria Łaskawiec-Żuławińska [1], Mateusz Grajek [3,*], Karolina Krupa-Kotara [4], Jarosław Markowski [5] and Ilona Korzonek-Szlacheta [1]

1 Department of Prevention of Metabolic Diseases, Faculty of Public Health in Bytom, Medical University of Silesia, 41902 Katowice, Poland; pgwizdak@sum.edu.pl (P.G.); d201081@365.sum.edu.pl (D.Ł.-Ż.); ikorzonek@sum.edu.pl (I.K.-S.)
2 Department of Basic Medical Sciences, Faculty of Public Health in Bytom, Medical University of Silesia, 41902 Katowice, Poland; pszlacheta@sum.edu.pl
3 Department of Public Health, Faculty of Public Health in Bytom, Medical University of Silesia, 41902 Katowice, Poland
4 Department of Epidemiology, Faculty of Public Health in Bytom, Medical University of Silesia, 41902 Katowice, Poland; kkrupa@sum.edu.pl
5 Department of Laryngology, Faculty of Medical Sciences in Katowice, Medical University of Silesia in Katowice, 40007 Katowice, Poland; jmarkowski@sum.edu.pl
* Correspondence: mgrajek@sum.edu.pl

Abstract: Background: The thyroid gland plays a crucial role in regulating metabolism and various bodily functions through hormone production. Women are particularly susceptible to thyroid disorders such as hypothyroidism and Hashimoto's disease, with associated symptoms affecting overall well-being. Prior research has inadequately addressed the influence of dietary habits and nutritional knowledge on thyroid health, especially in women. Objective: This study aimed to evaluate the dietary habits and nutritional awareness of women aged 18–45 with diagnosed thyroid disorders, emphasizing the effects of education level on knowledge and dietary practices. Material and Methods: A cross-sectional survey was conducted with 297 women diagnosed with thyroid conditions. The survey assessed demographics, comorbidities, hydration habits, and knowledge about nutrient intake critical for thyroid health. Chi-square tests, ANOVA, and correlation analyses were performed to evaluate associations. Results: Hypothyroidism and Hashimoto's disease were most prevalent among younger women (18–25 years). A significant association was observed between higher education and knowledge of protein and carbohydrate roles in managing thyroid health ($p < 0.01$). Women with higher educational backgrounds more frequently used healthier cooking methods and were more informed about beneficial nutrients, including vitamin D and omega-3. A chi-square test indicated that low water intake was significantly associated with comorbid conditions, including insulin resistance and cardiovascular disease ($p < 0.01$). Conclusions: Significant gaps remain in dietary knowledge, particularly concerning protein intake and nutrient–drug interactions, indicating a need for targeted dietary education. Women with higher education demonstrated greater dietary awareness, emphasizing the importance of tailored educational interventions to enhance thyroid disorder management.

Keywords: thyroid disorders; dietary habits; nutritional knowledge; women's health; education; vitamin D; obesity; inflammatory process

1. Introduction

The thyroid gland is a critical endocrine organ that produces key hormones, including thyroxine (T4) and triiodothyronine (T3), both of which play a pivotal role in regulating metabolism, cardiovascular and nervous system functions, and overall homeostasis. These hormones are essential in managing the metabolism of proteins, fats, and carbohydrates, while also influencing cholesterol levels and skin, hair, and nail health [1]. Disruptions in thyroid function, either through hormone overproduction or deficiency, can lead to significant metabolic imbalances, potentially resulting in conditions such as obesity, weight loss, or other metabolic disorders [2].

Women are disproportionately affected by thyroid disorders, with conditions like hypothyroidism, Hashimoto's thyroiditis, and Graves' disease being more prevalent in females compared to males [1]. This gender disparity is likely due to both hormonal fluctuations (such as those related to pregnancy and menopause) and immune system differences. Research has shown that autoimmune thyroid diseases, particularly Hashimoto's thyroiditis, are more common in women, emphasizing the need for gender-specific studies to better understand how these conditions manifest and can be managed in female populations [3].

Given that thyroid dysfunction often leads to associated comorbidities—such as insulin resistance, cardiovascular disease, and hyperlipidemia—it is crucial to understand how dietary management can impact these outcomes [3–5]. Hypothyroidism, for example, often results in slowed metabolism and increased risk of weight gain, making appropriate nutritional support an integral part of disease management. Similarly, hyperthyroidism, which leads to accelerated metabolism, requires careful dietary adjustments to prevent complications like postprandial hyperglycemia and insulin resistance [6,7].

Despite the critical role of diet in managing thyroid health, previous research has primarily focused on general dietary recommendations without sufficiently addressing the unique needs of women with thyroid disorders [8,9]. There is a significant gap in understanding how well women are equipped to manage their conditions through nutrition and whether they possess the necessary knowledge about dietary interventions, supplementation, and hydration [10]. This gap is particularly evident in the understanding of the role of essential nutrients like selenium, zinc, and vitamin D, which have been shown to play a key role in thyroid function, especially in autoimmune thyroid diseases like Hashimoto's [11–13].

Furthermore, the influence of educational background on dietary knowledge has been underexplored. Women with varying levels of education may have different access to, or understanding of, information related to thyroid health, which can affect their ability to manage their condition effectively [14]. This study aims to fill this gap by analyzing how educational level, dietary habits, and the use of dietary supplements such as vitamin D, selenium, and omega-3 fatty acids impact women with thyroid disorders [15].

A key factor in addressing thyroid health is hydration, yet its role in managing thyroid conditions remains under-examined [10,13]. Proper hydration is essential for overall metabolic function, yet many individuals, particularly women, may fail to meet recommended water intake levels, potentially exacerbating symptoms such as fatigue and constipation in thyroid disorders [1–5].

This study, focusing exclusively on women aged 18–45, addresses the current gap in research by evaluating the knowledge and dietary habits of this population in the context of thyroid disease management. By examining how factors such as educational background, dietary practices, and the use of supplements influence disease management, this research aims to provide a clearer picture of the unique challenges women face in managing thyroid disorders. Additionally, this study aims to uncover specific areas where targeted educational interventions can be implemented to improve knowledge and health outcomes.

By conducting a detailed analysis of how women manage their thyroid conditions through diet, hydration, and supplementation, this study contributes valuable insights into gender-specific needs in thyroid health management and underscores the importance of tailored dietary education programs.

The aim of this study is to assess dietary habits and knowledge regarding nutrition in the context of thyroid health among women with thyroid disorders, with a focus on the influence of educational attainment on these aspects.

2. Material and Methods

2.1. Study Group Characteristics

The study group consisted of 297 women, all diagnosed with thyroid disorders, and ranged in age from 18 to 45 years. Participants were recruited from GP practices and hospitals in Katowice, Poland, between March and August 2023, through a voluntary self-selection process. The participants were selected to provide a comprehensive understanding of how dietary habits and knowledge relate to the management of thyroid diseases.

Participation in this study was entirely voluntary and anonymous. Each participant provided informed consent after receiving detailed information about this study's purpose and procedures. All respondents were assured of their privacy, with no identifying information collected, ensuring the confidentiality of the data. This ethical approach ensured that the participants felt comfortable providing accurate information about their health and dietary habits.

The diverse demographic and health profiles of the participants made it possible to explore how various factors, including age, education, and comorbid conditions, affect dietary management and knowledge in women living with thyroid disorders.

All variables were complete for all the participants, as the questionnaire required responses to all questions.

2.2. Research Tool

The questionnaire used in this study was specifically designed to assess the dietary habits, nutritional knowledge, and supplementation practices of women diagnosed with thyroid disorders. It consisted of 31 questions divided into two main sections: a demographic profile and a dietary assessment. The dietary section included questions related to meal frequency, types of foods consumed, hydration practices, and the use of supplements such as vitamin D, selenium, and zinc. Additionally, the questionnaire evaluated the participants' knowledge of thyroid-friendly nutrients, cooking methods, and potential food–drug interactions.

To ensure the reliability and accuracy of the questionnaire, a comprehensive validation process was undertaken, which included both content validation and pilot testing.

In the content validation phase, a panel of five experts—including two dietitians specializing in thyroid health, an endocrinologist, and two academic researchers in nutrition and public health—reviewed the questionnaire. They assessed each item for its relevance, clarity, and comprehensiveness in relation to this study's objectives. Based on their feedback, modifications were made to improve ambiguous questions and to add additional items that ensured the full scope of thyroid-related dietary management was covered. The relevance of each item was rated on a 1 to 4 scale, where 4 indicated high relevance. The content validity index (CVI) was calculated for each item, with a threshold of 0.80 considered acceptable. The overall CVI for the questionnaire was 0.92, indicating a high level of agreement among the experts regarding the relevance of the questions to this study.

Following content validation, the questionnaire underwent pilot testing with a small sample of 20 women diagnosed with thyroid disorders. This phase assessed the questionnaire for clarity, ease of use, and participant understanding. The participants were asked to complete the questionnaire and provide feedback on any confusing or unclear items. The pilot testing indicated that the questionnaire was well understood, with an average completion time of 12 min. Based on the feedback, minor adjustments were made to improve the wording of two questions for enhanced clarity.

To further ensure the reliability of the questionnaire, Cronbach's alpha was calculated based on the responses from the pilot group, focusing on internal consistency. The dietary knowledge and habits section achieved a Cronbach's alpha of 0.85, demonstrating good internal consistency. Additionally, test–retest reliability was assessed by asking the same participants to complete the questionnaire again after two weeks. The correlation coefficient for test–retest reliability was 0.88, showing that the questionnaire produced stable and consistent results over time.

In the final version of the questionnaire, all feedback from both the expert panel and the pilot group was incorporated, ensuring that the tool was comprehensive, user-friendly, and reliable. The questionnaire covered a wide range of factors important to thyroid health, including dietary intake, hydration, supplementation, and patient knowledge. Its high validity and reliability ensured accurate and consistent assessment of the dietary habits and knowledge of the women participating in this study.

Data on educational attainment and comorbidities were collected through self-reports provided by the participants within the structured questionnaire. The questionnaire included specific items asking the participants to indicate their highest level of completed education and to list any comorbid conditions they had been diagnosed with, such as insulin resistance, cardiovascular diseases, and food allergies. This approach allowed for comprehensive demographic and health profiling of the study population.

2.3. Ethical Considerations

This study was conducted in accordance with ethical standards for research involving human participants. Prior to participation, all the respondents were informed about this study's objectives, procedures, and the voluntary nature of their involvement. Informed consent was obtained from each participant, and they were assured that their responses would remain anonymous and confidential. No personal identifying information was collected, ensuring the privacy of the participants.

Additionally, the participants were given the right to withdraw from this study at any time without providing a reason and without any negative consequences. This study adhered to the ethical guidelines outlined by the Declaration of Helsinki, ensuring that the rights, dignity, and well-being of the participants were prioritized throughout the research process. No interventions or manipulations were performed on the participants, as this study focused solely on gathering information through a questionnaire.

This study was conducted in accordance with the provisions of the Declaration of Helsinki and did not require the approval of the Bioethics Committee of the Silesian Medical University in Katowice (decision: PCN/CBN/0052/KB/187/22; 12 July 2022).

2.4. Statistical Methods

The data were compiled using Microsoft Excel 2024, and all calculations were performed using the software program Statistica 13.1. The results were analyzed to assess the knowledge and dietary habits of the participating women, with particular attention on how these habits align with current guidelines for managing thyroid conditions through diet.

A variety of statistical methods were employed to analyze the data. The chi-square test of independence was used to explore relationships between categorical variables, such as the prevalence of thyroid disorders across different age groups and the association between water intake and the presence of comorbid conditions. This test was crucial for identifying whether significant associations existed between variables by comparing observed and expected frequencies. The normality of continuous variables was verified with the Shapiro–Wilk test, and Pearson's correlation coefficient was employed for normally distributed data.

To examine differences in dietary knowledge and practices based on educational background, an analysis of variance (ANOVA) was conducted. This method allowed for the comparison of means between different educational levels and assessed whether there were significant differences in knowledge regarding the role of protein, carbohydrate choices, and cooking methods. The ANOVA also helped determine whether educational background influenced the frequency of healthier cooking methods, such as boiling and steaming, compared to frying.

A correlation analysis was performed to examine the relationship between continuous variables, specifically meal frequency and water intake. This analysis helped identify whether an increase in one variable, such as meal frequency, was associated with a corresponding increase or decrease in water intake.

In the validation phase of the questionnaire, Cronbach's alpha was calculated to assess the internal consistency of the dietary knowledge section, ensuring that the questions reliably measured the intended constructs. Additionally, test–retest reliability was assessed by having a pilot group complete the questionnaire twice over a period of time, allowing for the evaluation of consistency in responses across time.

For all statistical tests, the significance level was set at $p < 0.05$, ensuring that only statistically meaningful relationships were considered in the analysis.

3. Results

This study included 297 women aged between 18 and 45 years. The largest group of participants was aged 18–25 years (33%), followed by those aged 36–45 years (21%). Regarding educational background, the majority of the participants (64%) had obtained higher education, indicating a relatively high level of formal education. Additionally, 32% of the participants had completed secondary education, while a smaller percentage (4%) had vocational education. This diversity in educational levels provided a broad perspective on how knowledge of thyroid health and dietary management may differ across educational attainment (Table 1).

Table 1. Demographic characteristics of the study group (N = 297).

Demographic Characteristics	Category	N (%)
Age	18–25 years	98 (33%)
	26–35 years	65 (22%)
	36–45 years	62 (21%)
Education	Higher	190 (64%)
	Secondary	94 (32%)
	Vocational	13 (4%)

Among the women surveyed, 16% reported being diagnosed with hyperthyroidism. The highest prevalence of hyperthyroidism was observed in the 36–45 years age group (6%), while the lowest was in the 26–35 years age group (2%). The chi-square test did not show a statistically significant relationship between age and hyperthyroidism diagnosis ($p = 0.26$). Graves' disease was reported by 11% of the women in this study, with the highest prevalence (6%) found in the women aged 45 years and older. Hypothyroidism was the most common thyroid disorder among the women surveyed, with 60% reporting a diagnosis. The 18–25 years age group had the highest prevalence (23%). A chi-square test revealed a statistically significant relationship between age group and hypothyroidism diagnosis ($p < 0.05$), indicating that younger women are more likely to be diagnosed with hypothyroidism. Similarly, for Hashimoto's disease, the highest prevalence was in the 18–25 age group (11%), and a significant relationship was found between age group and Hashimoto's diagnosis ($p < 0.05$). Thyroid cancer was diagnosed in 16% of the women in this study, with the highest prevalence in the 36–45 years age group (7%). However, the relationship between age group and thyroid cancer diagnosis was not statistically significant ($p = 0.27$); see Table 2.

Among the women surveyed, 54% did not report any comorbid conditions. However, 22% had food allergies or intolerances, 21% were diagnosed with insulin resistance, and 15% reported cardiovascular diseases. Other conditions, including diabetes, were present in 15% of the respondents. A chi-square test revealed a significant association between low water intake (0.5–1 L/day) and the presence of comorbidities, such as insulin resistance and cardiovascular diseases ($p < 0.01$), indicating that women with lower water intake were more likely to have these conditions.

Table 2. Associations between thyroid diseases and the age of the study group (N = 297).

Thyroid Conditions	Category	N (%)	Statistics	
Hypothyroidism	Overall	178 (60%)	$\chi^2 = 4.12$	$p < 0.05$
	Age 18–25 years	68 (23%)		
Hashimoto's Disease	Overall	72 (24%)	$\chi^2 = 5.07$	$p < 0.05$
	Age 18–25 years	33 (11%)		
Thyroid Cancer	Overall	48 (16%)	$\chi^2 = 3.89$	$p = 0.27$
	Age 36–45 years	21 (7%)		

The majority of the women (66%) relied on the internet as their primary source of information about dietary management for thyroid diseases. Other sources included books (32%), dietitian consultations (31%), and scientific publications (20%). A statistically significant association was found between the use of dietitians as a primary source of information and higher dietary knowledge ($p < 0.01$), suggesting that the women who consulted dietitians were more informed about managing their thyroid conditions.

A significant portion of the women (64%) did not have knowledge about the importance of high-quality protein in managing hypothyroidism and Hashimoto's disease. Only 24% of the respondents indicated that increasing protein intake was important in these conditions. An ANOVA test showed a statistically significant difference in protein knowledge based on educational background ($p < 0.01$), with the women who had higher education more likely to understand the importance of protein in thyroid management. Similarly, 64% of the participants correctly identified whole-grain products as the best types of carbohydrates for individuals with thyroid diseases, while 31% were unsure. The ANOVA test also revealed a significant relationship between education and carbohydrate knowledge ($p < 0.05$).

Most of the women (76%) were aware of the beneficial role of omega-3 fatty acids in thyroid health, although 23% did not know which types of fats were the most beneficial for thyroid conditions. Vitamin D supplementation was practiced by 49% of the respondents, while only 2% used supplements like zinc and selenium. Additionally, 20% did not use any supplements. A chi-square test indicated a significant relationship between the type of thyroid disorder and vitamin D supplementation ($p < 0.05$), with the women diagnosed with Hashimoto's disease and hypothyroidism more likely to supplement with vitamin D compared to those with hyperthyroidism or thyroid cancer.

These results are shown in Table 3.

Table 3. Associations between dietary habits and supplementation of the study group (N = 297).

Dietary Habits and Supplementation	Category	N (%)	Statistics	
Protein's role	Awareness	71 (24%)	F = 7.45	$p < 0.01$
	Lack of awareness	190 (64%)		
Carbohydrate types	Whole-grain products	190 (64%)	F = 3.22	$p < 0.05$
	Uncertain	92 (31%)		
Vitamin D supplementation	Yes	146 (49%)	$\chi^2 = 3.89$	$p < 0.05$

When asked about the most beneficial vitamins and minerals for managing hypothyroidism and Hashimoto's disease, 47% of the women identified selenium, zinc, iron, iodine, and vitamin D as key nutrients. Additionally, 82% of the respondents correctly identified fish and seafood as primary sources of iodine. Only 28% of the women were aware that vitamin C enhances the absorption of levothyroxine, a common thyroid medication. Around 46% correctly identified goitrogenic foods, such as cruciferous vegetables (e.g., cabbage,

Brussels sprouts, and broccoli), which can interfere with thyroid function. However, 47% were unaware of which foods may inhibit thyroid hormone production.

More than half of the respondents (52%) reported consuming 4–5 meals per day, while only 6% consumed more than 5 meals daily. The majority of the women (90%) maintained intervals of more than two hours between meals. Only 34% of the participants consumed more than 2 L of water per day, while 14% reported a low water intake of 0.5–1 L daily. A correlation analysis found a positive relationship between meal frequency and water intake ($p < 0.05$), indicating that the women who consumed more meals per day also tended to have higher water intake.

Regarding specific food groups, 67% of the women consumed fish 1–2 times per week, and 47% consumed eggs with the same frequency. Whole-grain products were consumed regularly by 64% of the women, while 32% reported consuming legumes no more than twice a week. Vegetables were frequently consumed, with 57% eating them more than five times a week, and fruits were consumed with similar regularity, by 48% of the respondents. The most common cooking method among the women surveyed was boiling (25%), followed by steaming and frying (20%). An ANOVA test revealed a significant difference in cooking methods based on educational background ($p < 0.01$), with the women with higher education more likely to use healthier cooking methods, such as boiling and steaming, while frying was more common among those with vocational education.

In Table 4, a significant association was observed between education level and knowledge of the role of protein in thyroid health management ($p < 0.01$). It was shown that the women with higher education were more likely to know the importance of protein for thyroid health, highlighting the role of education in effective dietary management in patients with thyroid disease.

Table 4. Selected relationships demonstrated in the study (N = 297) *.

Sociomedical Data	Selected Relationship	N%	Statistics	
Age (18–25 years)	Hypothyroidism	23%	$\chi^2 = 4.12$	$p < 0.05$
Age (18–25 years)	Hashimoto's disease	11%	$\chi^2 = 5.07$	$p < 0.05$
Education (higher)	Knowledge of the role of protein in thyroid diseases	64%	F = 7.45	$p < 0.01$
Dietitian consultation	Dietary knowledge about thyroid diseases	31%	$\chi^2 = 8.21$	$p < 0.01$
Hashimoto's disease, hypothyroidism	Lack vitamin D supplementation	49%	$\chi^2 = 3.89$	$p < 0.05$

* χ^2 = chi-square test result; F = ANOVA analysis result; p = statistical significance level.

4. Discussion

The study results provide valuable insights into the dietary habits, hydration practices, and general knowledge of dietary management among women with thyroid diseases. These findings are essential for understanding both the strengths and gaps in patient awareness and the management of their condition, particularly through diet and lifestyle.

This study confirmed a higher prevalence of thyroid disorders among women, consistent with the findings of Vanderpump, who also indicated a higher incidence of these conditions in women [12]. Conditions such as hypothyroidism, Hashimoto's disease, and thyroid cancer were more common in the surveyed women. The largest group of women with hypothyroidism fell within the 18–25 age range, while hyperthyroidism and Graves' disease were more prevalent among the women aged 36–45. This age-related variation in the prevalence of different thyroid conditions highlights the need for age-specific educational and clinical interventions, which is also supported by the findings of Alevizaki et al. [16]. These findings align with the research of Carlé et al., who reported a higher incidence of thyroid disorders in women compared to men, due to hormonal and immune system differences [17].

This study revealed gaps in the participants' knowledge of dietary management for thyroid conditions. Only 24% of the women understood the importance of increasing protein intake in the management of hypothyroidism and Hashimoto's disease, while the majority (64%) were unaware of this key nutritional factor. This result is consistent with the research of Wiersinga et al., who emphasized the importance of protein in managing hypothyroidism [18]. Adequate protein intake is crucial for maintaining metabolic function, particularly in individuals with hypothyroidism, as protein supports muscle mass, regulates blood sugar, and maintains energy balance, which is also confirmed by Biondi and Cooper [19]. The lack of awareness on this topic suggests that many women may not be optimizing their diets for effective disease management.

A similar knowledge gap was observed regarding the types of carbohydrates beneficial for thyroid health. While 64% correctly identified whole grains as the most appropriate source, 31% of the participants were unsure. This finding aligns with Ittermann et al., who highlighted the importance of whole grains in regulating blood sugar levels in individuals with thyroid conditions [20].

Another key area was hydration practices. Only 34% of the participants consumed more than 2 L of water per day, while 14% had a daily intake of just 0.5–1 L. This result aligns with the findings of Pereira and Neves, who emphasize the importance of proper hydration in managing thyroid disorders [21]. Adequate hydration is essential for individuals with thyroid disorders, as water supports metabolic processes, aids digestion, and can alleviate symptoms of hypothyroidism, such as dry skin, fatigue, and constipation, which is also confirmed by Fenton and Silverberg [22].

Regarding general dietary habits, most participants consumed 4–5 meals per day at regular intervals of more than two hours between meals, which is consistent with dietary guidelines for stabilizing energy levels and metabolism in individuals with thyroid disorders, as also confirmed by Tan et al. [23]. However, a minority (6%) consumed more than five meals per day, which may indicate issues with overeating or insufficient portion control, in line with findings by Saponaro and Santini [24].

While 49% of the women reported taking vitamin D supplements, only 2% used essential minerals like selenium and zinc, despite their well-documented benefits in supporting thyroid function. This result is consistent with the findings of Nacamulli et al., who indicated that selenium supplementation can reduce thyroid antibody levels in patients with Hashimoto's disease [25]. Similarly, zinc plays a crucial role in supporting the immune system and hormone production, as confirmed by Prasad [26].

Moreover, this study revealed that many of the women were unaware of how vitamin C enhances the absorption of thyroid medications; only 28% were aware of this interaction. This finding aligns with Stott et al., who emphasized the importance of proper supplementation and awareness of nutrient–drug interactions in managing thyroid diseases [27].

The most commonly used cooking methods among the participants were boiling and steaming, both suitable for preserving the nutritional integrity of foods, especially in a thyroid-friendly diet. However, a significant number of the participants (20%) still used frying, which is less ideal as it can introduce inflammatory fats that may exacerbate thyroid disorder symptoms. This result aligns with the findings of Wanjek, who noted the negative effects of frying on thyroid health [28].

This study also found that most participants relied on the internet (66%) and social media (31%) as their primary sources of dietary information, while only 20% consulted scientific publications. This finding is consistent with research by Adams and Lomax, who highlighted the risks associated with relying on online sources for health information [29]. Given the complexities of managing thyroid disorders, it is crucial for patients to access evidence-based guidance, preferably from nutrition specialists, to ensure they are following appropriate dietary and lifestyle recommendations, as supported by Mettler and Zimmermann [30].

In conclusion, this study highlights several important findings regarding the dietary habits and knowledge of women with thyroid diseases. While many participants demonstrated an understanding of basic dietary principles, significant gaps remain in their knowledge of protein intake, hydration, and the role of key supplements in thyroid health. The findings suggest a need for targeted educational interventions to improve patients' dietary practices and awareness of nutrient–drug interactions. Additionally, there is a need to promote more reliable sources of dietary information, such as consultations with dietitians, to ensure that individuals with thyroid disorders receive accurate and beneficial guidance. This result is also consistent with research by the International Alliance of Patients' Organizations, which underscores the importance of access to reliable health information [31].

5. Strengths and Limitations

This study has several strengths that contribute to its significance. First, it focuses specifically on women, a group with a higher prevalence of thyroid disorders, providing a more targeted understanding of the challenges they face in managing their condition through diet. The relatively large sample size of 97 women offers a representative view of dietary habits and knowledge gaps among this population. Additionally, the use of a custom-designed questionnaire allowed for a detailed exploration of specific dietary behaviors, hydration practices, and supplementation habits relevant to thyroid health.

However, this study also has some limitations. The reliance on self-reported data may introduce bias, as participants might not accurately recall or report their dietary intake or hydration levels. Moreover, the cross-sectional design of this study captures data at a single point in time, which limits the ability to observe changes in dietary habits or the long-term impact of educational interventions. Another limitation is the use of online platforms for distributing the questionnaire, which may have restricted participation to women with internet access, potentially excluding those from lower socioeconomic backgrounds or rural areas. Finally, this study primarily relied on descriptive statistics without more advanced analysis to determine causal relationships between dietary habits and thyroid disease management outcomes.

It should be noted that educational level may have influenced the participants' knowledge of thyroid health, a potential confounding factor in the relationship between dietary habits and thyroid-related health outcomes. In addition, the presence of comorbidities, such as insulin resistance or cardiovascular disease, may have independently influenced the participants' dietary choices and thyroid disease management. Consideration of these confounding factors is important for understanding the full context of the results and limitations of this study.

6. Conclusions

This study showed that women aged 18–25 had the highest prevalence of hypothyroidism and Hashimoto's disease, indicating that younger women are more susceptible to these conditions. This highlights the importance of early education and intervention in managing thyroid disorders in this age group.

The women with higher education were significantly more likely to understand the role of protein and carbohydrates in managing thyroid health. This suggests a need for tailored educational efforts, particularly for the women with lower levels of education, to improve dietary management of thyroid disorders.

The women with lower water intake (0.5–1 L/day) were more likely to have comorbid conditions such as insulin resistance and cardiovascular diseases. Ensuring proper hydration is crucial for managing both thyroid health and related conditions.

The women with Hashimoto's disease and hypothyroidism were more likely to use vitamin D supplements, but the overall use of selenium and zinc, both essential for thyroid function, remains low. Increased awareness of these nutrients is necessary for comprehensive thyroid health management.

The women who consulted dietitians had significantly higher knowledge of dietary management compared to those relying on the internet or social media. This underscores the importance of promoting professional, evidence-based dietary advice for thyroid patients.

The women with higher education were more likely to use healthier cooking methods such as boiling and steaming, while frying was more common among those with vocational education. This highlights the need for educational interventions that encourage healthier cooking practices.

Given the gaps in knowledge and dietary practices, particularly among the women with lower education, there is a clear need for targeted educational programs that address diet, hydration, and supplementation for effective thyroid management.

Author Contributions: Conceptualization, P.G.; Methodology, P.S.; Software, P.G. and K.K.-K.; Validation, P.G.; Formal analysis, D.Ł.-Ż. and K.K.-K.; Investigation, P.G. and P.S.; Resources, P.S., M.G. and K.K.-K.; Data curation, D.Ł.-Ż. and M.G.; Writing—original draft, M.G.; Writing—review & editing, J.M. and I.K.-S.; Visualization, M.G.; Supervision, J.M. and I.K.-S.; Project administration, M.G. All authors have read and agreed to the published version of the manuscript.

Funding: This research received no external funding.

Institutional Review Board Statement: Not applicable.

Informed Consent Statement: Not applicable.

Data Availability Statement: The original contributions presented in the study are included in the article, further inquiries can be directed to the corresponding author.

Conflicts of Interest: The authors declare no conflict of interest.

References

1. LeClair, K.; Bell, K.J.L.; Furuya-Kanamori, L.; Doi, S.A.; Francis, D.O.; Davies, L. Evaluation of Gender Inequity in Thyroid Cancer Diagnosis: Differences by Sex in US Thyroid Cancer Incidence Compared with a Meta-analysis of Subclinical Thyroid Cancer Rates at Autopsy. *JAMA Intern Med.* **2021**, *181*, 1351–1358. [CrossRef] [PubMed]
2. American Thyroid Association. Thyroid Disease Facts and Statistics. Endocrine Society. 2020. Available online: https://www.endocrine.org (accessed on 20 August 2024).
3. Verywell Health. Thyroid Disease: Facts and Statistics. Verywell Health. 2021. Available online: https://www.verywellhealth.com (accessed on 20 August 2024).
4. JAMA Network. Autoimmune Thyroid Disease in Women: Prevalence and Risk Factors. JAMA Network. 2020. Available online: https://jamanetwork.com/journals/jama (accessed on 20 August 2024).
5. Fountoulakis, S.; Lee, C.J.; Kotsiou, A. Selenium and iodine: Critical for thyroid hormone synthesis. *J. Endocr. Res.* **2017**, *45*, 237–245.
6. Marwaha, R.K.; Tandon, N.; Garg, M.K.; Mehan, N. Impact of omega-3 fatty acids on thyroid health. *Nutr. J.* **2018**, *17*, 56–62.
7. Garber, J.R.; Cobin, R.H.; Gharib, H.; Hennessey, J.V. Clinical practice guidelines for hypothyroidism in adults: Cosponsored by the American Association of Clinical Endocrinologists and the American Thyroid Association. *Thyroid* **2012**, *22*, 1200–1235. [CrossRef]
8. Yilmaz, S.; Aykan, F.; Ozdemir, O.; Erdogan, M. Vitamin D deficiency in women with Hashimoto's thyroiditis. *Endocr. Pract.* **2020**, *26*, 423–430.
9. Chiovato, L.; Magri, F.; Carlé, A. Hydration and its impact on thyroid function. *Endocr. Rev.* **2018**, *39*, 145–160.
10. Kravets, I. Prevalence and gender disparities in thyroid diseases. *Lancet* **2021**, *397*, 45–52.
11. Hollowell, J.G.; Staehling, N.W.; Flanders, W.D.; Hannon, W.H. Serum TSH, T4, and thyroid antibodies in the United States population (1988 to 1994): National Health and Nutrition Examination Survey (NHANES III). *J. Clin. Endocrinol. Metab.* **2002**, *87*, 489–499. [CrossRef]
12. Vanderpump, M.P.J. The epidemiology of thyroid disease. *Br. Med. Bull.* **2011**, *99*, 39–51. [CrossRef]
13. Zimmermann, M.B.; Boelaert, K. Iodine deficiency and thyroid disorders. *Lancet Diabetes Endocrinol.* **2015**, *3*, 286–295. [CrossRef]
14. Pearce, E.N.; Andersson, M.; Zimmermann, M.B. Global iodine nutrition: Where do we stand in 2013? *Thyroid* **2013**, *23*, 523–528. [CrossRef] [PubMed]
15. Garmendia Madariaga, A.; Santos Palacios, S.; Guillén-Grima, F.; Galofré, J.C. The incidence and prevalence of thyroid dysfunction in Europe: A meta-analysis. *J. Clin. Endocrinol. Metab.* **2014**, *99*, 923–931. [CrossRef] [PubMed]
16. Alevizaki, M.; Saltiki, K.; Voigt, K. Age and sex-specific differences in the presentation and course of autoimmune thyroid disease *J. Endocrinol. Investig.* **2009**, *32*, 48–53.
17. Carlé, A.; Pedersen, I.B.; Knudsen, N.; Perrild, H.; Ovesen, L.; Rasmussen, L.B.; Laurberg, P. Epidemiology of subtypes of hyperthyroidism in Denmark: A population-based study. *Eur. J. Endocrinol.* **2014**, *170*, 649–659. [CrossRef]

18. Wiersinga, W.M.; Duntas, L.; Fadeyev, V.; Nygaard, B.; Vanderpump, M.P. 2012 ETA Guidelines: The Use of L-T4 and L-T3 in the Treatment of Hypothyroidism. *Eur. Thyroid J.* **2012**, *1*, 55–71. [CrossRef]
19. Biondi, B.; Cooper, D.S. The clinical significance of subclinical thyroid dysfunction. *Endocr. Rev.* **2010**, *31*, 149–194. [CrossRef]
20. Ittermann, T.; Khattak, R.M.; Nauck, M.; Völzke, H. Shift of the TSH reference range with improved iodine supply in adults: An analysis of NHANES 2001–2002. *Thyroid* **2014**, *24*, 552–558.
21. Pereira, T.D.; Neves, A.P. The importance of hydration in thyroid dysfunction: A review. *Eur. J. Clin. Nutr.* **2020**, *74*, 22–29.
22. Fenton, M.S.; Silverberg, A.B. Low serum sodium in hypothyroid patients. *Clin. Chem.* **2010**, *56*, 1834–1835.
23. Tan, Z.S.; Beiser, A.S.; Vasan, R.S. Thyroid function and the risk of Alzheimer's disease: The Framingham Study. *Arch. Intern. Med.* **2009**, *169*, 1199–1205.
24. Saponaro, F.; Santini, F. Current perspectives on the treatment of hypothyroidism. *Endocr. J.* **2016**, *45*, 315–328.
25. Nacamulli, D.; Mian, C.; Petricca, D.; Lazzarotto, F.; Barollo, S.; Pozza, D.; Masiero, S.; Faggian, D.; Plebani, M.; Girelli, M.E.; et al. Influence of physiological dietary selenium supplementation on the natural course of autoimmune thyroiditis. *Clin. Endocrinol.* **2010**, *73*, 535–539. [CrossRef] [PubMed]
26. Prasad, A.S. Zinc: Role in immunity, oxidative stress, and chronic inflammation. *Curr. Opin. Clin. Nutr. Metab. Care* **2009**, *12*, 646–652. [CrossRef] [PubMed]
27. Stott, D.J.; Rodondi, N.; Kearney, P.M.; Ford, I.; Westendorp, R.G.; Mooijaart, S.P.; Sattar, N.; Aubert, C.E.; Aujesky, D.; Bauer, D.C.; et al. Thyroid hormone therapy for older adults with subclinical hypothyroidism. *N. Engl. J. Med.* **2017**, *376*, 2534–2544. [CrossRef] [PubMed]
28. Wanjek, B. *Bad Medicine: Misconceptions and Misuses Revealed, from Distance Healing to Vitamin O*; John Wiley & Sons: Hoboken, NJ, USA, 2012.
29. Adams, S.; Lomax, M. Social media and its influence on nutrition and health behaviours. *Health Promot. Int.* **2017**, *32*, 953–962.
30. Mettler, S.; Zimmermann, M.B. Effects of excess iodine intake on thyroid hormone metabolism in healthy men. *Thyroid* **2010**, *20*, 293–294.
31. International Alliance of Patients' Organizations (IAPO). *Global Patient Empowerment Study*; IAPO: London, UK, 2021.

nutrients MDPI

Article

Short-Term Zinc Supplementation Stimulates Visceral Adipose Catabolism and Inflammation in Mice

Xiaohua Huang [1,2,3], Dandan Jiang [2], Yingguo Zhu [2], Zhengfeng Fang [2,4] and Bin Feng [2,3,*]

1 Meat Processing Key Laboratory of Sichuan Province, College of Food and Biological Engineering, Chengdu University, Chengdu 610106, China; hxh3028@163.com
2 Animal Nutrition Institute, Sichuan Agricultural University, Chengdu 611130, China; dandanjiang17@163.com (D.J.); yingguo_zhu@163.com (Y.Z.); zfang@sicau.edu.cn (Z.F.)
3 Key Laboratory of Animal Disease-Resistant Nutrition of Ministry of Education, Sichuan Agricultural University, Chengdu 611130, China
4 Key Laboratory for Food Science and Human Health, College of Food Science, Sichuan Agricultural University, Ya'an 625014, China
* Correspondence: fengbin@sicau.edu.cn

Abstract: Background: Zinc (Zn), a fundamental trace element in human biology, exhibits pivotal roles in sustaining vital physiological processes and regulating metabolic homeostasis. Insufficient zinc intake has been linked to deleterious consequences on growth, reproductive functions, metabolic activities, and immune responses in both humans and animals. Oral zinc supplementation is usually performed to meet zinc requirement. Previous studies have shown that long-term supplementation of zinc in mice impaired AKT signaling and induced adipocyte hypertrophy in visceral adipose tissue. Methods: The presented study was conducted to investigate the role and mechanism of short-term zinc supplementation on lipids metabolism. Zinc sulfate was supplemented in the drinking water of C57/BL6J male mice at 30 ppm or 90 ppm for one week. Water consumption, food intake, and body weight were analyzed, adipose tissue and serum profile of metabolites were investigated, and the key genes related to lipid metabolism were analyzed. Results: Short-term zinc supplementation decreased visceral adipose tissue weight and adipocyte size compared to the control group, but no difference was observed in food intake, water consumption, and body weight between the two groups. Further studies revealed that short-term zinc supplementation significantly increased the serum insulin level while decreasing the serum NEFA content. In addition, zinc supplementation increased the expression of *Atgl* and *Hsl* in the visceral adipose tissue compared with the control mice. Furthermore, the phosphorylation level of HSL and protein level of PPARg in the epididymal adipose tissue increased by zinc supplementation compared with the control mice. In comparison, the protein level of FASN was down-regulated by short-term zinc supplementation in the epididymal adipose tissue, although the expression of lipogenic genes was not changed. The expression of *F4/80* and *Tnfa* were increased in zinc-supplemented adipose tissue as compared with the control group. Conclusions: Our findings suggest that short-term zinc supplementation might reduce fat deposition by enhancing lipolysis in mice. Future studies could focus on the effect of intermittent zinc supplementation on fat reduction in both animal models and humans.

Keywords: zinc; adipose tissue; lipolysis; insulin; F4/80; inflammation

1. Introduction

Micronutrient deficiency may result in impaired health, including retarded growth, increased risk of acute infectious disease, birth defects, and even death [1,2]. Recently, epidemiological research has shown that people in a "vulnerable period" have a higher demand for micronutrients, especially vegetarians, infants, young children, adolescents, elderly individuals, and pregnant and lactating women [3,4]. Increasing the micronutrient intake of populations by diet supplementation or functional food has been shown to reduce the burden on the mother and morbidity and mortality of the child [5,6].

Zinc is recognized as a vital trace element that influences various physiological processes of humans and animals, including development, immunity, cognitive, reproductive, oxidative stress, and metabolism [7–9]. Remarkably, zinc deficiency aggravates obesity-related metabolic diseases, including hyperlipidemia, insulin resistance, inflammation, and hyperglycemia [10,11]. Zinc supplementation could improve blood glucose levels in individuals with diabetes, and it is more significant in elderly individuals [8,12]. On the other hand, zinc deficiency increases the concentrations of glucose, cholesterol, and triglycerides in streptozotocin (STZ)-induced diabetic rats [13], while increasing leptin production and stimulating macrophage recruitment in adipose tissue in obese mice [14]. Individuals fulfill their nutritional requirements by incorporating zinc-rich foods into their diet, including various types of meat (pork, chicken, lamb, etc.), vegetables (broccoli, carrots, spinach, etc.), and seafood options (kelp, salmon, oysters, etc.), as well as fruits and nuts [15]. However, despite the global accessibility of these foods, zinc deficiency remains prevalent, particularly in developing nations, indicating that malnutrition or other biological and physiological factors, not solely attributed to the country's economic status, contribute significantly to this widespread phenomenon [1]. Therefore, maintaining adequate zinc intake is crucial for optimal health. In addition, the bioavailability and functional outcomes of zinc are subject to the influence of the food matrix or the specific zinc formulation utilized in supplements. When administered in supplemental forms, zinc must traverse the intestinal barrier and be absorbed into the systemic circulation for effective utilization. Zinc sulfate ($ZnSO_4$) has traditionally been employed as a reference compound for zinc supplementation owing to its favorable bioavailability and cost-effectiveness.

Adipose tissue is the primary storage depot for lipids in the human body. Excessive accumulation of lipids in adipose tissue constitutes the morphological foundation of obesity. Meanwhile, the adipose tissue is also the main organ involved in energy metabolism. Imbalances in glucose and lipid metabolism within the body are the leading causes of metabolic disorders [16,17]. When energy intake is in excess, the adipose tissue synthesizes and stores triglycerides (TAG) using glucose, which is known as lipogenesis. Several key enzymes, including fatty acids synthetase (FASN), stearoyl CoA desaturase 1 (SCD1), and acetyl-CoA carboxylase (ACC1), are involved in the process of lipogenesis. The expression of these key enzymes is upregulated by sterol regulatory element binding transcription factor 1 (SREBF1) and peroxisome proliferator proliferator-activated receptor gamma (PPARg) [18]. On the contrary, TAG can be broken down, known as lipolysis, which is a crucial process in the body that involves the breakdown of stored fats into smaller components for energy production. During lipolysis, adipose triglyceride lipase (ATGL) and hormone-sensitive lipase (HSL) are activated. These enzymes work together to hydrolyze triglycerides into glycerol and fatty acids. Fatty acids can be then broken down into acetyl-CoA, which is finally used to synthesize ATP, the body's primary energy source [19].

Our previous study found that chronic zinc supplementation impaired AKT signaling and induced adipocyte hypertrophy in perirenal adipose tissue [20]. In the current study, we hypothesized that short-term zinc supplementation might benefit lipids metabolism status in adipose tissue. Mice were supplemented with $ZnSO_4$ in the drinking water for one week. The results showed that short-term zinc supplementation decreased visceral adipose tissue weight and adipocyte size compared to the control group and stimulated the expression of lipolytic genes and proteins in the visceral adipose tissue.

2. Materials and Methods

2.1. Animal Study

Animal protocols were reviewed and approved by the Animal Care and Use Committee of Sichuan Agricultural University (SICAU-2021-078, 8 March 2021). All animal studies were carried out under the guide of Use and Care of Laboratory Animals. Five-week-old

C57BL/6J mice were obtained from GenPharmatech (Chengdu, China). The mice were housed in a pathogen-free environment with a constant temperature of 22 °C and humidity of 60%. A total of 9 cages of male mice (2 mice per cage) were randomly split into 3 groups based on comparable average body weight at the age of 15 weeks with 3 cages (6 mice) per group. The control group received spring water (zinc content was less than 0.01 mg/L), while the other two groups were given 30 ppm or 90 ppm zinc-supplemented spring water (132.4 mg/L or 397.2 mg/L zinc sulfate heptahydrate (Z0251, Sigma, Shanghai, China)) for one week. All mice were provided a regular chow diet according to AIN93 (Dossy Experimental Animals Co., Ltd., Chengdu, China. Zinc content was 38.3 mg/kg) [20], and had free access to water. Water consumption and food intake were recorded. One week later, mice were weighed and euthanized with CO_2 under fed status. Blood was collected by cardiac exsanguination, and serum was stored at −20 °C for further analysis. Liver, perirenal, epididymal, and subcutaneous adipose tissues were removed from the body with delicate scissors. All tissue from the liver, perirenal, epididymal, and subcutaneous adipose tissues were collected, and non-fat tissue was removed from the adipose tissues. The liver was dried with absorbent tissue paper to remove residual blood. Fresh tissues were then weighed, flash-frozen in liquid nitrogen, and stored at −80 °C for further analysis.

2.2. Analysis of Serum Zinc Content

Zinc concentration in the serum was measured with a zinc detection kit (E011, Nanjing Jiancheng Bioengineering Institute, Nanjing, China) according to the manufacturer's instructions.

2.3. Analysis of the Metabolite Profiles and Hormone Concentration in Serum

The serum levels of glucose, high-density lipoprotein cholesterol (HDL-C), low-density lipoprotein cholesterol (LDL-C), triglycerides (TAG), non-esterified fatty acids (NEFA), and total cholesterol (TC) were measured on an automatic biochemical analyzer (7020, HITACHI, Tokyo, Japan) with the respective assay kits from Kehua Bio-Engineering (Shanghai, China). The serum insulin levels were measured using a mouse insulin ultrasensitive ELISA kit (80-INSMSU, ALPCO, Salem, MA, USA) according to the manufacturer's instructions.

2.4. Histology Staining

For H and E staining, adipose tissue samples were fixed, embedded in paraffin, and subsequently sectioned into 4 μm slices using a precision microtome (RM2016, Leica, Shanghai, China). The sections were rigorously dehydrated through graded alcohols, immersed in hematoxylin for 5 min, rinsed thoroughly with double distilled water (ddH_2O), and then counterstained with eosin for 2 min, followed by dehydrating again and mounting onto slides. Imaging software (NIS-Elements F3.2, v4.60, Nikon, Tokyo, Japan) was used to capture the images on a microscope (TS100, Nikon, Tokyo, Japan) with a CCD (DS-U3, Nikon, Tokyo, Japan).

Cell area measurements were conducted using the ImageJ software (v1.49, National Institutes of Health, Bethesda, MD, USA). In brief, 16 images obtained from four mice per group were analyzed, and within each image, 10 representative cells of the average cell size were selected for calculating the mean cell size. The cumulative cell area of 160 cells, spanning four mice in each tissue group, was then utilized to determine the average cell area for that specific group.

2.5. RNA Extraction and Real-Time PCR (qRT-PCR)

RNA extraction and real-time PCR procedures were conducted following previously documented protocols [21]. In summary, adipose tissues were grinded in liquid nitrogen, and 100 mg tissue was used for total RNA extraction using TRIzol reagent (Invitrogen, Shanghai, China). The quality of the extracted RNA was evaluated via agarose gel electrophoresis, and its concentration was measured using a spectrophotometer (NanoDrop 2000, Thermo Fisher Scientific, Shanghai, China). Subsequently, complementary DNA (cDNA) synthesis was achieved using a reverse-transcription PCR kit (RR047A, Takara, Dalian, China). Real-time PCR analysis was conducted with Power SYBR Green RT-PCR reagents (4367659, Thermo Fisher Scientific, Shanghai, China) on a quantitative-PCR machine (7900HT, ABI, Carlsbad, CA, USA). For each PCR reaction, the following reagent concentrations were optimized: 300 nM forward primer, 300 nM reverse primer, and 20 ng cDNA sample. The PCR conditions were as follows: 95 °C for 10 min for 1 cycle, followed by 40 cycles of 95 °C for 15 s and 60 °C for 1 min. B-actin (*Actb*) was used as the reference gene, and the $2^{-\Delta\Delta Ct}$ method was employed to quantify the levels of gene expression. The primer sequences utilized in this study are presented in Table 1.

Table 1. Primers for real-time quantitative PCR.

Genes	Forward	Reverse
Actb	GGCTGTATTCCCCTCCATCG	CCAGTTGGTAACAATGCCATGT
Acc1	CGGACCTTTGAAGATTTTGTCAGG	GCTTTATTCTGCTGGGTGAACTCTC
Acc2	GGAAGCAGGCACACATCAAGA	CGGGAGGAGTTCTGGAAGGA
Apob	TTGGCAAACTGCATAGCATCC	TCAAATTGGGACTCTCCTTTAGC
Atgl	CTGTGTGGAACCAAAGGACCTG	GCTACCCGTCTGCTCTTTCATC
Cd11c	CTGGATAGCCTTTCTTCTGCTG	GCACACTGTGTCCGAACTCA
Cd36	ATGGGCTGTGATCGGAACTG	GTCTTCCCAATAAGCATGTCTCC
Dgat1	TCCGTCCAGGGTGGTAGTG	TGAACAAAGAATCTTGCAGACGA
Fasn	GGCTCTATGGATTACCCAAGC	CCAGTGTTCGTTCCTCGGA
F4/80	TGACTCACCTTGTGGTCCTAA	CTTCCCAGAATCCAGTCTTTCC
Glut4	ACCGGATTCCATCCCACAAG	TCCCAACCATTGAGAAATGATGC
Hsl	TGAAGCCAAAGATGAAGTGAGAC	CTTGACTATGGGTGACGTGTAGAG
Il6	TAGTCCTTCCTACCCCAATTTCC	TTGGTCCTTAGCCACTCCTTC
Leptin	GAGACCCCTGTGTCGGTTC	CTGCGTGTGTGAAATGTCATTG
Mcp1	TTAAAAACCTGGATCGGAACCAA	GCATTAGCTTCAGATTTACGGGT
Mcp2	CCCTTCGGGTGCTGAAAAG	CCACTTCTGTGTGGGGTCTAC
Mgll	CGGACTTCCAAGTTTTTGTCAGA	GCAGCCACTAGGATGGAGATG
Pgc1a	TATGGAGTGACATAGAGTGTGCT	CCACTTCAATCCACCCAGAAAG
Plin	CGTGGAGAGTAAGGATGTCAATG	GGCTTCTTTGGTGCTGTTGTAG
Pparg	GGAAGACCACTCGCATTCCTT	TCGCACTTTGGTATTCTTGGAG
Scd1	CCTACGACAAGAACATTCAATCCC	CAGGAACTCAGAAGCCCAAAGC
Srebf1	AACTGCCCATCCACCGACTC	ATTGATAGAAGACCGGTAGCGC
Tnfa	GACCCTCACACTCAGATCATCTTCT	CCACTTGGTGGTTTGCTACGA

2.6. Western Blot Analysis

The protein extraction procedure was conducted according to previously documented protocols [21]. Briefly, adipose tissue powder was homogenized using a homogenizer in a protease inhibitor cocktail (4693116001, Roche, Mannheim, Germany) supplemented cell lysis buffer (Beyotime Biotechnology, Shanghai, China). Subsequently, 30 µg of total protein was separated on a polyacrylamide gel, followed by electro-transferring onto PVDF membranes. The PPARg (2443), pAKT S473 (4060), AKT (4691), and pHSL (3891) antibodies were purchased from Cell Signaling Technology (Shanghai, China); GAPDH (abs132004) antibody was obtained from Absin Biotechnology Company (Shanghai, China). After thorough washing, the membranes were incubated with appropriate horseradish peroxidase-conjugated secondary antibodies (7074 and 7076, Cell Signaling Technology) for 1 h. After additional washing, membranes were incubated with an ECL western blotting detection reagent (1705060, Bio-Rad, Hercules, CA, USA). Then, the protein signals were detected on a Molecular Imager ChemiDoc XRS+ System (Bio-Rad).

2.7. Statistical Analysis

The data were analyzed with SAS 9.3 software (SAS Institute Inc., Cary, NC, USA). Firstly, the univariate test was performed to examine the normality and homogeneity of data variances. For normally distributed data, an independent *t*-test was employed to compare differences between the two groups. One-way analysis of variance (ANOVA) procedure followed by Tukey's multiple range was applied to analyze the difference between the three groups. The results are presented as mean ± SE. *p*-Values less than 0.05 were considered statistically significant.

3. Results

3.1. Short-Term Zinc Supplementation Decreased Fat Deposition in Mice

To explore the effect of short-term zinc supplementation on lipids metabolism, 15-week-old C57/BL6J male mice were given spring water, 30 ppm, or 90 ppm zinc supplemented spring water for one week. The results showed that during the treatment, the food intake (Figure 1A) and water consumption (Figure 1B) were not changed by short-term zinc supplementation compared with the control mice. However, zinc supplementation significantly increased serum concentration of zinc compared to the control (Figure 1C).

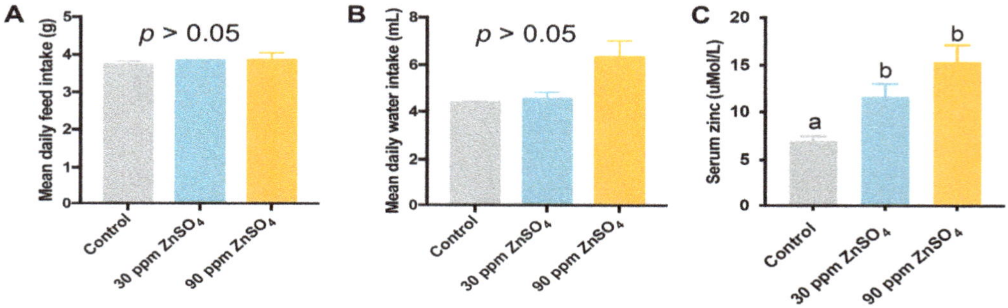

Figure 1. Food intake and water consumption of mice. A total of 18 mice were housed in 9 cages with 2 mice per cage. The initial food weight and water volume of every cage were recorded at the beginning of the study when the mice were 15 weeks of age. Seven days later, the left food weight and water volume were measured. Food intake and water consumption were calculated as follows: (initial food weight − left food weight)/7/2 or (initial water volume − left water volume)/7/2. (**A**) Weight of food intake (N = 3 per group). (**B**) Volume of water consumption (N = 3 per group). (**C**) Serum concentration of zinc (N = 6 per group). Data are shown as mean ± SE. ^{a,b} Columns with different superscript letters mean significant differences ($p < 0.05$).

Though the body weight and liver weight were not changed, the tissue weight and weight index of the perirenal and subcutaneous fats were decreased by short-term zinc supplementation compared to the control mice, while only the weight index of the epididymal fat decreased (Table 2).

Because 90 ppm zinc supplementation had a similar effect on the serum concentration of zinc and weight index of the white fat with 30 ppm zinc supplementation, the following indicators were analyzed only with 30 ppm zinc-supplemented group and control group. Hematoxylin and eosin (H and E) stain showed that the cell area of adipocytes in 30 ppm zinc-supplemented mice decreased by 50.6% and 52.4% in the perirenal and epididymal adipose tissues, respectively, compared with those in the control mice (Figure 2A–C). These data suggest that short-term zinc supplementation could decrease visceral adipose tissue weight and adipocyte size in mice.

Table 2. Effect of short-term zinc supplementation on body weight and tissue weight.

	Control	30 ppm $ZnSO_4$	90 ppm $ZnSO_4$	p Value
Body weight (g)	28.682 ± 0.282	29.100 ± 0.343	28.767 ± 0.388	0.6823
Tissue weight (g)				
Liver weight	1.282 ± 0.065	1.425 ± 0.025	1.353 ± 0.023	0.1364
Epididymal fat	0.428 ± 0.018	0.345 ± 0.033	0.348 ± 0.024	0.0704
Perirenal fat	0.158 ± 0.010 ^{a}	0.105 ± 0.015 ^{b}	0.114 ± 0.012 ^{b}	0.0248
Subcutaneous fat	0.304 ± 0.018 ^{a}	0.206 ± 0.024 ^{b}	0.232 ± 0.013 ^{b}	0.0099
Sum of white fat depots	0.890 ± 0.045 ^{a}	0.656 ± 0.064 ^{b}	0.695 ± 0.046 ^{b}	0.0184
Brown fat	0.062 ± 0.002	0.061 ± 0.001	0.052 ± 0.002	0.1930
Tissue weight index (%)				
Liver weight	4.469 ± 0.222	4.898 ± 0.075	4.707 ± 0.087	0.2080
Epididymal fat	1.493 ± 0.057 ^{a}	1.182 ± 0.108 ^{b}	1.214 ± 0.088 ^{b}	0.0496
Perirenal fat	0.549 ± 0.031 ^{a}	0.360 ± 0.051 ^{b}	0.398 ± 0.044 ^{b}	0.0200
Subcutaneous fat	1.057 ± 0.060 ^{a}	0.710 ± 0.084 ^{b}	0.808 ± 0.046 ^{b}	0.0076
Sum of white fat depots	3.099 ± 0.143 ^{a}	2.253 ± 0.217 ^{b}	2.421 ± 0.171 ^{b}	0.0135
Brown fat	0.214 ± 0.006	0.210 ± 0.019	0.182 ± 0.007	0.1590

Note: Tissue weight index = (tissue weight/body weight) × 100%. Data are presented as means ± SE. (N = 6 per group). ^{a,b} Values within a row with different superscript letters were significantly different ($p < 0.05$).

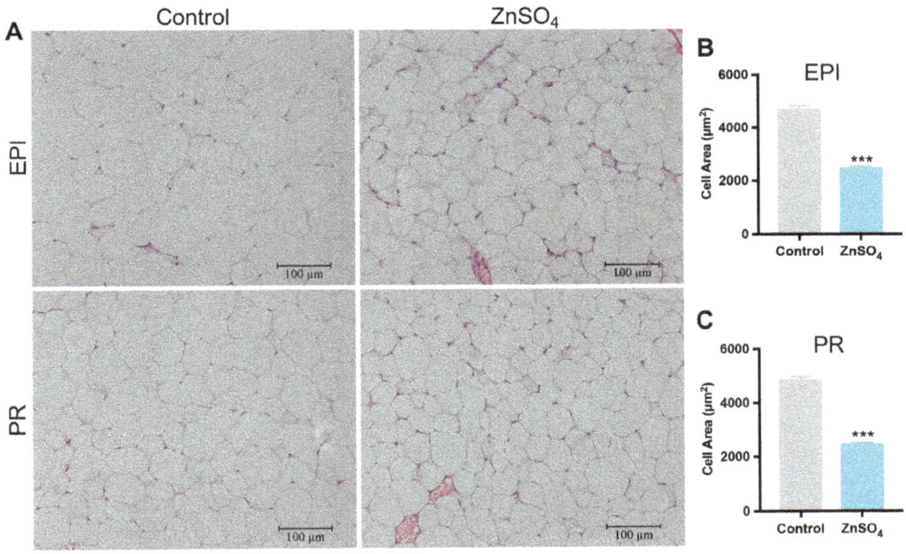

Figure 2. Histological analysis of adipose tissue. (**A**) H and E staining images for epididymal and perirenal adipose tissues; (**B**) Adipocyte cell area of epididymal adipose tissue (N = 160 cells from four mice for each group); (**C**) Adipocyte cell area of perirenal adipose tissue (N = 160 cells from four mice for each group). Bars equal to 100 µm. Data are shown as mean ± SE. $ZnSO_4$, 30 ppm zinc supplemented group. *** $p < 0.001$ $ZnSO_4$ vs. Control.

3.2. Short-Term Zinc Supplementation Induced Hyperglycemia and Hyperinsulinism in Mice

Blood glucose levels were then evaluated in the mice. The results revealed that zinc-supplemented mice showed higher blood glucose levels than the control mice under the fed state (Figure 3A). Moreover, serum insulin concentration significantly increased by short-term zinc supplementation compared with the control mice (Figure 3B). Further studies indicated that the gene expression of glucose transporter 4 (*Glut4*), the main regulator for glucose uptake in adipocyte, decreased in the epididymal adipose tissue of zinc-supplemented mice compared with the control group (Figure 3C). The expression of adipokine gene *Leptin* was increased in the epididymal adipose tissue of zinc-supplemented mice compared with that in the control mice (Figure 3C). However, the mRNA levels of *Glut4*, *Pgc1a*, and *Leptin* were not changed by zinc supplementation in perirenal or subcutaneous adipose tissues (Figure 3D,E). Western blot analysis showed that short-term zinc supplementation impaired the phosphorylation level of AKT at Ser473 in the epididymal adipose tissue compared with the control group (Figure 3F,G). These data suggest that short-term zinc supplementation might impair the insulin sensitivity of adipose tissue and thus induce hyperglycemia in mice.

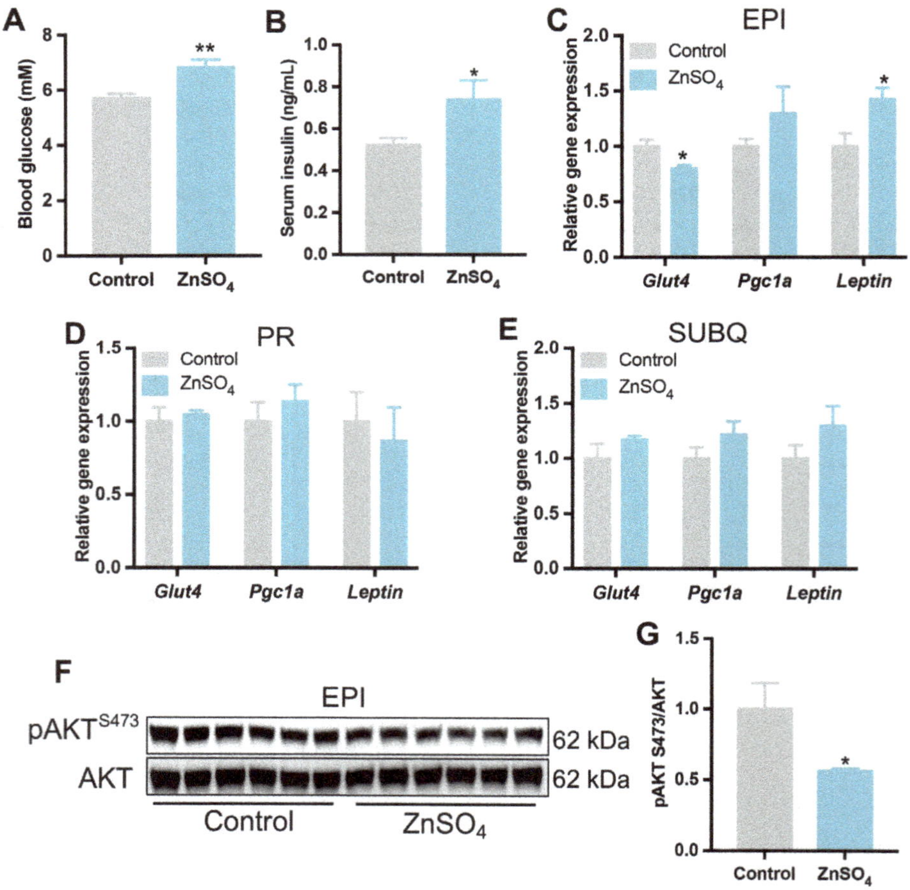

Figure 3. Short-term zinc supplementation induced hyperglycemia in mice. (**A**) Blood glucose level at harvest. (**B**) Insulin concentration in serum. Gene expression level of *Glut4*, *Pgc1a*, and *Leptin* in the epididymal (**C**), perirenal (**D**), and subcutaneous (**E**) adipose tissues. (**F**,**G**) Phosphorylation levels of AKT at S473 in the epididymal adipose tissue. (N = 6 for each group). Data are expressed as Mean ± SE. $ZnSO_4$, 30 ppm zinc supplemented group. * $p < 0.05$, ** $p < 0.01$.

3.3. Short-Term Zinc Supplementation Reduced Serum NEFA Concentration

Serum lipid profiles were then investigated, which indicated that serum NEFA concentration was lower in the zinc-supplemented mice than that of the control mice (Figure 4A).

However, serum levels of TAG, TC, LDL-C, and HDL-C were similar between the zinc-supplemented mice and control mice (Figure 4B–E). Of the fatty acid transporter genes, the mRNA level of fatty acid translocase (*Cd36*) was down-regulated in the epididymal, perirenal, and subcutaneous adipose tissues (Figure 4F–H), by zinc supplementation compared with control. These data suggest that short-term zinc supplementation might reduce circulating NEFA concentration by impairing fatty acid exportation in adipose tissue.

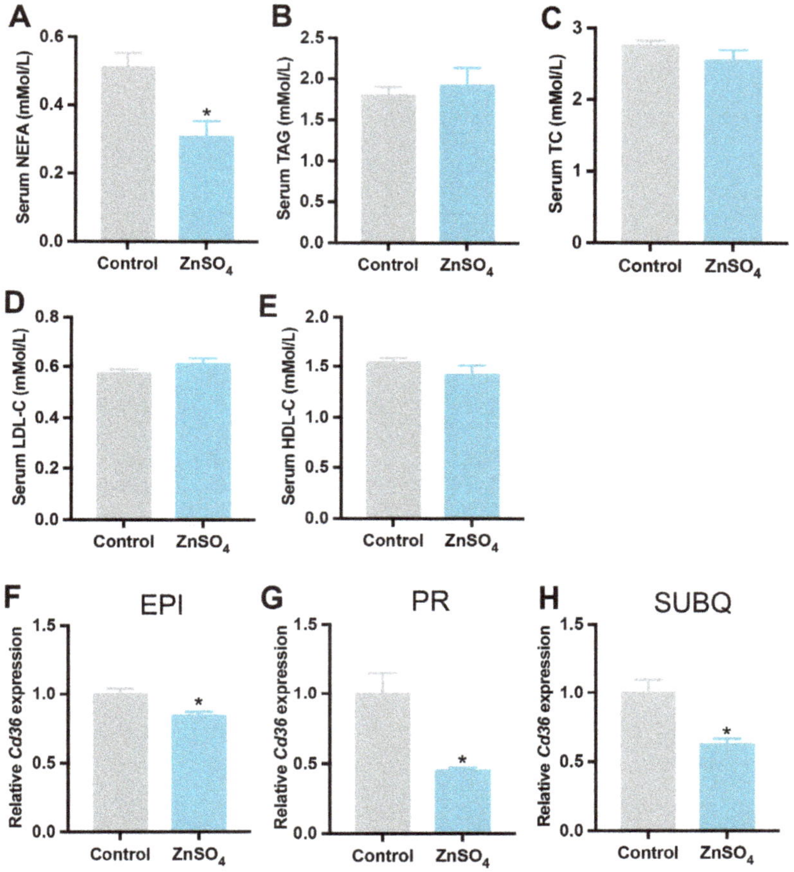

Figure 4. Lipid profiles in the serum. C57/BL6J male mice were supplemented with 30 ppm $ZnSO_4$ or control for one week. Serum and adipose tissues were harvested at fed state 7 days post zinc supplementation. (**A**) NEFA concentration in the serum. (**B**) TAG concentration in the serum. (**C**) TC concentration in the serum. (**D**) LDL-C concentration in the serum. (**E**) HDL-C concentration in the serum. The expression level of *Cd36* in the epididymal (**F**), perirenal (**G**), and subcutaneous (**H**) adipose tissues (N = 6 for each group). Data are expressed as mean ± SE. $ZnSO_4$, 30 ppm zinc supplemented group. * $p < 0.05$.

3.4. Short-Term Zinc Supplementation Impaired Lipogenesis in the Epididymal Adipose Tissue

The accumulation of triglycerides in the adipocytes can be affected by lipogenesis and lipolysis. The expression of lipogenic and lipolytic genes was then analyzed in adipose tissues. The results showed that the mRNA levels of acetyl-CoA carboxylase 2 (*Acc2*) and triglyceride synthetic gene *Dgat1* were increased, while the expression of sterol regulatory element binding transcription factor 1 (*Srebf1*), *Acc1*, *Fasn*, and *Scd1* were not changed by zinc supplementation in the epididymal adipose tissue compared with the control

(Figure 5A). The expression of *Srebf1*, *Acc1*, *Acc2*, *Fasn*, *Scd1*, and *Dgat1* was not changed by zinc supplementation in either perirenal or subcutaneous adipose tissue compared with the control mice (Figure 5B,C). Interestingly, the protein level of FASN was much lower in the epididymal adipose tissue of the zinc-supplemented mice than that of control mice (Figure 5D,E). These data demonstrate that short-term zinc supplementation might impair fatty acid synthesis by decreasing the protein expression of FASN in adipose tissue.

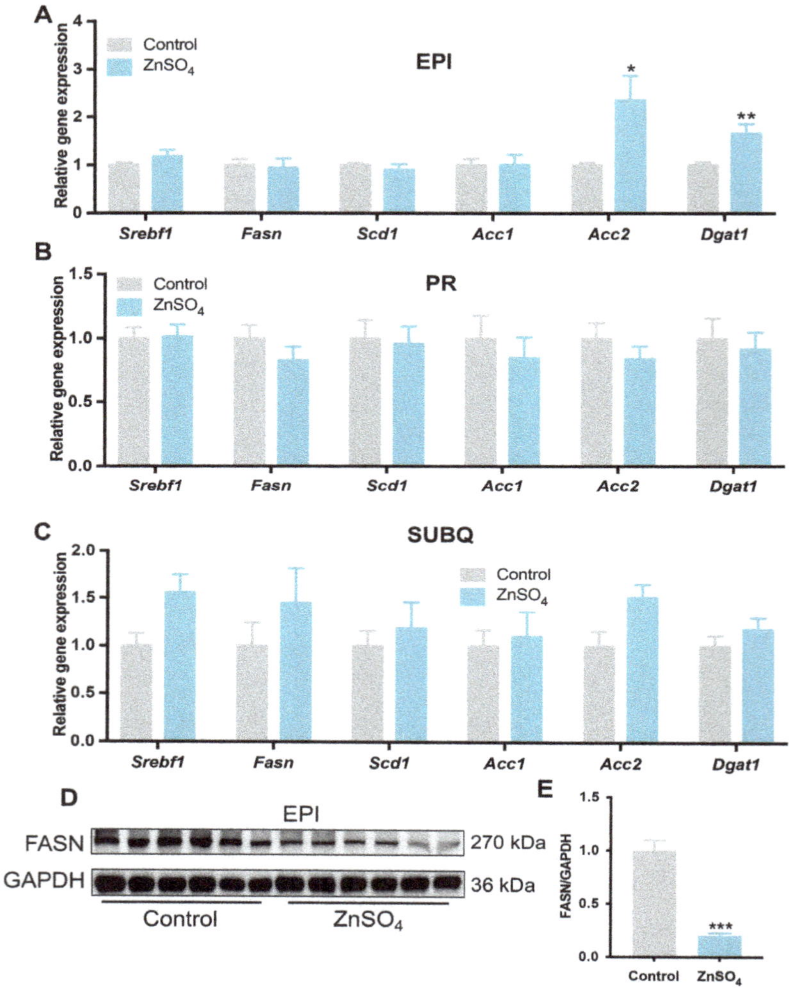

Figure 5. Lipid anabolism genes and protein expression in adipose tissues. The expression level of lipogenesis genes in the epididymal (**A**), perirenal (**B**), and subcutaneous (**C**) adipose tissues. (**D**,**E**) The protein levels of FASN in the epididymal adipose tissue (N = 6 for each group). Data are expressed as Mean ± SE. $ZnSO_4$, 30 ppm zinc supplemented group. * $p < 0.05$, ** $p < 0.01$, *** $p < 0.001$.

3.5. Short-Term Zinc Supplementation Promoted the Expressions of Lipolytic Genes and Proteins in the Epididymal Adipose Tissue

Next, the expressions of fat catabolism-related genes were analyzed in the visceral adipose tissue. The results suggested that the mRNA levels of the main enzymes related to lipid catabolism, including adipose triglyceride lipase (*Atgl*), hormone-sensitive lipase (*Hsl*), and *Pparg* were significantly higher in the epididymal adipose tissue of zinc-supplemented mice compared with the control mice; although no change was observed in the expression of monoglyceride lipase (*Mgll*), lipid droplet-associated protein (*Plin*), and apolipoprotein b (*Apob*) between the two groups (Figure 6A). Furthermore, as compared with the control mice, the zinc-supplemented mice had higher expression levels of *Hsl*, *Atgl*, *Mgll*, and *Pparg* in the subcutaneous adipose tissue, while the expression level of *Hsl* in the perirenal adipose tissue was decreased by short-term zinc supplementation, compared with the control mice (Figure 6B,C). Furthermore, the protein level of PPARg and phosphorylation level of HSL in the epididymal adipose tissue of zinc-supplemented mice were increased compared with those in the control mice, while the phosphorylation level of AMPKa tended to be decreased (Figure 6D–G). These data suggest that short-term zinc supplementation could promote lipolysis in epididymal and subcutaneous adipose tissues.

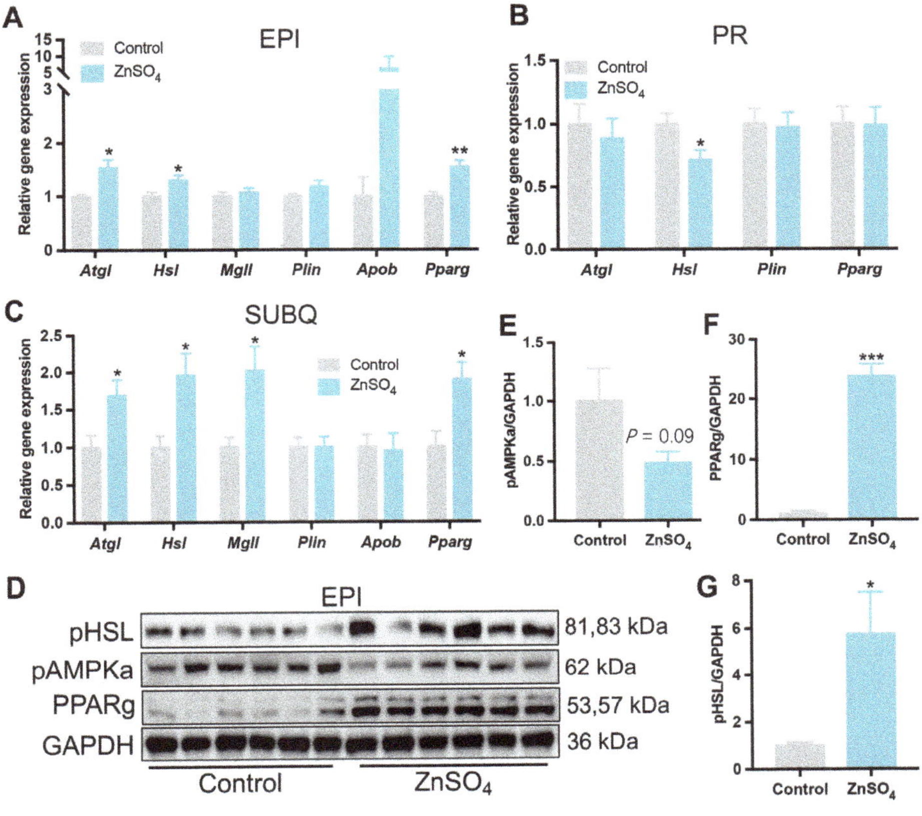

Figure 6. Expression of genes and proteins related to lipid metabolism in adipose tissues. (**A–C**) The expression levels of lipid metabolism-related genes in the epididymal (**A**), perirenal (**B**), and subcutaneous (**C**) adipose tissues. (**D–G**) Phosphorylation levels of HSL and AMPKa, and protein levels of PPARg in the epididymal adipose tissue (N = 6 for each group). Data are expressed as mean ± SE. $ZnSO_4$, 30 ppm zinc supplemented group. * $p < 0.05$, ** $p < 0.01$, *** $p < 0.001$.

3.6. Short-Term Zinc Supplementation Increased the Expressions of Inflammatory Genes in Adipose Tissue

Previous studies showed that long-term zinc supplementation induced the expression of inflammatory genes in adipose tissue [20]. Here, the expressions of marker genes for inflammatory cytokines and macrophages were analyzed in the adipose tissue. The results showed that short-term zinc supplementation increased the expression of *Tnfa* and *Il6* in both epididymal and perirenal adipose tissues compared with the control group (Figure 7A,B). The mRNA levels of *Mcp1* and *F4/80* were significantly increased in both epididymal and subcutaneous adipose tissues of zinc-supplemented mice as compared to the control mice (Figure 7A,C). However, the expression of *Cd11c* was not changed by zinc supplementation in either of the adipose tissues (Figure 7A–C). These data suggest that short-term zinc supplementation might induce inflammation in adipose tissue.

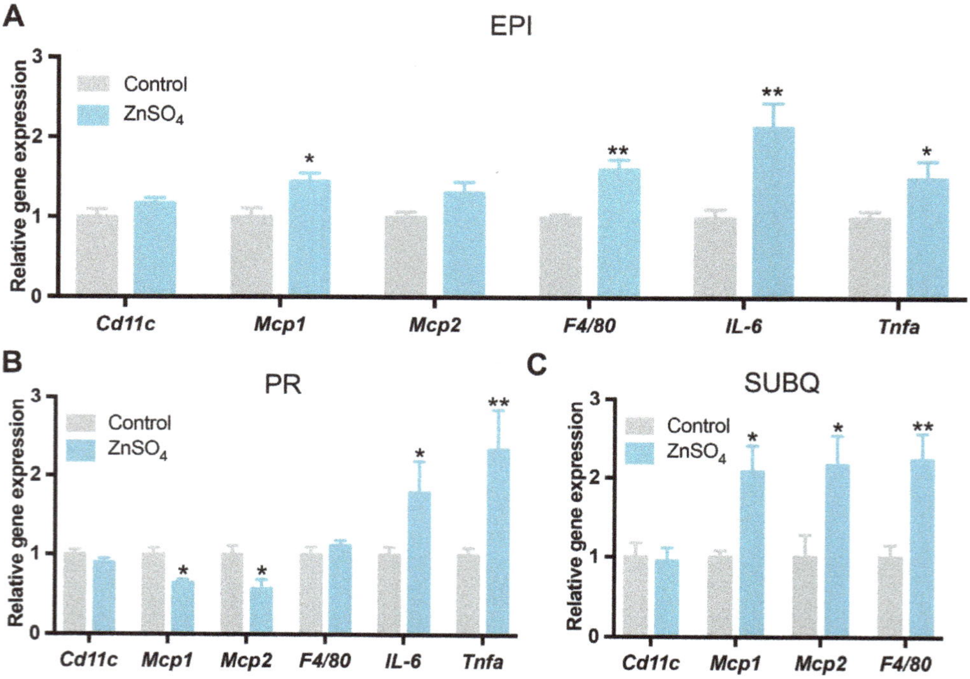

Figure 7. Expression of inflammatory genes in adipose tissues. (**A–C**) The expression level of inflammatory genes in the epididymal (**A**), perirenal (**B**), and subcutaneous (**C**) adipose tissues (N = 6 for each group). Data are expressed as mean ± SE. $ZnSO_4$, 30 ppm zinc supplemented group. * $p < 0.05$, ** $p < 0.01$.

4. Discussion

Micronutrient deficiency is harmful to health, such as impairing growth, increasing the risk of acute infectious disease, causing birth defects, and even death [1,2]. On the other hand, appropriate supplementation of microelement benefits human health [22]. The current study reported that short-term dietary zinc supplementation could regulate lipid metabolism in animals. Firstly, short-term zinc-supplemented mice exhibited lower adipose tissue weight and adipocyte size than the control mice. In addition, short-term zinc supplementation significantly increased blood glucose level and serum insulin concentration, while decreasing the serum NEFA content. Furthermore, the expression of lipolytic genes *Atgl* and *Hsl* were higher in the epididymal and subcutaneous adipose tissue of zinc-supplemented mice than those of the control mice. Meanwhile, the phosphorylation level of HSL and protein of PPARg were increased in the epididymal adipose tissue of zinc-supplemented mice compared with the control mice.

Previous studies found that chronic-zinc supplementation (30 ppm zinc in drinking water for 21 weeks) did not change body weight or food intake, but increased fat deposition in mice [20]. The current study showed that short-term zinc supplementation (30 ppm or 90 ppm zinc in drinking water for one week) did not change the food intake, water consumption, or body weight, but decreased fat deposition. Thus, the decreased adipose tissue weight and adipocyte size seemed not to depend on the feed intake in the current study. In another study with male C57BL/6J mice, zinc supplementation (60 ppm in diet) for 8 weeks did not change the body weight gain or fat weight index under either a normal-chow diet or a high-fat diet [23]. Furthermore, when prepubertal male Swiss mice (5-week-old) were fed with zinc-supplemented diet (25, 50, and 100 mg/kg) for 60 days, 50 and 100 mg/kg zinc-supplemented mice had higher body weight than the control mice [24]. The study by Rech et al. indicated that zinc supplementation (270 mg/kg zinc in the diet for 8 weeks) in Zucker diabetic fatty (ZDF) rats did not alter body weight, epididymal, or visceral fat weight [10]. The different animal models and different treatment durations between different studies might be the reason for the inconsistent results on body weight and fat weight under zinc supplementation. A recent study by Jiang et al. showed that thermogenic adipocytes could secret zinc ion to promote thermogenesis to reduce fat deposition [25]. Another study showed that a high-zinc diet-induced hepatic lipolysis [26]. Studies on humans showed that obese individuals had lower serum zinc concentrations than normal controls [27–29]. Thus, in clinic intermittent zinc supplementation might be applied to treat obese men and to reduce fat deposition. However, this should be further explored.

Insulin plays a pivotal role in maintaining glycemia. An increase in blood glucose levels prompts pancreatic islet cells to secrete more insulin, subsequently activating the PI3K/AKT signaling pathway and enhancing the uptake of glucose by adipose tissue and muscle, ultimately resulting in a reduction in the blood glucose level. However, when insulin sensitivity in peripheral tissues is compromised, the hypoglycemic effect of insulin is diminished, giving rise to insulin resistance and hyperinsulinemia [30]. Herein, we showed that zinc-supplemented mice exhibited higher blood glucose levels and serum insulin concentration compared to control mice. Further investigations revealed that short-term dietary zinc intake reduced the phosphorylation level of AKT and inhibited the expression of *Glut4* in the epididymal adipose tissue. GLUT4 plays a key role in maintaining whole-body glucose homeostasis in the adipose tissues and muscle [31,32]. The diminished GLUT4 delivery to the cell surface in the adipose tissue and muscle is the cause of the impairment of insulin-stimulated glucose transport [33,34]. Thus, the increase of blood glucose and insulin levels might be due to the impaired AKT signaling and the expression of *Glut4* in the epididymal adipose tissue of zinc-supplemented mice. Moreover, previous studies revealed that chronic high dose zinc supplementation diminished systemic insulin sensitivity and impaired AKT signaling in the perirenal adipose tissue [20]. Although variations existed in the duration of zinc treatment, these two studies indicated that both long-term and short-term zinc supplementation could impair AKT signaling in adipose tissue, through zinc-enhanced insulin sensitivity in vitro [20]. The increased expression of inflammatory genes in adipose tissue might be the reason for the impairment of insulin sensitivity. It is interesting to figure out the effect of intermittent zinc supplementation on insulin sensitivity and inflammatory gene expression in adipose tissue, which will be explored in future studies.

Fat deposition in the adipose tissue is regulated by lipolysis and lipogenesis [19]. In this study, short-term dietary zinc supplementation reduced adipose tissue weight. It was reported that FASN is the key enzyme for lipogenesis [35]. Here, the protein level of FASN was down-regulated in the epididymal adipose tissue of the zinc-supplemented mice, while the gene expression of *Fasn* was not changed. Previous studies have revealed that chronic high-dose zinc supplementation significantly decreased the protein expression of FASN in the perirenal adipose tissue [20]. Thus, both short-term and long-term zinc supplementation might suppress lipogenesis by reducing the protein expression of FASN in the adipose tissue. However, the exact mechanism for high-dose zinc suppression of FASN protein expression needs further study to be illustrated.

In the case of nutrient scarcity or increased energy requirements, animals undergo lipolysis, which breakdown TAG into NEFAs and glycerol in adipose tissue. ATGL and HSL are the major enzyme for lipolysis in the adipose tissue [36]. In the current study, the mRNA levels of *Atgl* and *Hsl* were significantly increased in the epididymal and subcutaneous adipose tissues of mice receiving short-term dietary zinc supplementation compared to the control mice. These genes are responsible for the central step of TAG lipolysis in the adipose tissue [37]. In this study, the phosphorylation level of HSL was upregulated in the epididymal adipose tissue of zinc-supplemented mice. Consequently, the decreased adipose deposition is probably dependent on lipolysis.

The current study also showed that short-term dietary zinc supplementation decreased the serum NEFA level. NEFA can be used for oxidation to meet the energy requirement of cells [38]. TAG is degraded into NEFA through lipolysis, and a proportion of circulating NEFA originates from visceral adipose tissue. CD36 plays an important role in both fatty acid uptake and the release in adipocytes [39]. Here, the mRNA levels of *Cd36* were decreased in the epididymal, perirenal, and subcutaneous adipose tissues of zinc-supplemented mice. At the same time, the expression of lipolytic gene *Atgl* and *Hsl* were increased by zinc supplementation in epididymal and subcutaneous adipose tissues, and the phosphorylation level of HSL in epididymal adipose tissue was increased, which means an increase in lipolysis in the adipose tissue by zinc supplementation. However, the plasma concentration of NEFA was decreased by zinc supplementation. Thus, short-term zinc supplementation might impair NEFA release by reducing the expression of *Cd36* in adipocytes, which could induce the accumulation of NEFA in adipocytes.

TNFa and IL6 are the main inflammatory factors in the adipose tissue [40]. In the current study, the expression of *Il6* and *Tnfa* was elevated in the visceral adipose tissue of zinc-supplemented mice, suggesting that short-term zinc supplementation may induce inflammation in adipose tissue, which might be a reason for the attenuation of AKT signaling in the epididymal adipose tissue of zinc-supplemented mice. Moreover, we also observed that short-term zinc supplementation stimulated the expression of *F4/80* and *Mcp1* in the epididymal adipose tissue. It has been reported that MCP1 is involved in the recruitment of macrophages in adipose tissue [41]. Macrophage infiltration into adipose tissue also contributes to adipose inflammation and insulin resistance [42]. Thus, the increased inflammation in zinc-supplemented adipose tissue might be due to the accumulation of NEFA in adipocytes and the infiltration of macrophages.

5. Conclusions

In summary, this study revealed that short-term zinc supplementation might reduce fat deposition by enhancing lipolysis in mice. Future studies could focus on the effect of intermittent zinc supplementation on fat reduction in both animal models and humans.

Author Contributions: B.F. and X.H. conceived and designed the experiments; X.H., D.J., Y.Z. and Z.F. performed the experiments; X.H. and B.F. analyzed the data; X.H. wrote the paper; B.F. revised the manuscript. All authors have read and agreed to the published version of the manuscript.

Funding: This study was supported by the National Natural Science Foundation of China (32272893) and the Sichuan Provincial Science and Technology Plan Project (2021NZZJ0016).

Institutional Review Board Statement: The study was conducted in accordance the Animal Care and Use Committee of Sichuan Agricultural University (protocol SICAU-2021-078, 8 March 2021).

Data Availability Statement: Data are available on request from the corresponding author.

Conflicts of Interest: The authors declare that they have no competing interests.

References

1. Stevens, G.A.; Beal, T.; Mbuya, M.N.N.; Luo, H.; Neufeld, L.M.; Global Micronutrient Deficiencies Research, G. Micronutrient deficiencies among preschool-aged children and women of reproductive age worldwide: A pooled analysis of individual-level data from population-representative surveys. *Lancet Glob. Health* **2022**, *10*, e1590–e1599. [CrossRef] [PubMed]
2. Black, R.E.; Victora, C.G.; Walker, S.P.; Bhutta, Z.A.; Christian, P.; de Onis, M.; Ezzati, M.; Grantham-McGregor, S.; Katz, J.; Martorell, R.; et al. Maternal and child undernutrition and overweight in low-income and middle-income countries. *Lancet* **2013**, *382*, 427–451. [CrossRef] [PubMed]
3. Wells, J.C.; Sawaya, A.L.; Wibaek, R.; Mwangome, M.; Poullas, M.S.; Yajnik, C.S.; Demaio, A. The double burden of malnutrition: Aetiological pathways and consequences for health. *Lancet* **2020**, *395*, 75–88. [CrossRef]
4. Foster, M.; Chu, A.; Petocz, P.; Samman, S. Effect of vegetarian diets on zinc status: A systematic review and meta-analysis of studies in humans. *J. Sci. Food Agric.* **2013**, *93*, 2362–2371. [CrossRef]
5. Bhutta, Z.A.; Das, J.K.; Rizvi, A.; Gaffey, M.F.; Walker, N.; Horton, S.; Webb, P.; Lartey, A.; Black, R.E.; The Lancet Nutrition Interventions Review Group; et al. Evidence-based interventions for improvement of maternal and child nutrition: What can be done and at what cost? *Lancet* **2013**, *382*, 452–477. [CrossRef] [PubMed]
6. Tam, E.; Keats, E.C.; Rind, F.; Das, J.K.; Bhutta, A.Z.A. Micronutrient supplementation and fortification interventions on health and development outcomes among children under-five in low- and middle-income countries: A systematic review and meta-analysis. *Nutrients* **2020**, *12*, 289. [CrossRef]
7. Prasad, A.S. Discovery of human zinc deficiency: Its impact on human health and disease. *Adv. Nutr.* **2013**, *4*, 176–190. [CrossRef]
8. Wang, X.; Wu, W.; Zheng, W.; Fang, X.; Chen, L.; Rink, L.; Min, J.; Wang, F. Zinc supplementation improves glycemic control for diabetes prevention and management: A systematic review and meta-analysis of randomized controlled trials. *Am. J. Clin. Nutr.* **2019**, *110*, 76–90. [CrossRef]
9. Rondanelli, M.; Miccono, A.; Lamburghini, S.; Avanzato, I.; Riva, A.; Allegrini, P.; Faliva, M.A.; Peroni, G.; Nichetti, M.; Perna, S. Self-Care for Common Colds: The Pivotal Role of Vitamin D, Vitamin C, Zinc, and Echinacea in Three Main Immune Interactive Clusters (Physical Barriers, Innate and Adaptive Immunity) Involved during an Episode of Common Colds-Practical Advice on Dosages and on the Time to Take These Nutrients/Botanicals in order to Prevent or Treat Common Colds. *Evid.-Based Complement. Altern. Med.* **2018**, *2018*, 5813095.
10. Rech, L.; Zahradka, P.; Taylor, C.G. Marginal Zinc Deficiency Promotes Pancreatic Islet Enlargement While Zinc Supplementation Improves the Pancreatic Insulin Response in Zucker Diabetic Fatty Rats. *Nutrients* **2024**, *16*, 1819. [CrossRef]
11. Mendes Garrido Abregu, F.; Caniffi, C.; Arranz, C.T.; Tomat, A.L. Impact of Zinc Deficiency During Prenatal and/or Postnatal Life on Cardiovascular and Metabolic Diseases: Experimental and Clinical Evidence. *Adv. Nutr.* **2022**, *13*, 833–845. [CrossRef] [PubMed]
12. Aslam, M.; Bashir, S.; Zeb, A. Effect of zinc supplementation on blood glucose level in different age groups of diabetes type 2. *Nutr. Health* **2023**, *29*, 599–605. [CrossRef] [PubMed]
13. Beloucif, A.; Kechrid, Z.; Bekada, A.M.A. Effect of zinc deficiency on blood glucose, lipid profile, and antioxidant status in streptozotocin diabetic rats and the potential role of sesame oil. *Biol. Trace Elem. Res.* **2022**, *200*, 3236–3247. [CrossRef] [PubMed]
14. Liu, M.J.; Bao, S.; Bolin, E.R.; Burris, D.L.; Xu, X.; Sun, Q.; Killilea, D.W.; Shen, Q.; Ziouzenkova, O.; Belury, M.A.; et al. Zinc deficiency augments leptin production and exacerbates macrophage infiltration into adipose tissue in mice fed a high-fat diet. *J. Nutr.* **2013**, *143*, 1036–1045. [CrossRef]
15. Saunders, A.V.; Craig, W.J.; Baines, S.K. Zinc and vegetarian diets. *Med. J. Aust.* **2013**, *199*, S17–S21. [CrossRef]
16. Perry, R.J.; Samuel, V.T.; Petersen, K.F.; Shulman, G.I. The role of hepatic lipids in hepatic insulin resistance and type 2 diabetes. *Nature* **2014**, *510*, 84–91. [CrossRef]
17. Sun, K.; Kusminski, C.M.; Scherer, P.E. Adipose tissue remodeling and obesity. *J. Clin. Investig.* **2011**, *121*, 2094–2101. [CrossRef]
18. Ferre, P.; Foufelle, F. Hepatic steatosis: A role for de novo lipogenesis and the transcription factor SREBP-1c. *Diabetes Obes. Metab.* **2010**, *12* (Suppl. S2), 83–92. [CrossRef]
19. McTernan, P.G.; Harte, A.L.; Anderson, L.A.; Green, A.; Smith, S.A.; Holder, J.C.; Barnett, A.H.; Eggo, M.C.; Kumar, S. Insulin and rosiglitazone regulation of lipolysis and lipogenesis in human adipose tissue in vitro. *Diabetes* **2002**, *51*, 1493–1498. [CrossRef]
20. Huang, X.; Jiang, D.; Zhu, Y.; Fang, Z.; Che, L.; Lin, Y.; Xu, S.; Li, J.; Huang, C.; Zou, Y.; et al. Chronic high dose zinc supplementation induces visceral adipose tissue hypertrophy without altering body weight in mice. *Nutrients* **2017**, *9*, 1138. [CrossRef]
21. Huang, X.; He, Q.; Zhu, H.; Fang, Z.; Che, L.; Lin, Y.; Xu, S.; Zhuo, Y.; Hua, L.; Wang, J.; et al. Hepatic leptin signaling improves hyperglycemia by stimulating MAPK Phosphatase-3 protein degradation via STAT3. *Cell. Mol. Gastroenterol. Hepatol.* **2022**, *14*, 983–1001. [CrossRef] [PubMed]
22. Smith, E.R.; Shankar, A.H.; Wu, L.S.; Aboud, S.; Adu-Afarwuah, S.; Ali, H.; Agustina, R.; Arifeen, S.; Ashorn, P.; Bhutta, Z.A.; et al. Modifiers of the effect of maternal multiple micronutrient supplementation on stillbirth, birth outcomes, and infant mortality: A meta-analysis of individual patient data from 17 randomised trials in low-income and middle-income countries. *Lancet Glob. Health* **2017**, *5*, e1090–e1100. [CrossRef] [PubMed]
23. Bolatimi, O.E.; Head, K.Z.; Luo, J.; Gripshover, T.C.; Lin, Q.; Adiele, N.V.; Watson, W.H.; Wilkerson, C.; Cai, L.; Cave, M.C.; et al. Can zinc supplementation attenuate high fat diet-induced non-alcoholic fatty liver disease? *Int. J. Mol. Sci.* **2023**, *24*, 1763. [CrossRef] [PubMed]

24. Onaolapo, O.J.; Jegede, O.R.; Adegoke, O.; Ayinde, M.O.; Akeredolu, O.M.; Onaolapo, A.Y. Dietary zinc supplement militates against ketamine-induced behaviours by age-dependent modulation of oxidative stress and acetylcholinesterase activity in mice. *Pharmacol. Rep.* **2020**, *72*, 55–66. [CrossRef] [PubMed]
25. Jiang, J.; Zhou, D.; Zhang, A.; Yu, W.; Du, L.; Yuan, H.; Zhang, C.; Wang, Z.; Jia, X.; Zhang, Z.N.; et al. Thermogenic adipocyte-derived zinc promotes sympathetic innervation in male mice. *Nat. Metab.* **2023**, *5*, 481–494. [CrossRef]
26. Xu, Y.C.; Zheng, H.; Hogstrand, C.; Tan, X.Y.; Zhao, T.; Song, Y.F.; Wei, X.L.; Wu, L.X.; Luo, Z. Novel mechanism for zinc inducing hepatic lipolysis via the HDAC3-mediated deacetylation of beta-catenin at lysine 311. *J. Nutr. Biochem.* **2023**, *121*, 109429. [CrossRef]
27. Hernandez-Mendoza, H.; Martinez-Navarro, I.; Hernandez-Ochoa, E.; Espinoza-Ruiz, M.; Lugo-Trampe, A.; Trujillo-Murillo, K.D.C.; Lopez-Garcia, M.A.; Rios-Lugo, M.J.; Chang-Rueda, C. Serum zinc levels are associated with obesity and low-density lipoprotein cholesterol in Mexican adults. *J. Trace Elem. Med. Biol.* **2022**, *73*, 127002. [CrossRef]
28. Gu, K.; Xiang, W.; Zhang, Y.; Sun, K.; Jiang, X. The association between serum zinc level and overweight/obesity: A meta-analysis. *Eur. J. Nutr.* **2019**, *58*, 2971–2982. [CrossRef]
29. Rios-Lugo, M.J.; Madrigal-Arellano, C.; Gaytan-Hernandez, D.; Hernandez-Mendoza, H.; Romero-Guzman, E.T. Association of Serum Zinc Levels in Overweight and Obesity. *Biol. Trace Elem. Res.* **2020**, *198*, 51–57. [CrossRef]
30. Samuel, V.T.; Shulman, G.I. Mechanisms for insulin resistance: Common threads and missing links. *Cell* **2012**, *148*, 852–871. [CrossRef]
31. Abel, E.D.; Peroni, O.; Kim, J.K.; Kim, Y.B.; Boss, O.; Hadro, E.; Minnemann, T.; Shulman, G.I.; Kahn, B.B. Adipose-selective targeting of the GLUT4 gene impairs insulin action in muscle and liver. *Nature* **2001**, *409*, 729–733. [CrossRef] [PubMed]
32. Zisman, A.; Peroni, O.D.; Abel, E.D.; Michael, M.D.; Mauvais-Jarvis, F.; Lowell, B.B.; Wojtaszewski, J.F.; Hirshman, M.F.; Virkamaki, A.; Goodyear, L.J.; et al. Targeted disruption of the glucose transporter 4 selectively in muscle causes insulin resistance and glucose intolerance. *Nat. Med.* **2000**, *6*, 924–928. [CrossRef] [PubMed]
33. Garvey, W.T.; Maianu, L.; Zhu, J.H.; Brechtel-Hook, G.; Wallace, P.; Baron, A.D. Evidence for defects in the trafficking and translocation of GLUT4 glucose transporters in skeletal muscle as a cause of human insulin resistance. *J. Clin. Investig.* **1998**, *101*, 2377–2386. [CrossRef] [PubMed]
34. Garvey, W.T.; Maianu, L.; Zhu, J.H.; Hancock, J.A.; Golichowski, A.M. Multiple defects in the adipocyte glucose transport system cause cellular insulin resistance in gestational diabetes. Heterogeneity in the number and a novel abnormality in subcellular localization of GLUT4 glucose transporters. *Diabetes* **1993**, *42*, 1773–1785. [CrossRef]
35. Maan, M.; Peters, J.M.; Dutta, M.; Patterson, A.D. Lipid metabolism and lipophagy in cancer. *Biochem. Biophys. Res. Commun.* **2018**, *504*, 582–589. [CrossRef]
36. Morak, M.; Schmidinger, H.; Riesenhuber, G.; Rechberger, G.N.; Kollroser, M.; Haemmerle, G.; Zechner, R.; Kronenberg, F.; Hermetter, A. Adipose triglyceride lipase (ATGL) and hormone-sensitive lipase (HSL) deficiencies affect expression of lipolytic activities in mouse adipose tissues. *Mol. Cell. Proteom.* **2012**, *11*, 1777–1789. [CrossRef]
37. Egan, J.J.; Greenberg, A.S.; Chang, M.K.; Wek, S.A.; Moos, M.C., Jr.; Londos, C. Mechanism of hormone-stimulated lipolysis in adipocytes: Translocation of hormone-sensitive lipase to the lipid storage droplet. *Proc. Natl. Acad. Sci. USA* **1992**, *89*, 8537–8541. [CrossRef]
38. Snorgaard, O.; Poulsen, G.M.; Andersen, H.K.; Astrup, A. Systematic review and meta-analysis of dietary carbohydrate restriction in patients with type 2 diabetes. *BMJ Open Diabetes Res. Care* **2017**, *5*, e000354. [CrossRef]
39. Daquinag, A.C.; Gao, Z.; Fussell, C.; Immaraj, L.; Pasqualini, R.; Arap, W.; Akimzhanov, A.M.; Febbraio, M.; Kolonin, M.G. Fatty acid mobilization from adipose tissue is mediated by CD36 posttranslational modifications and intracellular trafficking. *JCI Insight* **2021**, *6*, e147057. [CrossRef]
40. Ahmed, B.; Sultana, R.; Greene, M.W. Adipose tissue and insulin resistance in obese. *Biomed. Pharmacother.* **2021**, *137*, 111315. [CrossRef]
41. Feng, B.; Jiao, P.; Nie, Y.; Kim, T.; Jun, D.; van Rooijen, N.; Yang, Z.; Xu, H. Clodronate liposomes improve metabolic profile and reduce visceral adipose macrophage content in diet-induced obese mice. *PLoS ONE* **2011**, *6*, e24358. [CrossRef] [PubMed]
42. Xu, H.; Barnes, G.T.; Yang, Q.; Tan, G.; Yang, D.; Chou, C.J.; Sole, J.; Nichols, A.; Ross, J.S.; Tartaglia, L.A.; et al. Chronic inflammation in fat plays a crucial role in the development of obesity-related insulin resistance. *J. Clin. Investig.* **2003**, *112*, 1821–1830. [CrossRef] [PubMed]

nutrients

MDPI

Article

Nutritional Counseling and Mediterranean Diet in Adrenoleukodystrophy: A Real-Life Experience

Maria Rita Spreghini [1], Nicoletta Gianni [2], Tommaso Todisco [1,3], Cristiano Rizzo [4], Marco Cappa [5] and Melania Manco [2,*]

1 UOC of Endocrinology and Diabetology, Bambino Gesù Children's Hospital, Istituti di Ricovero e Cura a Carattere Scientifico, IRCCS, 00165 Rome, Italy; mariarita.spreghini@opbg.net (M.R.S.); tommaso.todisco@chuv.ch (T.T.)

2 Research Unit for Predictive and Preventive Medicine, Bambino Gesù Children's Hospital, Istituti di Ricovero e Cura a Carattere Scientifico, IRCCS, 00165 Rome, Italy; nicoletta.gianni@opbg.net

3 Service of Endocrinology, Diabetology and Metabolism, Lausanne University Hospital, 1011 Lausanne, Switzerland

4 UOC of Metabolic Diseases, Bambino Gesù Children's Hospital, Istituti di Ricovero e Cura a Carattere Scientifico, IRCCS, 00165 Rome, Italy; cristiano.rizzo@opbg.net

5 Research Unit for Innovative Therapies for Endocrinopathies, Bambino Gesù Children's Hospital, Istituti di Ricovero e Cura a Carattere Scientifico, IRCCS, 00165 Rome, Italy; marco.cappa@opbg.net

* Correspondence: melania.manco@opbg.net

Abstract: Background/Objectives: Adrenoleukodystrophy (X-ALD) is a metabolic disorder caused by dysfunctional peroxisomal beta-oxidation of very-long-chain fatty acids (VLCFAs). A VLCFA-restricted Mediterranean diet has been proposed for patients and carriers to reduce daily VLCFA intake. **Methods**: We retrospectively evaluated plasma VLCFAs in a cohort of 36 patients and 20 carriers at baseline and after 1 year of restricted diet. **Results**: At T1, compliant adult patients had significantly lower C26:0 levels [1.7 (1.2) vs. 2.5 μmol/L (1.7), $p < 0.05$], C26:0/C22:0 ratio [0.04 (0.02) vs. 0.06 (0.03), $p < 0.05$], and triglycerides [93 (56.5) vs. 128 mg/dL (109.5), $p < 0.05$] than non-compliant ones. C26:0 [2.4 (1.7) vs. 1.7 (1.2) μmol/L, $p < 0.05$], the C26:0/C22:0 ratio [0.06 (0.04) vs. 0.04 (0.02), $p < 0.05$], and cholesterol [173.5 (68.3) mg/dL vs. 157 (54) mg/dL, $p < 0.05$] were significantly reduced in compliant adult patients at T1 vs. baseline. As for carriers, the C26:0/C22:0 ratio was lower [0.02 (0.01) vs. 0.04 (0.009), $p < 0.05$] at T1 in compliant carriers, as compared to non-compliant ones. The C26:0/C22:0 [0.03 (0.02) vs. 0.02 (0.01) $p < 0.05$] and C24:0/C22:0 [1.0 (0.2) vs. 0.9 (0.3), $p < 0.05$] ratios were significantly decreased at T1 vs. T0. **Conclusions:** A VLCFA-restricted diet is effective in reducing plasma VLCFA levels and their ratios and must be strongly encouraged as support to therapy.

Keywords: adrenoleukodystrophy; X-ALD; hexacosanoic acid; C26:0; Mediterranean diet; nutritional therapy; VLCFA-restricted diet; very-long-chain fatty acids; VLCFAs

1. Introduction

Adrenoleukodystrophy (X-ALD, 300100) [1], a rare metabolic disorder with estimated incidence at 1/20,000 [2], causes dysfunction in peroxisomal beta-oxidation and, consequently, increased plasma levels of very-long-chain fatty acids (VLCFAs, longer than C22). X-ALD is determined by a mutation in the ABCD1 gene, on chromosome X (Xq28), encoding Adrenoleukodystrophy Protein (ALDP), a member of the ATP-binding cassette transport family, carrying VLCFAs into peroxisomes. More than 600 mutations have been identified to date, thus partially explaining variability in the clinical manifestation of the disease. VLCFAs accumulate in the central nervous system, adrenal cortex, and gonads, leading to adrenal dysfunction and central and peripheral demyelination [3,4].

The clinical features of X-ALD phenotypes are described in Supplementary Table S1; the early onset of pediatric X-linked Cerebral ALD (CerALD) and Adrenomyeloneuropathy

(AMN) are the prevalent phenotypes, accounting for 48% and 25% of patients, respectively. Male patients carrying a gene defect without clinical signs are referred to as presymptomatic individuals, being at increased risk of developing the overt disease [3,5–9]. Women, who are gene carriers, are symptomatic after the fourth decade with a mild form of the disease, presenting ataxia, leg pain, incontinence, mild myelopathy, or, rarely, severe myelopathy [3,4,9,10].

The loss of function of ALDP correlates with high plasma levels of VLCFAs in the brain and adrenal glands, as revealed by postmortem analysis [3,5,6]. VLCFAs are esterified to cholesterol, glycerophospholipids, phospholipids, and sphingolipids, likely triggering an impairment in cerebral white matter and changes in the integrity of myelin, with demyelination and a reduction in cerebrosides, and moreover determining an increased inflammation and oxidative stress, owing to increased levels of reactive oxygen species. Dyslipidemia with high levels of total cholesterol is due to the dysfunctional transport and metabolism of cholesterol esterified to VLCFAs and to the reduced availability of cholesterol for the synthesis of steroid hormones. In adrenal glands, the excess of VLCFAs determines an alteration of membrane microviscosity, leading to an attenuated response to the adrenocorticotropic hormone and hypoadrenalism [4,6,7,11].

Only a small fraction of VLCFAs derive from the diet, while the majority are synthetized endogenously by the elongase enzyme ELOVL1 (ELOngation of Very-Long-chain fatty acids-1) [6]. Both C26:0 (hexacosanoic acid) and C24:0 (tetracosanoic acid), and their ratios to C22:0 (docosanoic acid), are biomarkers for suspected disease. Their measurements are recommended for the following groups: men with non-autoimmune hypoadrenalism, with or without gait and/or behavioral disturbances; men with behavioral disturbances and areas of demyelination identified by magnetic resonance imaging; men with progressive paraparesis; and women with idiopathic progressive paraparesis and/or positive family history. The suspected disorder should be confirmed by a genetic test that identifies the mutated gene [3,7].

Neonatal screening allows for a more accurate estimate of the prevalence and early identification [3,12], this being pivotal for therapeutic intervention, as well as a VLCFA-restricted diet to reduce deleterious effects and alleviate symptoms, delay the onset of symptoms, and improve the prognosis in pre-symptomatic patients and carriers.

In this retrospective study, we aimed at investigating the impact of a real-life nutritional scenario with the effect of reducing plasma VLCFA levels with a restricted Mediterranean diet (the "Bambino" diet), in patients with X-ALD and female carriers.

2. Materials and Methods

To evaluate the efficacy of the VLCFA-restricted nutritional treatment, we retrospectively reviewed clinical and biochemical data from Electronic Health Records (EHRs) of all the patients affected by either Cer-ALD or AMN, as well as Addison-only patients and female carriers, referred to the Unit of Endocrinology and Diabetology of the Bambino Gesù Children's Hospital (OPBG, Ospedale Pediatrico Bambino Gesù), from January 2019 to July 2023. We included 56 subjects in total, distributed as follows: 26 male adults, 10 male children, and 20 female adults.

Patients and carriers underwent a nutritional evaluation and were prescribed with the VLCFA-restricted Mediterranean diet (the "Bambino" VLCFA-restricted Mediterranean diet). Oil and butter could not be included in the restricted diet due to their high content of C26:0 and fats, and were replaced with Aldixyl OiLife, a special oil composed of pure oleic acid (Aldixyl OiLife; Pharmaelle, Bologna, Italy) and administered as needed, with a suggested dose of 1 spoon/day, as part of the nutritional treatment.

All the patients were prescribed with a food for special medical purposes, consisting of a mixture of trioleic and trierucic acids in a ratio of 4:1, conjugated linoleic acids, and antioxidants (Aldixyl; Pharmaelle, Bologna, Italy) [13,14]. The prescribed dose was 0.25 to 0.75 mL/kg/day (a lower dose, 15 to 20 mL/day, was usually administered in pediatric patients).

The study baseline (T0) was the date of the restricted diet prescription, and T1 the follow-up visit after 1 year. Plasma VLCFA levels (C26:0, C26:0/C22:0 ratio, and

C24:0/C22:0 ratio), total cholesterol, and triglycerides were measured at T0 and T1. Compliance to nutritional treatment was evaluated with a 24 h Recall, compiled by dietitians (MRS and NG) during the in-person follow-up visit and with a 7-day Food Diary, compiled by the patient.

2.1. Anthropometrics and Biochemical Evaluation

Height was measured using a wall Holtain-stadiometer, and weight was measured with scales certified for medical use (90/384/EEC, SECA, Hamburg, Germany). Patients wore minimal clothing and no shoes during these measurements. The average of two measurements was used to calculate the Body Mass Index (BMI). All measures were taken to ensure the confidentiality of participants whose data were used.

Blood samples were collected in EDTA tubes, and plasma was stored at −20 °C for up to six months. Plasma VLCFAs (from C22:0 to C26:0) were identified and quantified by gas chromatography/mass spectrometry, after extraction and derivatization to methyl esters, using a control plasma, certified ERNDIM (European Research Network for evaluation and improvement of screening, Diagnosis and treatment of Inherited disorders of Metabolism). Plasma C26:0 levels are expressed in µmol/L. C26:0/C22:0 ratio (hexacosanoic to docosanoic acid) and C24:0/C22:0 ratio (tetracosanoic to docosanoic acid) were then calculated, being crucial for therapeutic monitoring and diagnosis of X-ALD in carriers. The normal ranges are the following: hexacosanoic acid 0.010–0.900 µmol/L, C26:0/C22:0 ratio 0.006–0.020, and C24:0/C22:0 ratio 0.470–1.270 [15].

Total cholesterol and triglyceride levels were assessed using colorimetric kits (modular systems P/S Can 433; Roche Diagnostic/Hitachi, Monza, Italy).

2.2. The "Bambino" VLCFA-Restricted Mediterranean Diet

We specifically designed the "Bambino" diet to limit VLCFA intake to less than 3 mg/day (the lowest threshold identified in Van Duyn et al.'s study [16]), while minimizing total fatty acid consumption. Due to the lack of national or international reference data for C26:0 content in foods, we relied on C26:0 content values reported in the studies by Van Duyn et al. (1984) [16] and Kawahara et al. (1988) [17]. We set the C26:0 cut-off at 0.150 mg/100 g serving, excluding any foods with C26:0 content equal to or higher than this threshold [16,17]. Both studies referred to gross weight or baked weight, depending on the food.

We also set a maximum fat content limit of 2 g/100 g of the edible portion. We excluded foods with fat content equal to or higher than the cut-off, as reported in Italian or US databases [18–20]. For foods not listed in the CREA (Consiglio per la Ricerca in Agricoltura e l'Analisi dell'Economia Agraria, 2019) database, we referred to either the BDA (Banca Dati di Composizione degli Alimenti per Studi Epidemiologici in Italia, 2022) or the USDA (U.S. Department of Agriculture, 2019) databases, particularly for items like spices [18–20].

The food categories we analyzed included cereals, meat and meat-derived products, fish, legumes, eggs, milk and yogurt, dairy products, fruits, vegetables, spices, nuts and seeds, fried foods, pre-packaged and precooked foods, sweets, beverages, and seasonings. When discrepancies among the databases were observed, we considered the highest value and excluded the product if it met or exceeded the set limits.

In our food analysis, we adopted a non-brand-specific scenario. Many of the brand names referenced in the US and Japanese studies by Van Duyn et al. and Kawahara et al. [16,17] are not available in Italy. As a result, we considered only the available nutritional value, without considering or comparing the brand. Nutritional variations between Italian brands were not considered; instead, we used the average nutritional values from non-branded items as reported in food composition databases. Neither were differences in nutrient content due to agriculture, cultivation, and environmental factors considered.

After a thorough analysis, we selected foods based on the Mediterranean dietary pattern, which aligns with the eating habits of most of our patients and their families, as reported in their Food Diaries. We categorized these foods into three groups: non-allowed foods, foods with hidden fats, and allowed foods. We listed the foods in Supplementary Tables S2–S4. Among "non-allowed foods" group, we included both category products with elevated levels of C26:0 (≥0.150 mg/100 g) and the ones with high content of fats (≥ 2 g/100 g) (Supplementary Table S2). The "foods with hidden fats" category contains items that appear not to have high fat or VLCFA content, but are made from forbidden ingredients such as eggs, cheese, olive oil, milk, and some sausages (Supplementary Table S3). The "allowed foods" group consists of items all featuring low levels of fatty acids and/or low amount of VLCFAs, including non-whole-grain cereals, lean cuts of meat, certain types of fish, and legumes (peeled when possible). Foods made with skim milk (0% fat) are also included. Fruits, vegetables, spices, and beverages not mentioned in the other categories are permitted, with regard to the peel and seeds of plant foods, which must be removed before eating, as they can be sources of C26:0 [16,21] (Supplementary Table S4). Although white bread, rice, and other white flour products are relatively high in C26:0, we categorized them as "allowed foods" because they play an important role in the Mediterranean diet. By allowing their consumption, we ensure an adequate variety of dietary pattern and a balanced intake of all the nutrients, making the restricted diet more feasible for patients and carriers. Replacing these foods with alternative cereals or pseudocereals was not reasonable, as these alternatives contain higher levels of fatty acids.

2.3. Evaluation of Adherence to the Diet

Patients were provided with lists of allowed and non-allowed foods, along with general advice on consuming certain items, to ensure the best compliance to the diet. Rather than prescribing specific portions, patients were educated to make appropriate food choices based on their physiological needs and their hunger and satiety cues, conforming to the nutritional education therapy.

At the follow-ups, all patients received nutritional counseling. We assessed their adherence to the dietary protocol using two methods: the 24-h Recall and the Food Diary. The 24-h Recall is an assessment tool based on a short interview performed by the operator, in which the participants recall foods and drinks they consumed in the previous 24 h. The Food Diary is a 3- or 7-day assessment tool, self-administered by subjects, and records all meals eaten during that period. It is potentially affected by a minor risk of bias, such as inaccuracies due to false memories. Both tools aim at evaluating patients' eating habits and their compliance with the prescribed nutritional therapy.

2.4. Statistical Analysis

Data were reported as median and Interquartile Range (IQR). We used a non-parametric based-ranks test to assess the significant differences between groups. Comparisons were carried out using the Wilcoxon test. A p-value at 5% was considered statistically significant. Data analysis was performed using the R for Windows statistical software, version 3.0.3.

3. Results

A total of 78 EHRs were re-evaluated, with data from 56 subjects—36 affected males and 20 female carriers—ultimately included in the analysis. We excluded patients with severe dysphagia requiring enteral nutrition ($n = 8$) and those who did not complete the follow-up at T1 ($n = 14$). The studied cohort was aged from 4.8 to 72 years [men: 39.1 (22.3); children: 8.9 (8.1); carriers: 51.7 (14.3)]. At the follow-up (T1), participants were re-evaluated after a median period of 13 months (range: 8 to 33 months). Table 1 presents the anthropometric and biochemical measurements at the baseline (T0) and T1 for the three groups: adults, children, and female carriers.

Table 1. Median and IQR [median (IQR)] of plasma VLCFA levels (C26:0, C26:0/C22:0 ratio, C24:0/C22:0 ratio), total cholesterol and triglycerides, at T0 and T1, divided by compliant and non-compliant to nutritional treatment, for each sample group.

	Patients				Carriers	
	Children		Adults		Female Carriers	
	T0	*T1*	*T0*	*T1*	*T0*	*T1*
Age, Years Old	8.9 (8.1)		39.1 (22.3)		51.7 (14.3)	
Weight, kg	32.6 (21)	33.3 (18.4)	72.2 (13)	72 (7.1)	61.7 (20.4)	61.6 (17)
Height, cm	139.6 (32.4)	146.3 (33.3)	173 (10.6)	173 (11.1)	158 (5.5)	158 (4.9)
BMI, kg/m^2	17.6 (2.9)	17.3 (2.3)	23.5 (4.3)	23.8 (4.8)	24.7 (4.9)	22.3 (5.5)
C26:0, µmol/L	1.4 (0.9)	1.6 (0.7) vs. 1.5 (0.7)	2.4 (1.7)	1.7 (1.2) vs. 2.5 (1.7) * b ** b	1.4 (0.7)	1.2 (0.7) vs. 1.8 (0.2)
C26:0/C22:0 ratio	0.04 (0.03)	0.03 (0.03) (A) 0.04 (0.008) (B)	0.06 (0.04)	0.04 (0.02) (A) 0.06 (0.03) (B) * b ** e	0.03 (0.02)	0.02 (0.01) (A) 0.04 (0.009) (B) * d ** a
C24:0/C22:0 ratio	1.2 (0.2)	1.1 (0.08) (A) 1.2 (0.2) (B)	1.4 (0.3)	1.2 (0.5) (A) 1.4 (0.4) (B)	1.0 (0.2)	0.9 (0.3) (A) 1.2 (0.3) (B) ** b
Cholesterol, mg/dL	141.5 (32.8)	123 (29) (A) 149 (35) (B)	173.5 (68.3)	157 (54) (A) 193 (39) (B) ** b	191 (50.8)	183 (40.5) (A) 186.5 (12.8) (B)
Triglycerides, mg/dL	63 (19.8)	62 (81) (A) 70.5 (19.5) (B)	103 (63.8)	93 (56.5) (A) 128 (109.5) (B) * c	77.5 (37)	74 (48.3) (A) 84.5 (79.8) (B)
Adherence to Nutritional Treatment, %	50%		53.9%		80%	

(A): compliant to nutritional treatment; (B): non-compliant to nutritional treatment; * significant comparison between compliant and non-compliant at T1; ** significant comparison between compliant at T1 and T0. [a]: $p = 0.001$; C 26:0: hexacosanoic acid; [b]: $p = 0.01$; C26:0/C22:0 ratio: hexacosanoic acid/docosanoic acid ratio; [c]: $p = 0.02$; C24:0/C22:0 ratio: tetracosanoic acid/docosanoic acid ratio; [d]: $p = 0.03$; [e]: $p = 0.05$.

Fifty percent of the adult and child patients adhered the nutritional advice (fourteen out of twenty-six adults and five out of ten children), while in carriers, we observed a higher percentage of adherence ($n = 16$; 80%).

Total C26:0 [1.7 (1.2) vs. 2.5 µmol/L (1.7), $p < 0.05$], C26:0/C22:0 ratio [0.04 (0.02) vs. 0.06 (0.03), $p < 0.05$] and triglycerides [93 (56.5) vs. 128 mg/dL (109.5) mg/dL, $p < 0.05$] at T1 were significantly lower in compliant adult patients as compared to non-compliant ones (Figure 1A–C). Total C26:0 [2.4 (1.7) vs. 1.7 (1.2) $p < 0.05$], C26:0/C22:0 ratio [0.06 (0.04) vs. 0.04 (0.02) $p < 0.05$] and total cholesterol [173.5 mg/dL (68.3) vs. 157 mg/dL (54), $p < 0.05$] were significantly decreased at T1 as compared to baseline in compliant individuals (Figure 2A–C).

In the group of child patients, we did not find a significant difference in C26:0 levels or in the ratio following the nutritional intervention, regardless of the treatment compliance.

In the 20 carriers, there was a significant reduction in C26:0/C22:0 ratio [0.02 (0.01) vs. 0.04 (0.009), $p < 0.05$] at T1 in those who adhered to the nutritional treatment as compared to those who did not (Figure 3), and in the C26:0/C22:0 ratio [0.03 (0.02) vs. 0.02 (0.01), $p < 0.05$] and C24:0/C22:0 ratio [1.0 (0.2) vs. 0.9 (0.3), $p < 0.05$], comparing longitudinal values at T0 versus those at T1 (Figure 4A,B).

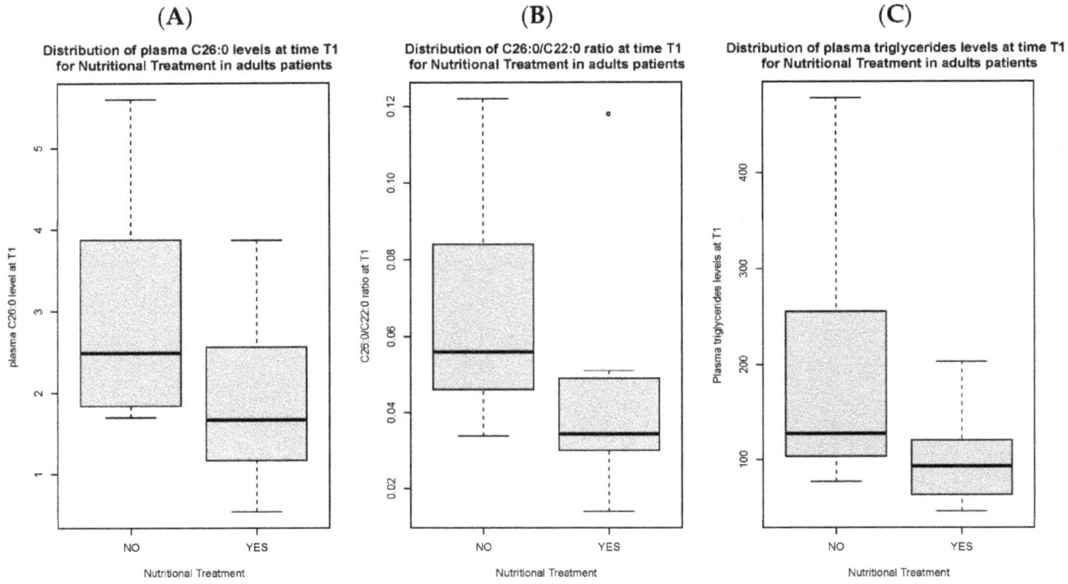

Figure 1. Distribution of C26:0 plasma levels (**A**), C26:0/C22:0 ratio (**B**) and triglycerides (**C**) at time T1 in adult patients compliant to nutritional treatment as compared to non-compliant patients. Figure 1B presents an outlier.

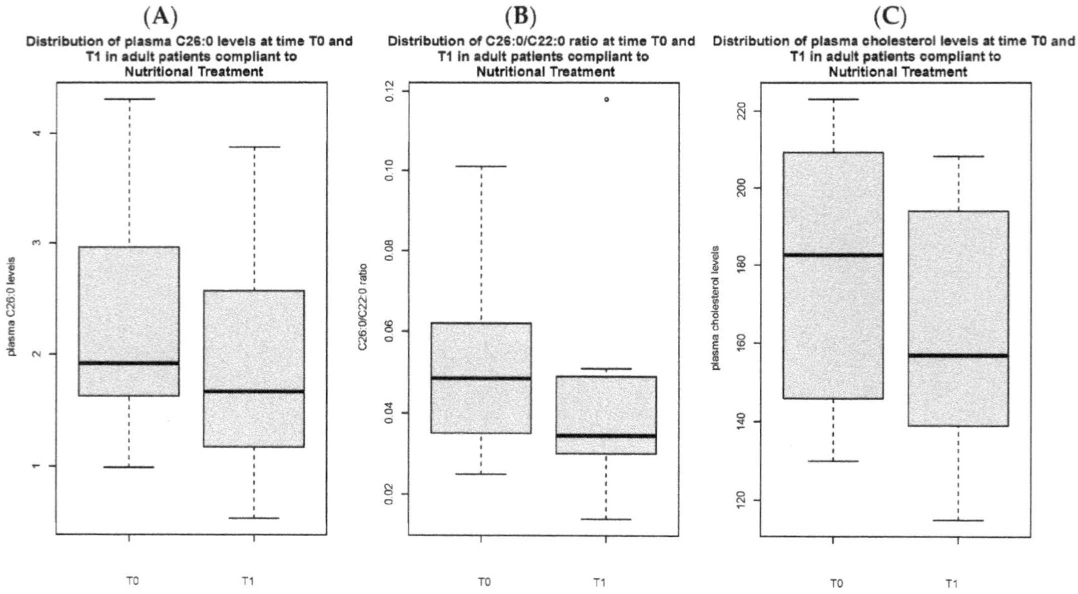

Figure 2. Distribution of C26:0 plasma levels (**A**), C26:0/C22:0 ratio (**B**) and cholesterol (**C**) in adult patients compliant to nutritional treatment, comparing T0 vs. T1. Figure 2B presents an outlier.

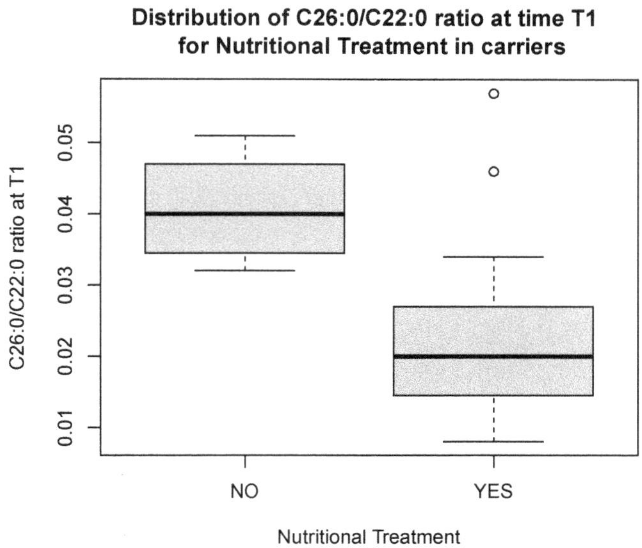

Figure 3. Distribution of C26:0/C22:0 ratio at time T1 in carriers compliant to nutritional treatment as compared to non-compliant carriers. The figure presents outliers.

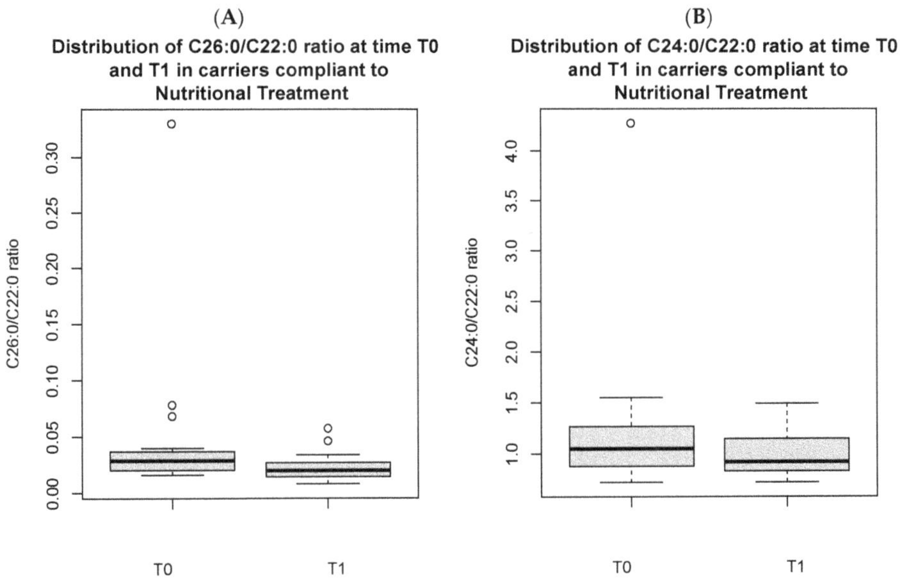

Figure 4. Distribution of C26:0/C22:0 ratio (**A**) and C24:0/C22:0 ratio (**B**) in carriers compliant to nutritional treatment, comparing T0 vs. T1. Both the figures present outliers.

Analysis of patients' and carriers' Food Diaries demonstrated a drastic average reduction in C26:0 intake from 12.1 mg/day to 2.5 mg/day, with the same caloric content of about 1400–1500 kcal/day distributed over five meals per day in those who adhered to the restricted diets as compared with those who did not adhere. Total fatty acid intake decreased from about 20% to about 6% with daily oil seasoning not included (Table 2).

Table 2. Example of restricted Mediterranean diet vs. unrestricted one.

Restricted Diet			Unrestricted Diet		
Meal	**Foods**	**C26:0 (mg)**	**Meal**	**Foods**	**C26:0 (mg)**
Breakfast	200 g Skim Milk	0.1 [16]	Breakfast	150 g Ordinary Liquid Milk	0.5715 [17]
	Sandwich Made with 50 g Italian Bread and 50 g Lean Ham	0.35 (Bread) [16] + 0.036 (Ham) [16]		30 g Cereals	0.066 [16]
Snack	Fruit: 150 g Peeled and Deseed Strawberries and 150 g Watermelon	0.096 (Strawberries) [16] + 0.24 (Watermelon) [16]	Snack	150 g Peaches w/peel + 30 g Sweet Almonds	0.1365 (Peaches) [16] + 0.7548 (Almonds) [17]
Lunch	100 g rice [22]	0.7506 [16]	Lunch	80 g Rice [22]	0.5994 [16]
	200 g Skinned Chicken Breast [22]	0.108 [16]		180 g Skinned Turkey Breast [22]	0.3212 [16]
	200 g p.d. Tomatoes	0.142 [16]		200 g Lettuce Hearts	0.74 [16]
Snack	150 g Skim Milk	0.075 [16]	Snack	125 g Yoghurt	2.0125 [17]
	150 g Peeled Apples	0.0675 [16]		150 g Banana	1.17 [16]
Dinner	120 g Egg Whites	0.0084 [16]	Dinner	200 g Fresh Cod [22]	0.96 [16]
	200 g Peeled Carrots	0.122 [16]		200 g Cabbage	4.38 [16]
	60 g Italian Bread	0.42 [16]		50 g Italian Bread	0.35 [16]
Total: 2.5 mg/day C26:0; 8.7 g/day total fatty acids (5.6%) [18]; 1404 kcal/day (daily seasoning oil not included) [18]			Total: 12.1 mg/day C26:0; 33.40 g/day total fatty acids (20.0%) [18];1.501 kcal/day (daily seasoning oil not included) [18]		

4. Discussion

The "Bambino" diet prescribed in our study is designed on a Mediterranean pattern, with special attention to those foods high in C26:0 and fats. While some of the pillars of the Mediterranean diet are missing in the "Bambino" diet, it maintains a considerable variety. It still incorporates a wide range of fruits, vegetables, and legumes, with only a few exceptions, as well as cereals. Notably, bread, rice, and products made from white flour are included, although having high levels of VLCFAs, thus aligning with the Mediterranean diet's principles and ensuring an adequate food variety and nutrient intake. Nevertheless, the consumption of extra virgin olive oil is excluded from the diet, being replaced by the X-ALD specifically formulated oil.

The "Bambino" diet restricted in VLCFAs leads to a significant improvement of biochemical markers of X-ALD disease, such as plasma C26:0 content and its ratios. A reduction in circulating levels of both biomarkers was evident in patients and carriers complying with the diet restrictions.

We considered "non-compliant" patients the ones declaring not to follow general advice and/or eating non-allowed foods and/or using common seasoning oils. Analysis of the Food Diaries and 24-h Recalls indicated good adherence to the diet, with nearly 62.5% of participants being compliant. However, among both children and adults, the lowest compliance was registered, with a major consumption of non-allowed foods (industrial and processed items, such as sweets, meat-derived products, sweetened beverages), while joining parties and social events. Thus, recreational events were the major obstacle to adherence.

In adult patients who adhered to the diet, we found lower triglyceride levels compared to their non-compliant counterparts at T1. While, in the adult male group, total cholesterol was lower at T1, no differences were found in total cholesterol and triglycerides levels in carriers.

Given that a small portion of VLCFAs originate from exogenous sources [8,16,17,23], the "Bambino" VLCFA-restricted diet serves as an adjunctive treatment. It results in the significant restriction of C26:0 intake (up to 2.5 mg/day in the sample diet reported here). Importantly, we also observe a decrease in total fat intake (6%, with daily oil seasoning not included (see Table 2)). The adequate seasoning oil (Aldixyl OiLife), composed of pure oleic acid, is then part of the nutritional treatment, due to high levels of VLCFAs in common seasonings.

When comparing the food for a special medical purpose (Aldixyl; Pharmaelle, Bologna, Italy) to the first oil therapy, "Lorenzo's Oil" [24,25], the mixture of triolein and trierucin in a 4:1 ratio, conjugated linoleic acids and antioxidants (alpha-lipoic acid, L-glutathione reduced, vitamin E), proposed by Cappa et al. [13,14], permits crossing of the blood–brain barrier. It has been demonstrated to inhibit the accumulation of VLCFAs, interfere with elongases to prevent synthesis of VLCFAs, and increase the activity of ALDP and consequently beta-oxidation in peroxisomes. Thanks to the addition of antioxidants, inflammation and oxidative stress are also reduced. In a two-month study involving five women with no or moderate symptoms, this newly formulated oil caused significant outcomes: a decrease in total cholesterol and LDL (Low-Density Lipoprotein) cholesterol, an increase in HDL (High-Density Lipoprotein) cholesterol, a decrease in plasma C26:0 levels and C26:0/C22:0 ratio, and an increase in the C22:6/C20:5 (docosahexaenoic acid/eicosapentaenoic acid) ratio, markers of peroxisomal beta-oxidation. Levels of interleukin-6 were significantly reduced in plasma and cerebrospinal fluid [13,14].

The average C26:0 intake in our study was comparable to levels reported in previous studies on restricted diets [16,17,23,26]. The diet described by Van Duyn et al. in 1984 [16] aimed to limit daily C26:0 intake to 3 mg, far below the usual intake of VLCFAs in the US diet [16]. A higher threshold of 10 mg/day was suggested by Kawahara et al. (1988) [17] and Rizzo et al. (1987) [26]. However, these studies were conducted with small sample sizes, with Van Duyn et al. including only seven patients [16]. Adherence to the diet and VLCFA intake restrictions were assessed using Food Diaries. This study did not find significant changes in plasma C26:0 levels, nor amelioration in the clinical prognosis of the disease [16]. Despite this, they claimed the potential benefit of dietary restriction of VLCFAs, combined with other therapies. Patients were followed for a time frame varying from 4 to 24 months and the possibility of poor adherence to the diet was mentioned by the authors as a plausible explanation for the lack of significant difference in C26:0 plasma level following the diet with respect to the baseline [16]. Another relevant study by Moser et al. [23] included 36 individuals and analyzed plasma VLCFA levels in patients following a diet restricted to 3 mg/day of C26:0, along with the administration of glyceryl trioleate oil, for a range period that varied from 60 days to 1.5 years. This cohort encompassed all the phenotypes of X-ALD and also included asymptomatic individuals and carriers. The authors revealed lower plasma VLCFA levels in 25 patients out of 36. They also observed improvement in peripheral nerve function in two individuals [23]. When considering the rarity of the disease, the "Bambino" study sample size appears to be larger, and our findings are in keeping with those of Moser et al.'s study (1987) [23].

Few studies addressing the putative role of VLCFA-restricted diets in the management of X-ALD demonstrate that these attempts are feasible for different dietary patterns, such as US and Japanese diets [16,17]. We worked on it, demonstrating that this attempt is also feasible using the Mediterranean diet, according to our lists of foods, based on C26:0 and total fat content. Given the lack of comprehensive data on food content of these nutrients in the major nutrient databases, we provided a ready-to-use list of allowed and non-allowed foods for patients, carriers, and caregivers, providing them with education on their consumption.

To build up these lists, we set the cut-off of C26:0 at 0.150 mg/100 g serving, following the guidelines by Van Duyn et al. (1984) [16] and Kawahara et al. (1988) [17], who are the only sources of information on VLCFA content in foods available in the scientific literature. They performed lipid analyses on food samples and documented C26:0 intake with the diet in the general populations of the US and Japan, reporting daily intakes of 15–40 mg and 12–36 mg, respectively [16,17]. Van Duyn et al. (1984) [16] found that in cereals, the higher the fiber content, the higher the C26:0 content was found to be; in particular, whole-grain products have the higher content, as compared to their non-whole-grain counterparts. They also identified elevated C26:0 levels in white bread, white flour, and homemade Italian bread [16]. Kawahara et al. (1988) [17] confirmed most of the findings from the earlier study. The two studies [16,17] differed slightly from each other, both confirming high levels in white bread and also detecting high levels in white rice. Among protein sources, fish, meat, legumes, and non-skimmed dairy products exhibited the highest VLCFA content. For fruits and vegetables, the C26:0 content per 100 g of edible portion was mostly low, except for the varieties listed in Supplementary Table S2. Notably, significant differences were observed between peeled and unpeeled fruits, with the outer coverings of plant foods containing the highest VLCFA levels. Therefore, the fruits and vegetables indicated in Supplementary Table S2 should be avoided, even if unpeeled. Additionally, spices, lard, and oils are known to have extremely high levels of C26:0 [16,17,21]. Unfortunately, current freely available databases do not report C26:0 content, making it challenging to assess VLCFA levels in foods.

It is worth noting that Van Duyn et al. (1984) [16] also suggested limiting total fat intake per day. Indeed, when they compared an average US diet with a diet restricted to VLCFAs of the same caloric content, they observed that a large decrease in C26:0 intake was paralleled by a moderate decrease in total daily fat intake (from 31% to 12%). Thus, we excluded high-fat foods, defining "high-fat" as a fatty acid content greater than (or equal to) 2 g/100 g of edible portion, as resulted from the analysis of food composition databases of CREA, BDA, and USDA [18–20]. Looking at products not listed in the previous studies [16,17], we found that pseudocereals with higher fiber content, certain fatty meats, spices, nuts and seeds, fried foods, precooked industrial products, sweets, and various beverages had the highest fat content [18–20]. We found moderate discrepancies in fat content values between databases for some foods, possibly due to non-standard experimental procedures.

Due to the large exclusion of foods, as general nutritional advice, we strongly recommended to our patients to eat fresh foods and avoiding prolonged cooking, to ensure adequate intake of all micronutrients and limit the risk of vitamin and mineral deficits.

The growing scientific interest in the nutritional management of X-ALD has been enhanced by the establishment of newly developed screening programs for X-ALD. Indeed, a nutritional intervention to reduce the VLCFA content in the diet may be beneficial to delay the onset of neurologic symptoms or improve them in patients and carriers. Neonatal screening for X-ALD should be an important tool to make the diagnosis earlier and consequently start nutritional intervention as soon as possible [27–31].

Our study is a retrospective but real-life investigation. Ethical concerns make it very complicated to carry out a randomized controlled trial in X-ALD, a rare condition [8], which also accounts for the small number of patients evaluated in the present study, despite our sample size being larger when compared to previous studies [16,23].

When studying nutritional treatment in X-ALD, a limit is caused by the lack of up-to-date national or international dietary tables that provide detailed information on the nutritional values of interest, first of all C26:0, in common Italian foods. Moreover, discrepancies in C26:0 content reported by previous studies [16,17] stem from the different techniques used for analysis. To support future research, it is crucial to standardize VLCFA measurement methods and analytical procedures to ensure consistent and comparable data. Despite this, the findings from the OPBG experience, along with the great compliance to the nutritional treatment shown by the enrolled individuals, has proven it possible to lower consumption by up to 2.5 mg/day, following our sample of a restricted diet, together with a total fat intake decrease too. The great amount of allowed foods according to the Mediterranean pattern makes it sustainable, as proven by data on adherence revealed by statistical analysis. However, more frequent in-person follow-up visits could have improved compliance. Age-specific behavioral strategies to comply with the diet must be elaborated in children.

As a limit of our study, we did not analyze data on micronutrient intake, which could occur in patients following a restricted diet, due to the exclusion of some foods. So, future research is necessary to eventually standardize specific supplementation. Due to the slight differences between the "Bambino" diet and a standard Mediterranean diet, the long-term effects, cost, and accessibility are comparable. The continuous monitoring of patients during the routine follow-ups will provide more data on the persistence in time of the biochemical improvements, along with the improvement in clinical symptoms.

Raising awareness of X-ALD is crucial among the general population and healthcare providers and stakeholders. We strongly recommend food labeling with C26:0 content and annotation of its content, measured by a gold-standard technique, in the main food databases.

5. Conclusions

The "Bambino" VLCFA-restricted diet caused a dramatic improvement of the X-ALD biomarkers, along with a reduced intake of VLCFAs and fats from the diet. In the "Bambino" cohort, adherence to the diet was good, thus proving that nutritional treatment must be considered a pivotal strategy. It is crucial in presymptomatic patients and in subjects who do not have a severe form of disease to delay the onset of symptoms and to improve the prognosis [8,16,24,32]. The possibility of drawing up this nutritional pattern and all subjects receiving the same nutritional protocol and general advice, reducing the risk of bias, represent strengths of our study.

Supplementary Materials: The following supporting information can be downloaded at: https://www.mdpi.com/article/10.3390/nu16193341/s1, Table S1: Phenotypes of X-ALD. Clinical features; Table S2: Non-allowed foods, according to C26:0 and total fatty acids content; Table S3: Foods with hidden fats; Table S4: Allowed foods, according to C26:0 and total fatty acids content.

Author Contributions: Conceptualization, M.R.S., M.C. and M.M.; data curation, N.G. and T.T.; writing—original draft preparation, N.G. and M.M.; blood sample analysis, C.R.; all authors, writing—review and editing; supervision, M.C. and M.M. All authors have read and agreed to the published version of the manuscript.

Funding: The study was supported by the Italian Ministry of Health with "current research funds". Dr. Manco also received Italian Ministry of Health FPRC 5xmille2023 funds. The funder had no role in the design of the study; in the collection, analyses, or interpretation of data; in the writing of the manuscript; or in the decision to publish the results.

Institutional Review Board Statement: The study was conducted in accordance with the Declaration of Helsinki, and approved by the OPBG Ethics Committee (2058, 1 June 2018).

Informed Consent Statement: Informed consent was obtained from adult patients and caregivers for minors. Children older than 6 years of age provided written consent.

Data Availability Statement: The data described in the study are not publicly available due to privacy and ethical restrictions but may be made available upon request.

Acknowledgments: We are greatly in debt to Nicola Spreghini for his support in statistical analysis.

Conflicts of Interest: The authors report that there are no competing interests to declare.

Abbreviations

ALDP	Adrenoleukodystrophy Protein
AMN	Adrenomyelomeuropathy
BDA	Banca Dati di composizione degli Alimenti
BMI	body mass index Kg/m^2
CerALD	Cerebral disease ALD
CREA	Consiglio per la Ricerca in agricoltura e l'analisi dell'Economia Agraria
EHRs	Electronic Health Records
ELOVL1	ELOngation of Very Long chain fatty acids-1
LO	Lorenzo's Oil
OPBG	Ospedale Pediatrico Bambino Gesù
USDA	U.S. Department of Agriculture
VLCFAs	Very-Long-Chain Fatty Acids
X-ALD	Adrenoleukodystrophy

References

1. OMIM. Online Mendelian Inheritance in Man. 1986. Available online: https://omim.org/entry/300100?search=x.ald&highlight=%22x%20ald%22,xald (accessed on 2 June 2024).
2. OMAR. Osservatorio Malattie Rare. Available online: https://www.osservatoriomalattierare.it/malattie-rare/adrenoleucodistrofia/16693-adrenoleucodistrofia-la-terapia-di-supporto-dietetico-e-un-ausilio-fondamentale (accessed on 2 June 2024).

3. EnEngelen, M.; Kemp, S.; de Visser, M.; van Geel, B.M.; Wanders, R.J.; Aubourg, P.; Poll-The, B.T. X-linked adrenoleukodystrophy (X-ALD): Clinical presentation and guidelines for diagnosis, follow-up and management. *Orphanet J. Rare Dis.* **2012**, *7*, 51. [CrossRef] [PubMed]
4. Cappa, M.; Todisco, T.; Bizzarri, C. X-linked Adrenoleukodystrophy and primary adrenal insufficiency. *Front. Endocrinol.* **2023**, *14*, 1309053. [CrossRef] [PubMed]
5. Van Geel, B.M.; Assies, J.; Wanders, R.J.; Barth, P.G. X linked adrenoleukodystrophy: Clinical presentation, diagnosis, and therapy. *J. Neurol. Neurosurg. Psychiatry* **1997**, *63*, 4–14. [CrossRef] [PubMed]
6. Turk, B.R.; Theda, C.; Fatemi, A.; Moser, A.B. X-linked adrenoleukodystrophy: Pathology, pathophysiology, diagnostic testing, newborn screening and therapies. *Int. J. Dev. Neurosci.* **2020**, *80*, 52–72. [CrossRef] [PubMed]
7. Di Biase, A.; Avellino, C.; Salvati, S. Adrenoleucodistrofia: Genetica, fenotipi, patogenesi e trattamento [Adrenoleukodystrophy: Genetics, phenotypes, pathogenesis and treatment]. *Ann. Dell'istituto Super. Sanità* **1999**, *35*, 185–192.
8. De Luis Román, D.A.; Bellido Guerrero, D.; García Luna, P.P. Trastornos congénitos del metabolismo de los lípidos: Adrenoleucodistrofia. Dieta controlada en ácidos grasos [Congenital disorders of lipid metabolism: Adrenoleukodystrophy. Fatty acid controlled diet]. In *Dietoterapia, Nutrición Clínica y Metabolismo [Diet Therapy, Clinic Nutrition and Metabolism]*, 3rd ed.; de Santos, E.D., Peláez, R.B., Casas, N.V., Eds.; SEEN (Sociedad Española de Endocrinología y Nutrición): Madrid, Spain, 2017; pp. 293–299.
9. Raymond, G.V.; Moser, A.B.; Fatemi, A. X-Linked Adrenoleukodystrophy, Synonym: X-ALD. In *Gene Reviews*; University of Washington: Seattle, WA, USA, 1999.
10. Schirinzi, T.; Vasco, G.; Aiello, C.; Rizzo, C.; Sancesario, A.; Romano, A.; Favetta, M.; Petrarca, M.; Paone, L.; Castelli, E.; et al. Natural history of a cohort of ABCD1 variant female carriers. *Eur. J. Neurol.* **2019**, *26*, 326–332. [CrossRef] [PubMed]
11. Whitcomb, R.W.; Linehan, W.M.; Knazek, R.A. Effects of long-chain, saturated fatty acids on membrane microviscosity and adrenocorticotropin responsiveness of human adrenocortical cells in vitro. *J. Clin. Investig.* **1988**, *81*, 185–188. [CrossRef] [PubMed]
12. OMAR. Osservatorio Malattie Rare. Available online: https://www.osservatoriomalattierare.it/malattie-rare/adrenoleucodistrofia (accessed on 3 April 2023).
13. Cappa, M.; Bizzarri, C.; Petroni, A.; Carta, G.; Cordeddu, L.; Valeriani, M.; Vollono, C.; De Pasquale, L.; Blasevich, M.; Banni, S. A mixture of oleic, erucic and conjugated linoleic acids modulates cerebrospinal fluid inflammatory markers and improve somatosensorial evoked potential in X-linked adrenoleukodystrophy female carriers. *J. Inherit. Metab. Dis.* **2012**, *35*, 899–907. [CrossRef] [PubMed]
14. Cappa, M. Adrenoleucodistrofia: L'evoluzione della terapia [Adrenoleukodystrophy: Evolution of therapy]. *L'Endocrinologo* **2022**, *23*, 168–175. [CrossRef]
15. Caruso, U.; Fowler, B.; Erceg, M.; Romano, C. Determination of very-long-chain fatty acids in plasma by a simplified gas chromatographic-mass spectrometric procedure. *J. Chromatogr.* **1991**, *562*, 147–152. [CrossRef] [PubMed]
16. Van Duyn, M.A.; Moser, A.E.; Brown, F.R., 3rd; Sacktor, N.; Liu, A.; Moser, H.W. The design of a diet restricted in saturated very long-chain fatty acids: Therapeutic application in adrenoleukodystrophy. *Am. J. Clin. Nutr.* **1984**, *40*, 277–284. [CrossRef] [PubMed]
17. Kawahara, K.; Iwasaki, Y.; Harada, Y.; Maruyama, Y.; Ono, S.; Kanja, K.; Sasai, H.; Nanba, S.; Nakano, H.; Kusaka, T. Hexacosanoate contents in Japanese common foods. *J. Nutr. Sci. Vitaminol.* **1988**, *34*, 633–639. [CrossRef] [PubMed]
18. Consiglio per la Ricerca in Agricoltura e L'analisi dell'Economia Agraria (CREA), by Marletta L, Camilli, E. Available online: https://www.alimentinutrizione.it/sezioni/tabelle-nutrizionali (accessed on 20 July 2023).
19. Food Composition Database for Epidemiological Studies in Italy (Banca Dati di Composizione degli Alimenti per Studi Epidemiologici in Italia—BDA), by Gnagnarella P, Salvini S, Parpinel, M. Versione 1. Available online: https://bda.ieo.it/ (accessed on 20 July 2023).
20. U.S. Department of Agriculture (USDA), Agricultural Research Service. *FoodData Central*. Available online: https://fdc.nal.usda.gov/ (accessed on 20 July 2023).
21. Kolattukudy, P.E.; Croteau, R.; Brown, L. Structure and Biosynthesis of Cuticular Lipids: Hydroxylation of Palmitic Acid and Decarboxylation of C(28), C(30), and C(32) Acids in Vicia faba Flowers. *Plant Physiol.* **1974**, *54*, 670–677. [CrossRef] [PubMed]
22. Bowman, S.A.; Martin, C.L.; Carlson, J.L.; Clemens, J.C.; Lin, B.H.; Moshfegh, A.J. *Food Intakes Converted to Retail Commodities Databases 2003-08: Methodology and User Guide*; U.S. Department of Agriculture, Agricultural Research Service: Beltsville, MD, USA; U.S. Department of Agriculture, Economic Research Service: Washington, DC, USA. Available online: https://www.ars.usda.gov/ARSUserFiles/80400530/pdf/ficrcd/Documentation%20FICRCD%202003_08.pdf (accessed on 18 May 2023).
23. Moser, A.B.; Borel, J.; Odone, A.; Naidu, S.; Cornblath, D.; Sanders, D.B.; Moser, H.W. A new dietary therapy for adrenoleukodystrophy: Biochemical and preliminary clinical results in 36 patients. *Ann. Neurol.* **1987**, *21*, 240–249. [CrossRef] [PubMed]
24. Rizzo, W.B.; Leshner, R.T.; Odone, A.; Dammann, A.L.; Craft, D.A.; Jensen, M.E.; Jennings, S.S.; Davis, S.; Jaitly, R.; Sgro, J.A. Dietary erucic acid therapy for X-linked adrenoleukodystrophy. *Neurology* **1989**, *39*, 1415–1422. [CrossRef] [PubMed]
25. Raymond, G.; Engelen, M.; Kemp, S. Lorenzo's Oil. Available online: https://adrenoleukodystrophy.info/treatment-options/lorenzos-oil (accessed on 18 May 2023).
26. Rizzo, W.B.; Phillips, M.W.; Dammann, A.L.; Leshner, R.T.; Jennings, S.S.; Avigan, J.; Proud, V.K. Adrenoleukodystrophy: Dietary oleic acid lowers hexacosanoate levels. *Ann. Neurol.* **1987**, *21*, 232–239. [CrossRef] [PubMed]

27. Bonaventura, E.; Alberti, L.; Lucchi, S.; Cappelletti, L.; Fazzone, S.; Cattaneo, E.; Bellini, M.; Izzo, G.; Parazzini, C.; Bosetti, A.; et al. Newborn screening for X-linked adrenoleukodystrophy in Italy: Diagnostic algorithm and disease monitoring. *Front. Neurol.* **2023**, *13*, 1072256. [CrossRef] [PubMed]
28. Salzman, R.; Kemp, S. Newborn Screening. Available online: https://adrenoleukodystrophy.info/clinical-diagnosis/ald-newborn-screening (accessed on 10 June 2023).
29. Albersen, M.; van der Beek, S.L.; Dijkstra, I.M.E.; Alders, M.; Barendsen, R.W.; Bliek, J.; Boelen, A.; Ebberink, M.S.; Ferdinandusse, S.; Goorden, S.M.I.; et al. Sex-specific newborn screening for X-linked adrenoleukodystrophy. *J. Inherit. Metab. Dis.* **2023**, *46*, 116–128. [CrossRef] [PubMed]
30. Barendsen, R.W.; Dijkstra, I.M.E.; Visser, W.F.; Alders, M.; Bliek, J.; Boelen, A.; Bouva, M.J.; van der Crabben, S.N.; Elsinghorst, E.; van Gorp, A.G.M.; et al. Adrenoleukodystrophy Newborn Screening in the Netherlands (SCAN Study): The X-Factor. *Front. Cell Dev. Biol.* **2020**, *8*, 499. [CrossRef] [PubMed]
31. SCAN-Studie. Screening op Adrenoleukodystrofie (ALD) in Nederland via de Hielprik [Screening for Adrenoleukodystrophy (ALD) in the Netherlands via the Heel Prick]. Available online: https://scanstudie.nl (accessed on 25 June 2023).
32. Moser, H.W.; Borel, J. Dietary management of X-linked adrenoleukodystrophy. *Annu. Rev. Nutr.* **1995**, *15*, 379–397. [CrossRef] [PubMed]

nutrients

MDPI

Article

Impact of Dietary Patterns on Metabolic Syndrome in Young Adults: A Cross-Sectional Study

Jingwen Liu †, Wenfeng Lu †, Qingyun Lv, Yaqi Wang, Xueying Xu, Yuan He, Hairong Chang, Yue Zhao, Xiaonan Zhang *, Xiaoying Zang * and Hua Zhang *

School of Nursing, Tianjin Medical University, No. 22 Qixiangtai Road, Heping District, Tianjin 300070, China; liujingwen0929@163.com (J.L.); lwf980426@163.com (W.L.); 18722595311@163.com (Q.L.); wangyaqi1229@163.com (Y.W.); xuxueying2022@163.com (X.X.); heyuan1632@163.com (Y.H.); r1307375960@163.com (H.C.); yuezhao35@hotmail.com (Y.Z.)

* Correspondence: zhangxiaonan19@163.com (X.Z.); xiaoyingzang@163.com (X.Z.); snzhanghua@tmu.edu.cn (H.Z.)

† These authors contributed equally to this work.

Abstract: Metabolic syndrome has become a significant public health concern. This study aims to investigate the impact of dietary patterns on metabolic syndrome in young adults and how physical activity modulates this effect. A cross-sectional study was conducted at a health management center in Tianjin, China, from September 2022 to March 2023. Participants aged 18–35 years were recruited using convenience sampling. Dietary intake was assessed using a validated food frequency questionnaire. Logistic regression models evaluated associations between these patterns and metabolic syndrome, adjusting for potential confounders. Among 442 participants, four dietary patterns were identified: Legume–Nut, Alcohol–Meat, Sugar–Processed, and Egg–Vegetable. The Legume–Nut dietary pattern was associated with a higher risk of metabolic syndrome (OR = 2.63, 95% CI: 1.08–6.37), while the Egg–Vegetable dietary pattern was associated with a lower risk (OR = 0.26, 95% CI: 0.10–0.70). No significant associations were found for the Sugar–Processed and Alcohol–Meat patterns. Subgroup analysis revealed that the Legume–Nut pattern increased the risk of metabolic syndrome among those with irregular physical activity, whereas the Egg–Vegetable pattern decreased the risk. These findings highlight the significant influence of dietary patterns on the risk of metabolic syndrome in young adults and the modifying effect of regular physical activity, underscoring the need for targeted dietary and lifestyle interventions to prevent metabolic syndrome in this population.

Keywords: metabolic syndrome; dietary patterns; young adults; cross-sectional study; subgroup analysis

1. Introduction

Metabolic syndrome is a clinical syndrome characterized by abdominal obesity, hypertriglyceridemia, low high-density lipoprotein cholesterol, elevated blood pressure, and increased fasting blood glucose [1]. Metabolic syndrome is a disease with severe public health consequences, reducing patients' quality of life, increasing morbidity and mortality, and leading to substantial healthcare expenses [2]. The prevalence of metabolic syndrome in young populations has reached 17.4% [3], and its incidence is rising in both developed and developing countries [4]. However, current research on metabolic syndrome predominantly focuses on physiological, biochemical, clinical, and metabolic factors [5].

Existing research indicates that metabolic syndrome is closely associated with diet [6]. Specifically, carbohydrate intake is associated with metabolic syndrome, and reducing carbohydrate intake can be an effective treatment for metabolic syndrome [7]. Excessive sugar intake directly correlates with hyperglycemia and its complications [8]. Protein intake plays a crucial role in muscle protein synthesis and the maintenance of muscle mass [9], thereby impacting overall body metabolism [10]. Furthermore, diets high in fats, particularly excessive intake of saturated fatty acids, are closely linked to several core symptoms of metabolic syndrome [11].

In recent years, many studies have focused on the health status of young people, particularly metabolic syndrome [12,13]. However, most studies have not focused on dietary patterns, or the classifications of dietary patterns have been overly simplistic [14]. Studying dietary patterns offers a comprehensive assessment of the impact of food combinations on health, avoiding the biases that may arise from focusing solely on individual nutrients [15]. For instance, a study investigating the relationship between total sugar intake and metabolic syndrome in middle-aged individuals suggests that the focus should not solely be on the total sugar intake but rather on the contribution of sugar intake to the overall energy intake [16]. Furthermore, the impact of consuming isoenergetic amounts of sugar on health varies depending on the original form of the food, highlighting the limitations of focusing exclusively on individual nutrients. By analyzing overall dietary patterns, researchers can gain deeper insights into the complex interactions between different foods and their collective effects on metabolism, providing a scientific basis for effective public health strategies [17]. Another advantage of using dietary patterns in research is the ability to identify potential synergistic effects within the diet and their long-term impact on overall health [18]. Additionally, the research indicates that combining dietary therapy with physical activity can effectively improve the metabolic status in patients with metabolic syndrome [19]. Studying the health status of young people is particularly significant because their eating habits and lifestyles are relatively easier to modify. This makes them more amenable to interventions and facilitates the formulation of effective public health strategies.

Hence, this study aimed to explore the impact of dietary patterns on metabolic syndrome in young adults and how physical activity modulates this effect, providing new strategies for the prevention and treatment of metabolic syndrome.

2. Material and Methods

2.1. Data Sources and Study Population

Participants were recruited using a convenience sampling method from a tertiary hospital's health management center in Tianjin, China, from September 2022 to March 2023. Inclusion criteria were as follows: (a) age within the young adult range: 18–35 years; (b) provided informed consent and voluntary participation; (c) normal cognitive and communication abilities, confirmed through a brief interview during the recruitment process where participants were asked to describe their daily routines and respond to basic questions to ensure clear and coherent communication, and no history of mental disorders as per their medical records; (d) underwent a physical examination in the past year. Exclusion criteria included the inability to complete the questionnaire due to severe illness. Before participation, all participants were thoroughly informed about the study's purpose and procedures, and their understanding was confirmed through verbal confirmation to ensure they comprehended the study's aims and requirements. Participants were also given detailed instructions on how to complete the study questionnaires. Each questionnaire took approximately 25–30 min to complete. A total of 442 samples were included in this study (Figure 1).

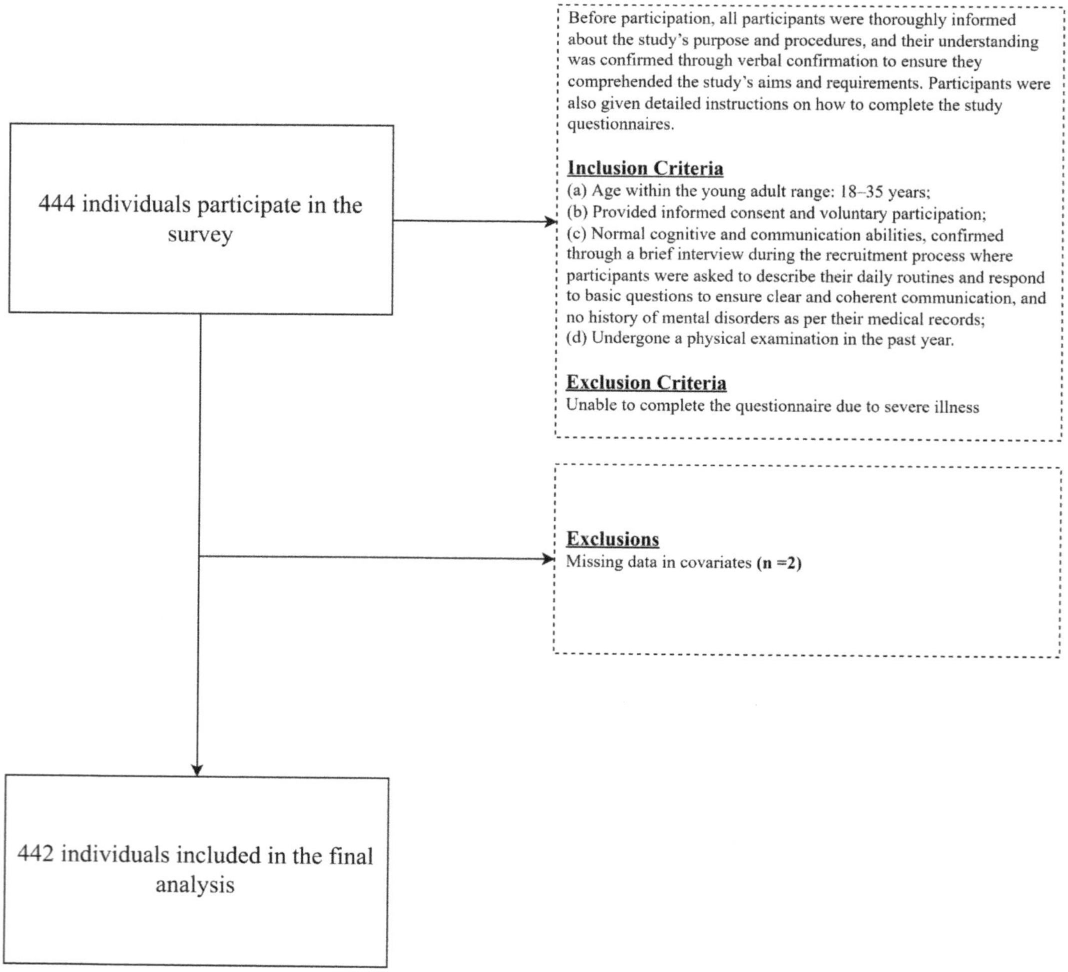

Figure 1. Flowchart of the sample selection process.

2.2. Metabolic Syndrome

Metabolic syndrome is diagnosed when any three of the following criteria are present [20]: (1) increased waist circumference, with standards varying by population and country (≥90 cm for Chinese men, ≥85 cm for women) [21]; (2) elevated triglyceride levels: over 150 mg/dL (1.7 mmol/L), or under treatment for reduced triglycerides; (3) reduced HDL cholesterol: below 50 mg/dL (1.3 mmol/L) or under treatment to increase HDL-C; (4) elevated blood pressure: systolic ≥130 mmHg and/or diastolic ≥85 mmHg, or history of hypertension under treatment; (5) raised fasting blood glucose: ≥100 mg/dL (5.6 mmol/L), or history of diabetes under treatment.

The disease is self-reported by the participant, or detected during a physical examination in the last year, or determined with relevant therapeutic medications. All diseases are diagnosed by a doctor, and all anthropometric measurements are con-ducted by nurses according to standard procedures.

2.3. Food Frequency Questionnaire (FFQ)

Dietary information was assessed using a validated semi-quantitative food frequency questionnaire (FFQ) [22–24], which has been widely employed in various studies [25]. Considering the consumption volume and frequency of food items among the surveyed population, the food items were selected based on the nutritional composition and dietary habits of Chinese individuals. This study included 17 major food groups and 189 food items within these groups, encompassing whole grains, tubers, beans and bean products, nuts, fresh vegetables, fruits, white meat, red meat, processed meat, aquatic and seafood products, milk and dairy products, eggs, cooking oils, fried food, sugary beverages, alcoholic beverages, and pastries (Supplementary Material Table S3). Participants were required to recall their frequency of intake and portion size consumed during the past 12 months.

Frequency options include: (1) every day; (2) 4–6 times a week; (3) 1–3 times a week; (4) 1–3 times a month; (5) almost never/never. During face-to-face interviewing, the portion size of each food item was estimated using food models and standard serving sizes (e.g., one standardized portion of cooked rice is one small bowl, weighing approximately 100 g). Then, the frequency of intake and portion size were used to calculate the amount of each food item consumed on average. These data were converted to grams or milliliters per day. Food intake was normalized for each participant. This study was normalized using a z-score, which is food intake minus the mean divided by the standard deviation. Finally, we calculated the factor scores of each participant for each pattern by summing the intake of food groups weighted by their factor loadings and grouped them into quartiles for further analysis, with Q1 corresponding to the lowest quartile of dietary pattern score. A higher quartile indicates more consistency with the pattern being calculated. Total energy intake was estimated through the FFQ. The intake of energy was calculated by using the China Food Composition Database (Version 6) [26].

2.4. Physical Activity

The World Health Organization recommends that for young adults, the most suitable physical activity should meet the following requirements: Heart rate should be maintained at 50–85% of maximum heart rate [27]. During moderate-intensity exercise, breathing should increase but still allow for conversation, while during high-intensity exercise, breathing should become rapid and conversation difficult, with a perceived exertion of 5–8 on a 10-point scale. Post-exercise, muscles should feel adequately fatigued but not overly sore, and joints should be free from abnormal pain. Body temperature should rise moderately with increased sweating. Each exercise session should last at least 30 min. Throughout and following the exercise, individuals should feel pleasant and satisfied, with adequate hydration and timely nutritional supplementation. The entire process should be safe and free from injury. In this study, the frequency of physical activity is categorized as monthly or less, weekly, 2–3 times per week, and 4+ times per week. 'Monthly or less' is defined as irregular physical activity, while 'Weekly', '2–3 times per week', and '4+ times per week' are defined as regular physical activity.

2.5. Statistical Analysis

Initially, the suitability of using factor analysis is determined by the Kaiser–Meyer–Olkin (KMO) measure and Bartlett's test of sphericity. The results of the KMO test yielded a value of 0.771, and Bartlett's test of sphericity showed a p-value < 0.001, indicating that factor analysis is appropriate for this study (Supplementary Material Table S1). Based on the correlation matrix and orthogonal varimax rotation, dietary patterns were derived. Eigenvalues greater than 1 were considered, and the scree plot indicated that the eigenvalues began to level off after the fourth principal component, contributing a smaller proportion of the total variance. Taking into account the specifics of this study, four factors were retained, extracting four distinct dietary patterns, with a cumulative variance contribution rate of 54.457% (Supplementary Materials Table S2 and Figure S1).

To present participant characteristics, continuous variables were expressed as mean values with standard deviations (SD), and categorical variables were presented as number and percentages. Characteristics between groups were compared using Chi-square tests, *t*-test, analysis of variance (ANOVA), or the Kruskal–Wallis rank sum test, as appropriate. To assess the relationship between dietary patterns and the incidence of metabolic syndrome, logistic regression models were established, providing odds ratios (OR) with their 95% confidence intervals (95% CI). Model 1 is an unadjusted crude model. Model 2 includes adjustments for age, gender, residence, education level, occupation, and marital status, while Model 3 further adjusts for age, gender, residence, education level, occupation, marital status, smoking, alcohol consumption, exercise frequency, sleep duration, height, weight and total energy intake (kcal/d). Subgroup analysis was performed according to physical activity frequency and also performed to detect potential modifiers.

Statistical analysis was carried out using IBM Statistical Package SPSS version 26.0 (SPSS Inc., Chicago, IL, USA) and R version 4.3.1 Forest plots were generated using the 'forestplot' package. Statistical significance in the analysis was indicated by a two-tailed *p*-value of less than 0.05.

This study was carried out in accordance with the Checklist for Strengthening The Reporting of Observational Studies in Epidemiology (STROBE), and it was approved by the medical ethics committee of Tianjin Medical University in China (TMuhMEC2022021) in 29 August 2022.

Each participant was given a written consent form before involvement in the study. The participants were briefed about the schedule, venue, and duration of data collection, along with the potential advantages and risks associated with their participation. Post obtaining informed consent, the principal investigator gathered data from both the medical records and the participants themselves. This methodology was followed until the targeted sample size was accomplished. The lead researcher meticulously read out the questions to all participants.

3. Results

3.1. Establishment of Dietary Patterns

This study identified four dietary factors through factor analysis, with factor load distributions shown in Table 1. Foods with a factor loading absolute value greater than 0.4 were considered to have a significant relationship with the respective factor. We independently named the dietary patterns based on their included food types, resulting in the following names: Legume–Nut dietary pattern, Alcohol–Meat dietary pattern, Sugar–Processed dietary pattern, and Egg–Vegetable dietary pattern. Each dietary pattern includes the following types of food as shown in Table 1.

Table 1. Factor load in food groups of dietary patterns.

	Sugar–Processed	Alcohol–Meat	Legume–Nut	Egg–Vegetable
Whole Grains	−0.053	0.124	−0.007	0.439
Tubers	0.042	−0.057	0.729	−0.037
Beans and Bean Products	−0.080	−0.001	0.787	−0.019
Nuts	0.158	0.181	0.678	0.002
Fresh Vegetables	−0.102	−0.483	0.334	0.483
Fruits	0.208	−0.326	0.504	0.238
White Meat	0.042	−0.033	0.623	0.276
Red Meat	0.121	0.517	0.277	0.420
Processed Meats	0.446	0.503	0.343	0.058
Aquatic and Seafood	0.207	0.304	0.536	−0.025
Milk and Dairy Products	0.376	−0.547	0.170	0.309

Table 1. *Cont.*

	Sugar–Processed	Alcohol–Meat	Legume–Nut	Egg–Vegetable
Eggs	0.069	−0.068	0.144	0.706
Cooking Oils	0.134	0.037	−0.09	0.753
Fried Foods	0.566	0.560	0.106	0.100
Sugary Beverages	0.836	0.072	0.054	−0.006
Alcoholic Beverages	0.139	0.765	0.028	0.161
Pastries	0.791	0.044	0.077	0.048

Factor loadings of food groups in each dietary pattern were identified using factor analysis. The color gradation denotes the strength and direction of the correlation between the food groups and dietary patterns. Dark green indicates a relatively high correlation between the food group and the corresponding dietary pattern, with factor loadings greater than or equal to 0.4. Light green signifies a moderate correlation, with factor loadings less than 0.4 but greater than or equal to 0.2. Orange denotes a relatively low correlation, with factor loadings less than 0.2. The type of food with a black border is the composition of the food group.

3.2. Characteristics of Participants Grouped by Dietary Patterns

In this study, the average age of the participants was 24.79 ± 4.97 years, with a distribution showing more males (54.52%) than females (45.48%). Most participants resided in urban areas (70.14%), had completed a bachelor's degree or higher (53.17%), and were unmarried (72.40%). A significant portion of the participants were non-smokers (75.11%) and non-drinkers (46.60%). Most of the participants were employed as mental workers (38.01%), and the majority exercised weekly (23.98%) (Table 2). The characteristics of study participants across quartile categories of the dietary pattern scores were shown in Table 3. Participants in the highest quartile (Q4) of the Legume–Nut dietary pattern were notably younger, predominantly male, taller, and heavier compared to those in the lowest quartile (Q1). They also were more likely to reside in rural areas and engage in more frequent physical activity. In contrast, individuals in the highest quartile of the Egg–Vegetable dietary pattern tended to be older, more likely to be male, taller, and heavier, with a higher likelihood of being married, smoking, and drinking. Additionally, those in the highest quartile of the Sugar–Processed dietary pattern were heavier, with higher rates of smoking and drinking, and engaged in less frequent physical activity. For the Alcohol–Meat dietary pattern, individuals in the highest quartile were older, predominantly male, taller, heavier, and more likely to smoke and drink.

Table 2. Characteristics of participants, Tianjin, China, 2022–2023.

Characteristics	Total	MetS− N (%) or M ± SD	MetS+ N (%) or M ± SD	*p* χ^2 or *t*
Sample size (N)	442	361	81	
Age (years)	24.79 ± 4.97	24.53 ± 4.75	25.95 ± 5.74	0.041
Gender				0.344
Male	241 (54.52)	193	48	
Female	201 (45.48)	168	33	
Height (cm)	171.97 ± 8.50	171.98 ± 8.33	171.93 ± 9.26	0.958
Weight (kg)	66.12 ± 14.07	64.64 ± 13.17	72.70 ± 16.06	<0.001
Waist circumference (cm)	77.56 ± 12.77	74.58 ± 10.75	90.83 ± 12.68	<0.001
Hypertension				<0.001
No	363 (82.13)	358	5	
Yes	79 (17.87)	3	76	

Table 2. *Cont.*

Characteristics	**Total**	**MetS−** *N* **(%) or M ± SD**	**MetS+** *N* **(%) or M ± SD**	***p*** χ^2 **or** t
Diabetes				<0.001
No	377 (85.29)	354	23	
Yes	65 (14.71)	7	58	
Dyslipidemia				<0.001
No	380 (85.97)	357	23	
Yes	62 (14.03)	4	58	
Residence				0.196
Urban area	310 (70.14)	258	52	
Suburban area	132 (29.86)	103	29	
Educational level completed				0.012
Junior middle school or below	113 (25.57)	85	28	
High school or Vocational school	94 (21.27)	72	22	
Bachelor's degree or Junior college or above	235 (53.17)	204	31	
Marital status				0.017
Unmarried	320 (72.40)	270	50	
Married	122 (27.60)	91	31	
Smoking status				0.012
No	332 (75.11)	280	52	
Yes	110 (24.89)	81	29	
Drinking status				0.666
No	206 (46.60)	170	36	
Yes	236 (53.39)	191	45	
Occupation type				0.042
Physical workers	168 (38.01)	135	33	
Mental workers	60 (13.57)	43	17	
Students	214 (48.42)	183	31	
Sleep duration (hours/day)	7.27 ± 0.82	7.47 ± 0.90	7.38 ± 1.05	0.452
Physical activity frequency				0.005
Monthly or less	202 (45.70)	152	50	
Weekly	106 (23.98)	97	9	
2–3 times per week	95 (21.49)	80	15	
4+ times per week	39 (8.82)	32	7	

Statistical tests: Continuous variables were analyzed using independent *t*-tests or Welch's *t*-test. Categorical variables were analyzed using Chi-square tests. MetS− indicates no metabolic syndrome. MetS+ indicates the presence of metabolic syndrome.

Table 3. Characteristics of participants grouped by dietary patterns.

Characteristics	N	Sugar-Processed		p	Alcohol-Meat		p	Legume-Nut		p	Egg-Vegetable		p
		Q1 (n = 111)	Q4 (n = 110)		Q1 (n = 111)	Q4 (n = 110)		Q1 (n = 111)	Q4 (n = 110)		Q1 (n = 111)	Q4 (n = 110)	
Age (years)	24.79 ± 4.97	25.41 ± 4.78	24.51 ± 5.32	0.100	23.42 ± 4.77	26.15 ± 4.82	0.001	24.28 ± 4.89	23.99 ± 4.73	0.033	21.86 ± 3.79	25.61 ± 5.01	<0.001
Gender				0.215			<0.001			0.006			<0.001
Male	241	55	65		26	87		49	67		27	84	
Female	201	56	45		85	23		62	43		84	26	
Height (cm)	171.97 ± 8.50	171.49 ± 8.24	171.78 ± 8.47	0.708	167.26 ± 7.75	174.87 ± 6.78	<0.001	169.28 ± 8.67	174.51 ± 7.52	<0.001	166.93 ± 6.76	175.67 ± 8.08	<0.001
Weight (kg)	66.12 ± 14.07	62.37 ± 13.74	71.16 ± 15.20	<0.001	58.10 ± 14.63	75.80 ± 12.66	<0.001	63.64 ± 14.82	68.89 ± 14.83	0.039	57.06 ± 12.32	74.39 ± 13.19	<0.001
Waist circumference (cm)	77.56 ± 12.77	75.41 ± 10.02	81.42 ± 16.75	0.002	72.92 ± 11.72	82.30 ± 15.69	<0.001	77.00 ± 11.87	76.37 ± 15.23	0.339	72.83 ± 11.65	79.25 ± 15.21	<0.001
Hypertension				0.028			0.110			0.131			0.750
No	363	96	80		97	83		99	87		94	87	
Yes	79	15	30		14	27		12	23		17	23	
Diabetes				0.767			0.134			0.919			0.896
No	377	96	91		100	90		96	92		95	96	
Yes	65	15	19		11	20		15	18		16	14	
Dyslipidemia				0.010			0.407			0.361			0.488
No	380	103	85		99	91		100	92		91	94	
Yes	62	8	25		12	19		11	18		20	16	
Residence				0.740			0.821			0.034			0.901
Urban area	310	80	75		77	74		80	68		80	77	
Suburban area	132	31	35		34	36		31	42		31	33	
Educational level completed				0.004			0.498			0.046			0.446
Junior middle school or below	168	15	35		23	30		29	20		22	27	
High school or Vocational school	60	25	26		30	21		28	23		20	25	
Bachelor's degree or Junior college or above	214	71	49		58	59		54	67		69	58	
Marital status				0.565			0.065			0.363			<0.001
Unmarried	320	76	81		86	70		83	84		101	77	
Married	122	35	29		25	40		28	26		10	33	
Smoking status				0.037			<0.001			0.220			<0.001
No	332	90	80		102	55		88	86		102	74	
Yes	110	21	30		9	55		23	24		9	36	

Table 3. *Cont.*

Characteristics	N	Sugar-Processed		p	Alcohol-Meat		p	Legume-Nut		p	Egg-Vegetable		p
		Q1 (n = 111)	Q4 (n = 110)		Q1 (n = 111)	Q4 (n = 110)		Q1 (n = 111)	Q4 (n = 110)		Q1 (n = 111)	Q4 (n = 110)	
Drinking status				0.050			<0.001			0.173			<0.001
No	206	60	45		75	25		57	56		83	42	
Yes	236	51	65		36	85		54	54		28	68	
Occupation type				0.445			0.098			0.006			<0.001
Physical workers	168	40	43		33	49		41	30		25	43	
Mental workers	60	18	17		14	16		10	16		8	17	
Students	214	53	50		64	45		60	64		78	50	
Sleep duration (hours/day)	7.45 ± 0.93	7.62 ± 0.79	7.28 ± 1.07	0.055	7.57 ± 0.87	7.31 ± 1.07	0.208	7.42 ± 0.92	7.47 ± 1.19	0.879	7.60 ± 1.14	7.42 ± 0.88	0.323
Physical activity frequency				0.004			0.268			0.005			0.073
Monthly or less	202	40	63		44	54		56	41		40	50	
Weekly	106	30	15		30	22		28	20		33	20	
2-3 times per week	95	26	22		30	20		23	29		31	26	
4+ times per week	39	15	10		7	14		4	20		7	14	
Metabolism Syndrome				0.023			0.071			0.125			0.996
No	361	16	31		97	82		99	86		91	89	
Yes	81	95	79		14	28		12	24		20	21	
Energy intake (kcal/day)	1908.5 ± 710.8	1443.0 ± 364.4	2632.4 ± 875.5	<0.001	1424.5 ± 402.7	2622.0 ± 879.5	<0.001	1694.2 ± 580.2	2341.2 ± 967.5	<0.001	1475.0 ± 527.8	2462.9 ± 852.5	<0.001

Statistical tests: Continuous variables were analyzed using ANOVA or Kruskal–Wallis rank sum test. Categorical variables were analyzed using Chi-square tests.

3.3. Relationship between Dietary Patterns and Metabolic Syndrome

Table 4 outlines the relationship between dietary patterns and metabolic syndrome, presenting the results of the logistic regression analyses of the associations between various dietary patterns and metabolic syndrome. We constructed three covariate models to elucidate these associations, each with different levels of adjustment. After adjusting for multiple covariates, the fully adjusted logistic regression model indicates that participants in the top quartile of the Legume–Nut pattern scores had greater OR for metabolic syndrome (OR = 2.63; 95%CI: 1.08–6.37; $p = 0.033$) than those in the bottom quartile. Compared with those in the bottom quartile, participants in the top quartile of the Egg–Vegetable pattern scores had a lower OR for metabolic syndrome (OR = 0.26; 95%CI: 0.10–0.70; $p = 0.007$). In addition, no significant impact was observed of the Sugar–Processed dietary pattern on metabolic syndrome, nor of the Alcohol–Meat dietary pattern on metabolic syndrome ($p > 0.05$).

Table 4. Relationship between dietary patterns and metabolic syndrome.

Cluster		Model 1 OR, 95% CI	*p*	Model 2 OR, 95% CI	*p*	Model 3 OR, 95% CI	*p*
Sugar–Processed	Q1	ref		ref		ref	
	Q4	2.33 (1.19, 4.57)	0.014	2.24 (1.11, 4.51)	0.025	1.10 (0.42, 2.84)	0.849
Alcohol–Meat	Q1	ref		ref		ref	
	Q4	2.37 (1.17, 4.79)	0.017	2.49 (1.13, 5.51)	0.024	1.29 (0.45, 3.70)	0.635
Legume–Nut	Q1	ref		ref		ref	
	Q4	2.30 (1.09, 4.88)	0.029	2.48 (1.14, 5.40)	0.022	2.63 (1.08, 6.37)	0.033
Egg–Vegetable	Q1	ref		ref		ref	
	Q4	1.07 (0.55, 2.12)	0.837	1.22 (0.57, 2.64)	0.608	0.26 (0.10, 0.70)	0.007

Statistical test: analyzed using logistic regression models.

3.4. Subgroup Analysis in Physical Activity Frequency and Metabolic Syndrome

As shown in Figure 2, within the irregular physical activity group, participants in the top quartile of the Legume–Nut pattern scores had greater OR for metabolic syndrome (OR = 4.15; 95%CI: 1.20–14.42; $p = 0.025$) than those in the bottom quartile. Compared with those in the bottom quartile, participants in the top quartile of the Egg–Vegetable pattern scores had a lower OR for metabolic syndrome (OR = 0.13; 95%CI: 0.03–0.68; $p = 0.015$). In addition, no significant association was observed between Sugar–Processed and Alcohol–Meat patterns and metabolic syndrome ($p > 0.05$). In the regular physical activity group, no significant association was observed between dietary patterns and metabolic syndrome ($p > 0.05$).

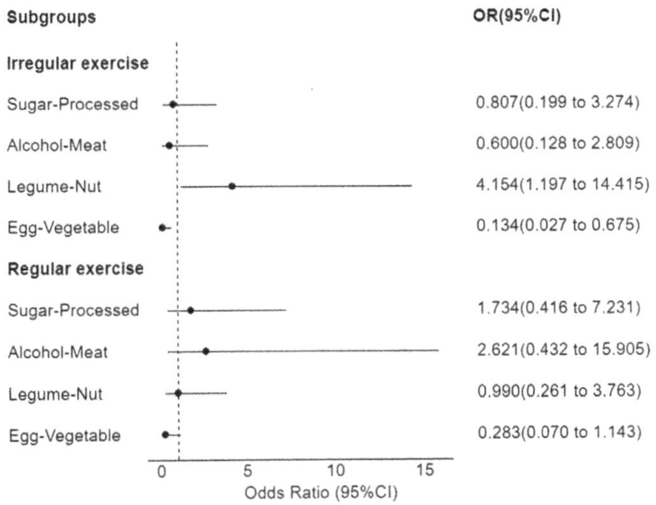

Figure 2. Subgroup analysis based on physical activity frequency.

4. Discussion

This study examined the impact of dietary patterns and exercise frequency on metabolic syndrome among younger demographics. Four dietary patterns were identified: the Legume–Nut dietary pattern, the Alcohol–Meat dietary pattern, the Sugar–Processed dietary pattern, and the Egg–Vegetable dietary pattern. The findings revealed that the Legume–Nut dietary pattern is associated with a higher risk of metabolic syndrome, whereas the Egg–Vegetable dietary pattern is associated with a lower risk. The Sugar–Processed dietary pattern and the Alcohol–Meat dietary pattern did not show a significant association with the risk of metabolic syndrome. In the subgroup analyses, similar results were observed among young adults with irregular physical activity. However, among young adults with regular physical activity, no significant association was observed between dietary patterns and metabolic syndrome.

This study identified four major dietary patterns: the Sugar–Processed dietary pattern, characterized by high consumption of sugary beverages, pastries, and fried foods, featuring high levels of carbohydrates (especially refined sugars) and unhealthy fats; the Alcohol–Meat dietary pattern, primarily consisting of red meat, processed meats, and alcoholic beverages, noted for its high protein and fat content, particularly saturated fats and alcohol; the Legume–Nut dietary pattern, which includes significant intake of beans and bean products, nuts, and aquatic and seafood, rich in plant proteins, unsaturated fats, fibers, and micronutrients; and the Egg–Vegetable dietary pattern, marked by the high consumption of eggs, fresh vegetables, whole grains, and white meat, offering high-quality proteins, low fats, abundant fibers, vitamins, minerals, and carbohydrates. Research by the Chinese Center for Disease Control and Prevention's Nutrition and Health Institute on dietary patterns among Chinese youth aged 18–35 indicates that these patterns can be categorized into the Traditional Rice dietary pattern, the Traditional Pasta dietary pattern and the High-Quality Protein dietary pattern [28]. These three dietary patterns closely resemble the Egg–Vegetable dietary pattern, the Legume–Nut dietary pattern, and Alcohol–Meat dietary pattern identified in this study. However, the earlier research did not include fried foods, sugary beverages, and pastries, which in our study, constituted the Sugar–Processed dietary pattern.

In the course of investigating the relationship between dietary patterns and metabolic syndrome, our findings indicate that the Legume–Nut dietary pattern is linked to an increased risk of developing metabolic syndrome. Conversely, the Egg–Vegetable dietary pattern appears to be associated with a reduced risk of metabolic syndrome. The Legume–Nut dietary pattern encompasses a variety of foods including tubers, beans, nuts, fruits, white meat, aquatic products, and seafood. This diet is predominantly characterized by a high intake of carbohydrates and sugars, particularly from potatoes and certain legumes that naturally contain significant amounts of starches and sugars [29,30]. In the absence of sufficient dietary fiber to mitigate these effects, there is an elevated risk of rapid blood sugar spikes [31]. Although fruits are abundant in natural fructose [32], excessive consumption can also result in fluctuations in blood sugar levels [33]. Nuts and seafood, while providing beneficial unsaturated fats that support cardiovascular health [34,35], can contribute to weight gain and negatively impact metabolic health if consumed in excess without adequate physical activity [36]. Furthermore, processed white meat and seafood may contain elevated levels of sodium [37], posing potential risks to blood pressure regulation and thereby increasing the likelihood of metabolic syndrome [38]. The Egg–Vegetable dietary pattern appears to be associated with a reduced risk of metabolic syndrome. This finding is consistent with a longitudinal study conducted in Australia, which examined the dietary patterns of young adults and their risk of metabolic syndrome and insulin resistance [14]. The Australian study found that a dietary pattern rich in fruits, vegetables, and whole grains was associated with a lower risk of metabolic syndrome. Similarly, our Egg–Vegetable dietary pattern is primarily composed of whole grains, fresh vegetables, eggs, and cooking oils, further supporting the beneficial effects of such diets on metabolic health. The high fiber content present in whole grains and fresh vegetables contributes to slower digestion and absorption, which in turn stabilizes blood sugar levels and promotes a prolonged sense of satiety, thereby preventing the overconsumption of high-calorie foods [39]. Eggs serve as a valuable source of protein, essential for muscle maintenance and overall metabolic health [40]. Additionally, healthy cooking oils, such as olive oil, provide essential monounsaturated fatty acids and small quantities of polyunsaturated fatty acids, both of which support normal cholesterol levels and cardiovascular health [41]. Furthermore, the low sugar content inherent in this dietary pattern aids in the prevention of metabolic disorders, including insulin resistance and diabetes [42].The Sugar–Processed dietary pattern and the Alcohol–Meat dietary pattern did not show a significant association with the risk of metabolic syndrome. This lack of association may be attributed to the relatively simplistic composition of these dietary patterns and the insufficient diversity of nutrient intake, which limits data variability. Furthermore, the absence of potential synergistic or interactive effects between foods may render the Sugar–Processed dietary pattern and the Alcohol–Meat dietary pattern less influential on metabolic syndrome compared to the Legume–Nut dietary pattern and the Egg–Vegetable dietary pattern, which are more nutritionally comprehensive and feature a greater variety of foods. Therefore, single or simplified food combinations may not be sufficiently relevant in studies addressing complex health conditions such as metabolic syndrome [17].

Subsequent subgroup analyses revealed that among individuals with irregular physical activity, the Legume–Nut dietary pattern is associated with a higher risk of metabolic syndrome, whereas the Egg–Vegetable dietary pattern is associated with a lower risk. In populations with irregular physical activity, the basal metabolic rate tends to be relatively low due to insufficient physical exertion, leading to reduced energy expenditure [43]. This condition makes the high-carbohydrate and high-fat content of the Legume–Nut pattern prone to causing an energy surplus and weight gain, further exacerbating the risk of metabolic syndrome [44]. Additionally, irregular physical activity may result in decreased insulin sensitivity and impaired blood glucose regulation [45]. Conversely, the Egg–Vegetable dietary pattern, characterized by its high fiber and low sugar content, can help improve blood glucose and insulin responses [46]. Thus, among those with irregular physical activity, this dietary pattern can effectively reduce the risk of metabolic syndrome. For individuals who engage in regular physical activity, the various physiological benefits conferred by physical activity, including an increased basal metabolic rate and improved glucose uptake by muscles and insulin sensitivity [45,47,48], may offset the potential negative impacts of different dietary patterns on the risk of metabolic syndrome. Furthermore, regular physical activity exhibits anti-inflammatory and antioxidant effects, which can effectively mitigate chronic inflammation and oxidative stress induced by unhealthy diets [49]. Consequently, the influence of dietary patterns on metabolic syndrome risk may be less pronounced in individuals who engage in regular physical activity. These findings underscore the importance of regular physical activity in preventing metabolic syndrome, demonstrating its efficacy across diverse dietary contexts.

The findings of this study highlight the crucial role of dietary patterns in preventing metabolic syndrome among young adults. Specifically, young adults, particularly those who do not engage in regular physical activity, should opt for high-fiber, low-sugar dietary patterns such as the Egg–Vegetable dietary pattern and avoid high-fat, high-carbohydrate dietary patterns like the Legume–Nut dietary pattern. This approach can aid in weight management, improve blood sugar regulation, and ultimately reduce the risk of metabolic syndrome. By adopting these healthier dietary habits and incorporating regular physical activity, young adults can enhance their overall metabolic health, decrease the likelihood of developing chronic conditions, and promote long-term well-being.

Nevertheless, it is crucial to recognize the specific limitations of this study. First, the sample size of this research is relatively small. This constraint may lead to insufficient statistical power, potentially affecting the generalizability and extendibility of the findings. Second, this cross-sectional study effectively reflects the status at a specific point in time but does not establish temporal sequence or causality, which may limit causal inference. Future research should include a larger sample size and employ a longitudinal design to track changes over time within the same cohort, thereby better assessing causal relationships and long-term effects. Additionally, this study may not have completely controlled for all potential confounding variables, such as socioeconomic status and genetic factors, which could influence the interpretation of the results.

5. Conclusions

In summary, this study revealed a significant association between dietary patterns and the risk of metabolic syndrome among younger demographics. These findings underscore the importance of balanced dietary choices as preventive strategies against metabolic syndrome. Future research should consider a larger sample size and a longitudinal design to better ascertain causal relationships and long-term effects. This approach can provide more definitive guidance for dietary recommendations to improve metabolic health in young adults.

Supplementary Materials: The following supporting information can be downloaded at: https://www.mdpi.com/article/10.3390/nu16172890/s1, Table S1: KMO and Bartlett's Test of Sphericity; Table S2: Eigenvalues and Variance Explained by Each Component in Principal Component; Table S3: Description of the food items included in the food categories; Figure S1: Appropriateness of Factor Analysis Scree Plot.

Author Contributions: J.L. contributed to the writing—review and editing, and the writing of the original draft of the study. W.L. was involved in the investigation and the data curation, review and editing. Q.L., Y.W., X.X. and H.C. were each responsible for the methodology. Y.H. performed the formal analysis. Y.Z., in addition to contributing to the writing—review and editing and supervision, was involved in the conceptualization of the study. X.Z. (Xiaonan Zhang) and H.Z. contributed to review and editing, supervision, and conceptualization. X.Z. (Xiaoying Zang) also contributed to the writing—review and editing, supervision, and conceptualization. X.Z. (Xiaoying Zang) assumes final responsibility for the decision to submit for publication. All authors have read and agreed to the published version of the manuscript.

Funding: This work was supported by the National Natural Science Foundation of China (72304206); Humanities and Social Science Fund of Ministry of Education of China (23YJAZH189); Nursing Special Development Fund of Tianjin Medical University (2022XKZX-04).

Institutional Review Board Statement: The study was conducted in accordance with the Declaration of Helsinki, and approved by the Medical Ethics Committee of Tianjin Medical University in China (TMuhMEC2022021) in 29 August 2022.

Informed Consent Statement: Informed consent was obtained from all subjects involved in the study.

Data Availability Statement: The original contributions presented in the study are included in the article/Supplementary Materials, further inquiries can be directed to the corresponding author.

Acknowledgments: The authors are grateful to all participants of this study and all surveyors who collected the data.

Conflicts of Interest: The authors declare no conflict of interest.

References

1. Chen, J.; Gu, D.; Chen, C.S.; Wu, X.; Hamm, L.L.; Muntner, P.; Batuman, V.; Lee, C.H.; Whelton, P.K.; He, J. Association between the metabolic syndrome and chronic kidney disease in Chinese adults. *Nephrol. Dial. Transplant.* **2007**, *22*, 1100–1106. [CrossRef] [PubMed]
2. Nóbrega da Silva, V.; Goldberg, T.B.L.; Silva, C.C.; Kurokawa, C.S.; Fiorelli, L.N.M.; Rizzo, A.; Corrente, J.E. Impact of metabolic syndrome and its components on bone remodeling in adolescents. *PLoS ONE* **2021**, *16*, e0253892. [CrossRef] [PubMed]
3. Lin, K.M.; Chiou, J.Y.; Kuo, H.W.; Tan, J.Y.; Ko, S.H.; Lee, M.C. Associations Between Unhealthy Lifestyle Behaviors and Metabolic Syndrome by Gender in Young Adults. *Biol. Res. Nurs.* **2019**, *21*, 173–181. [CrossRef] [PubMed]
4. Wang, F.; Wu, S.; Song, Y.; Tang, X.; Marshall, R.; Liang, M.; Wu, Y.; Qin, X.; Chen, D.; Hu, Y. Waist circumference, body mass index and waist to hip ratio for prediction of the metabolic syndrome in Chinese. *Nutr. Metab. Cardiovasc. Dis.* **2009**, *19*, 542–547. [CrossRef] [PubMed]

5. Expert Panel on Detection, Evaluation, and Treatment of High Blood Cholesterol in Adults. Executive Summary of The Third Report of The National Cholesterol Education Program (NCEP) Expert Panel on Detection, Evaluation, And Treatment of High Blood Cholesterol In Adults (Adult Treatment Panel III). *JAMA* **2001**, *285*, 2486–2497. [CrossRef] [PubMed]
6. Sperling, L.S.; Mechanick, J.I.; Neeland, I.J.; Herrick, C.J.; Despres, J.P.; Ndumele, C.E.; Vijayaraghavan, K.; Handelsman, Y.; Puckrein, G.A.; Araneta, M.R.; et al. The CardioMetabolic Health Alliance: Working Toward a New Care Model for the Metabolic Syndrome. *J. Am. Coll. Cardiol.* **2015**, *66*, 1050–1067. [CrossRef] [PubMed]
7. Tal, B.; Sack, J.; Yaron, M.; Shefer, G.; Buch, A.; Ben Haim, L.; Marcus, Y.; Shenkerman, G.; Sofer, Y.; Shefer, L.; et al. Increment in Dietary Potassium Predicts Weight Loss in the Treatment of the Metabolic Syndrome. *Nutrients* **2019**, *11*, 1256. [CrossRef] [PubMed]
8. Stanhope, K.L. Sugar consumption, metabolic disease and obesity: The state of the controversy. *Crit. Rev. Clin. Lab. Sci.* **2016**, *53*, 52–67. [CrossRef] [PubMed]
9. Koopman, R.; van Loon, L.J. Aging, exercise, and muscle protein metabolism. *J. Appl. Physiol. (1985)* **2009**, *106*, 2040–2048. [CrossRef]
10. Savkur, R.S.; Philips, A.V.; Cooper, T.A. Aberrant regulation of insulin receptor alternative splicing is associated with insulin resistance in myotonic dystrophy. *Nat. Genet.* **2001**, *29*, 40–47. [CrossRef]
11. Nazli, S.A.; Loeser, R.F.; Chubinskaya, S.; Willey, J.S.; Yammani, R.R. High fat-diet and saturated fatty acid palmitate inhibits IGF-1 function in chondrocytes. *Osteoarthr. Cartil.* **2017**, *25*, 1516–1521. [CrossRef] [PubMed]
12. Lee, H.; Rhee, T.M.; Park, H.E.; Han, K.; Choi, S.Y. Association between Cumulative Metabolic Risk Exposure and Cardiovascular Disease: A Nationwide Cohort of Over 3.6 Million Young Adults. *Eur. J. Prev. Cardiol.* **2024**, *00*, 1–13. [CrossRef] [PubMed]
13. Masrouri, S.; Shapiro, M.D.; Khalili, D.; Hadaegh, F. Impact of coronary artery calcium on mortality and cardiovascular events in metabolic syndrome and diabetes among younger adults. *Eur. J. Prev. Cardiol.* **2024**, *31*, 744–753. [CrossRef] [PubMed]
14. Ushula, T.W.; Mamun, A.; Darssan, D.; Wang, W.Y.S.; Williams, G.M.; Whiting, S.J.; Najman, J.M. Dietary patterns and the risks of metabolic syndrome and insulin resistance among young adults: Evidence from a longitudinal study. *Clin. Nutr.* **2022**, *41*, 1523–1531. [CrossRef] [PubMed]
15. Murakami, K.; Livingstone, M.B.E.; Shinozaki, N.; Sugimoto, M.; Fujiwara, A.; Masayasu, S.; Sasaki, S. Food Combinations in Relation to the Quality of Overall Diet and Individual Meals in Japanese Adults: A Nationwide Study. *Nutrients* **2020**, *12*, 327. [CrossRef] [PubMed]
16. Seo, E.H.; Kim, H.; Kwon, O. Association between Total Sugar Intake and Metabolic Syndrome in Middle-Aged Korean Men and Women. *Nutrients* **2019**, *11*, 2042. [CrossRef] [PubMed]
17. Michels, K.B.; Schulze, M.B. Can dietary patterns help us detect diet-disease associations? *Nutr. Res. Rev.* **2005**, *18*, 241–248. [CrossRef] [PubMed]
18. Godoy-Izquierdo, D.; Ogallar, A.; Lara, R.; Rodriguez-Tadeo, A.; Arbinaga, F. Association of a Mediterranean Diet and Fruit and Vegetable Consumption with Subjective Well-Being among Adults with Overweight and Obesity. *Nutrients* **2021**, *13*, 1342. [CrossRef] [PubMed]
19. Pelczynska, M.; Mikulska, A.A.; Czyzewska, K.; Bogdanski, P.; Grzelak, T. The Association of Serum Circulating Neuropeptide Q and Chemerin Levels with Cardiometabolic Risk Factors among Patients with Metabolic Syndrome. *Biomolecules* **2021**, 11. [CrossRef]
20. Alberti, K.G.; Eckel, R.H.; Grundy, S.M.; Zimmet, P.Z.; Cleeman, J.I.; Donato, K.A.; Fruchart, J.C.; James, W.P.; Loria, C.M.; Smith, S.C., Jr. Harmonizing the metabolic syndrome: A joint interim statement of the International Diabetes Federation Task Force on Epidemiology and Prevention; National Heart, Lung, and Blood Institute; American Heart Association; World Heart Federation; International Atherosclerosis Society; and International Association for the Study of Obesity. *Circulation* **2009**, *120*, 1640–1645. [CrossRef]
21. Ouyang, X.; Lou, Q.; Gu, L.; Ko, G.T.; Mo, Y.; Wu, H.; Bian, R. Anthropometric parameters and their associations with cardio-metabolic risk in Chinese working population. *Diabetol. Metab. Syndr.* **2015**, *7*, 37. [CrossRef]
22. Lee, M.S.; Pan, W.H.; Liu, K.L.; Yu, M.S. Reproducibility and validity of a Chinese food frequency questionnaire used in Taiwan. *Asia Pac. J. Clin. Nutr.* **2006**, *15*, 161–169.
23. Zhao, W.; Hasegawa, K.; Chen, J. The use of food-frequency questionnaires for various purposes in China. *Public Health Nutr.* **2002**, *5*, 829–833. [CrossRef]
24. Hotz, C.; Abdelrahman, L. Simple methods to obtain food listing and portion size distribution estimates for use in semi-quantitative dietary assessment methods. *PLoS ONE* **2019**, *14*, e0217379. [CrossRef]
25. Huang, Q.; Jiang, H.; Zhang, J.; Jia, X.; Huang, F.; Wang, H.; Zhang, B.; Wang, L.; Gu, M.; Huang, Y.; et al. Dietary Patterns Are Associated With Multi-Dimensional Cognitive Functions Among Adults Aged 55 and Older in China. *Front. Nutr.* **2022**, *9*, 806871. [CrossRef]
26. Wang, X.; Li, T.; Li, H.; Li, D.; Wang, X.; Zhao, A.; Liang, W.; Xiao, R.; Xi, Y. Association of Dietary Inflammatory Potential with Blood Inflammation: The Prospective Markers on Mild Cognitive Impairment. *Nutrients* **2022**, *14*, 2417. [CrossRef]
27. Bull, F.C.; Al-Ansari, S.S.; Biddle, S.; Borodulin, K.; Buman, M.P.; Cardon, G.; Carty, C.; Chaput, J.P.; Chastin, S.; Chou, R.; et al. World Health Organization 2020 guidelines on physical activity and sedentary behaviour. *Br. J. Sports Med.* **2020**, *54*, 1451–1462. [CrossRef]

28. Zhao, B.; Li, W.; Liu, M.; Chen, L.; Jiang, H.; Li, F.; Wang, S.; Ding, N.; Wang, L.; Wang, Z. [Change of dietary pattern among young people aged 18-35 in China from 1989 to 2018]. *Wei Sheng Yan Jiu* **2024**, *53*, 419–434. [CrossRef]
29. Rippe, J.M.; Angelopoulos, T.J. Sugars, obesity, and cardiovascular disease: Results from recent randomized control trials. *Eur. J. Nutr.* **2016**, *55*, 45–53. [CrossRef]
30. Sievenpiper, J.L. Low-carbohydrate diets and cardiometabolic health: The importance of carbohydrate quality over quantity. *Nutr. Rev.* **2020**, *78*, 69–77. [CrossRef]
31. Reynolds, A.N.; Akerman, A.P.; Mann, J. Dietary fibre and whole grains in diabetes management: Systematic review and meta-analyses. *PLoS Med.* **2020**, *17*, e1003053. [CrossRef] [PubMed]
32. Chang, H.H.; Lin, I.C.; Wu, C.W.; Hung, C.Y.; Liu, W.C.; Wu, C.Y.; Cheng, C.L.; Wu, K.L.H. High fructose induced osteogenic differentiation of human valve interstitial cells via activating PI3K/AKT/mitochondria signaling. *Biomed. J.* **2022**, *45*, 491–503. [CrossRef] [PubMed]
33. Basciano, H.; Federico, L.; Adeli, K. Fructose, insulin resistance, and metabolic dyslipidemia. *Nutr. Metab.* **2005**, *2*, 5. [CrossRef] [PubMed]
34. Salas-Salvado, J.; Garcia-Arellano, A.; Estruch, R.; Marquez-Sandoval, F.; Corella, D.; Fiol, M.; Gomez-Gracia, E.; Vinoles, E.; Aros, F.; Herrera, C.; et al. Components of the Mediterranean-type food pattern and serum inflammatory markers among patients at high risk for cardiovascular disease. *Eur. J. Clin. Nutr.* **2008**, *62*, 651–659. [CrossRef] [PubMed]
35. Zibaeenezhad, M.J.; Ghavipisheh, M.; Attar, A.; Aslani, A. Comparison of the effect of omega-3 supplements and fresh fish on lipid profile: A randomized, open-labeled trial. *Nutr. Diabetes* **2017**, *7*, 1. [CrossRef] [PubMed]
36. Swinburn, B.; Sacks, G.; Ravussin, E. Increased food energy supply is more than sufficient to explain the US epidemic of obesity. *Am. J. Clin. Nutr.* **2009**, *90*, 1453–1456. [CrossRef] [PubMed]
37. Balakrishna, Y.; Manda, S.; Mwambi, H.; van Graan, A. Identifying Nutrient Patterns in South African Foods to Support National Nutrition Guidelines and Policies. *Nutrients* **2021**, *13*, 3194. [CrossRef]
38. Jackson, S.L.; Cogswell, M.E.; Zhao, L.; Terry, A.L.; Wang, C.Y.; Wright, J.; Coleman King, S.M.; Bowman, B.; Chen, T.C.; Merritt, R.; et al. Association Between Urinary Sodium and Potassium Excretion and Blood Pressure Among Adults in the United States: National Health and Nutrition Examination Survey, 2014. *Circulation* **2018**, *137*, 237–246. [CrossRef] [PubMed]
39. Dahl, W.J.; Stewart, M.L. Position of the Academy of Nutrition and Dietetics: Health Implications of Dietary Fiber. *J. Acad. Nutr. Diet.* **2015**, *115*, 1861–1870. [CrossRef]
40. Arentson-Lantz, E.; Clairmont, S.; Paddon-Jones, D.; Tremblay, A.; Elango, R. Protein: A nutrient in focus. *Appl. Physiol. Nutr. Metab.* **2015**, *40*, 755–761. [CrossRef]
41. Torres-Moreno, M.; Torrescasana, E.; Salas-Salvado, J.; Blanch, C. Nutritional composition and fatty acids profile in cocoa beans and chocolates with different geographical origin and processing conditions. *Food Chem.* **2015**, *166*, 125–132. [CrossRef] [PubMed]
42. Rinaldi, A.E.; Gabriel, G.F.; Moreto, F.; Corrente, J.E.; McLellan, K.C.; Burini, R.C. Dietary factors associated with metabolic syndrome and its components in overweight and obese Brazilian schoolchildren: A cross-sectional study. *Diabetol. Metab. Syndr.* **2016**, *8*, 58. [CrossRef] [PubMed]
43. Brown, M.A.; Green, B.P.; James, L.J.; Stevenson, E.J.; Rumbold, P.L. The Effect of a Dairy-Based Recovery Beverage on Post-Exercise Appetite and Energy Intake in Active Females. *Nutrients* **2016**, *8*, 355. [CrossRef]
44. Keeble, M.; Adams, J.; Vanderlee, L.; Hammond, D.; Burgoine, T. Associations between online food outlet access and online food delivery service use amongst adults in the UK: A cross-sectional analysis of linked data. *BMC Public Health* **2021**, *21*, 1968. [CrossRef]
45. Holloszy, J.O. Exercise-induced increase in muscle insulin sensitivity. *J. Appl. Physiol. (1985)* **2005**, *99*, 338–343. [CrossRef]
46. Finelli, C.; Tarantino, G. Is there any consensus as to what diet or lifestyle approach is the right one for NAFLD patients? *J. Gastrointestin. Liver Dis.* **2012**, *21*, 293–302.
47. Sjodin, A.M.; Forslund, A.H.; Westerterp, K.R.; Andersson, A.B.; Forslund, J.M.; Hambraeus, L.M. The influence of physical activity on BMR. *Med. Sci. Sports Exerc.* **1996**, *28*, 85–91. [CrossRef] [PubMed]
48. Hawley, J.A.; Lessard, S.J. Exercise training-induced improvements in insulin action. *Acta Physiol.* **2008**, *192*, 127–135. [CrossRef]
49. Kasapis, C.; Thompson, P.D. The effects of physical activity on serum C-reactive protein and inflammatory markers: A systematic review. *J. Am. Coll. Cardiol.* **2005**, *45*, 1563–1569. [CrossRef]

nutrients MDPI

Article

Sub-Optimal Paternal Diet at the Time of Mating Disrupts Maternal Adaptations to Pregnancy in the Late Gestation Mouse

Afsaneh Khoshkerdar [1], Nader Eid [1], Vipul Batra [1], Nichola Baker [1], Nadine Holmes [2], Sonal Henson [2], Fei Sang [2], Victoria Wright [2], Jane McLaren [3], Kevin Shakesheff [3], Kathryn J. Woad [4], Hannah L. Morgan [1] and Adam J. Watkins [1,*]

[1] Lifespan and Population Health, School of Medicine, University of Nottingham, Nottingham NG7 2UH, UK; afsaneh.khoshkerdar@nottingham.ac.uk (A.K.); nader.eid@nottingham.ac.uk (N.E.); vipul.batra@nottingham.ac.uk (V.B.); nicky.baker@nottingham.ac.uk (N.B.); hannah.morgan@nottingham.ac.uk (H.L.M.)

[2] Deep Seq, School of Life Sciences, Queen's Medical Centre, University of Nottingham, Nottingham NG7 2UH, UK; nadine.holmes1@nottingham.ac.uk (N.H.); sonal.henson1@nottingham.ac.uk (S.H.); fei.sang2@nottingham.ac.uk (F.S.); victoria.wright@nottingham.ac.uk (V.W.)

[3] Regenerative Medicine and Cellular Therapies, School of Pharmacy, University of Nottingham, Nottingham NG7 2UH, UK; jane.mclaren@nottingham.ac.uk (J.M.)

[4] School of Veterinary Medicine and Science, University of Nottingham, Loughborough LE12 5RD, UK; katie.woad@nottingham.ac.uk

* Correspondence: adam.watkins@nottingham.ac.uk; Tel.: +44-(0)115-8230699

Abstract: Pregnancy represents a stage during which maternal physiology and homeostatic regulation undergo dramatic change and adaptation. The fundamental purpose of these adaptations is to ensure the survival of her offspring through adequate nutrient provision and an environment that is tolerant to the semi-allogenic foetus. While poor maternal diet during pregnancy is associated with perturbed maternal adaptations during pregnancy, the influence of paternal diet on maternal well-being is less clearly defined. We fed C57BL/6 male mice either a control (CD), low protein diet (LPD), a high fat/sugar Western diet (WD) or the LPD or WD supplemented with methyl donors (MD-LPD and MD-WD, respectively) for a minimum of 8 weeks prior to mating with C57BL/6 females. Mated females were culled at day 17 of gestation for the analysis of maternal metabolic, gut, cardiac and bone health. Paternal diet had minimal influences on maternal serum and hepatic metabolite levels or gut microbiota diversity. However, analysis of the maternal hepatic transcriptome revealed distinct profiles of differential gene expression in response to the diet of the father. Paternal LPD and MD-LPD resulted in differential expression of genes associated with lipid metabolism, transcription, ubiquitin conjugation and immunity in dams, while paternal WD and MD-WD modified the expression of genes associated with ubiquitin conjugation and cardiac morphology. Finally, we observed changes in maternal femur length, volume of trabecular bone, trabecular connectivity, volume of the cortical medullar cavity and thickness of the cortical bone in response to the father's diets. Our current study demonstrates that poor paternal diet at the time of mating can influence the patterns of maternal metabolism and gestation-associated adaptations to her physiology.

Keywords: paternal diet; maternal health; foetal programming; cardio-metabolic health

1. Introduction

During pregnancy, maternal physiology undergoes dramatic and dynamic adaptations, affecting almost every organ system including the cardiovascular, metabolic, microbiome and skeletal systems [1]. These maternal adaptations, which have the fundamental purpose of supporting the development of her offspring, initiate shortly after conception and continue throughout early neonatal life. Inappropriate maternal adaptation(s) can impair a range of processes including transport of nutrients via the placenta, utero-placental blood flow and the development and growth of the foetus. Such impairments can also result in the development of conditions such as gestational diabetes and pre-eclampsia, both major causes of maternal and foetal mortality [2,3]. Furthermore, gestational diabetes has been associated with foetal overgrowth, while pre-eclampsia results in growth restriction [4–6], both of which can impact the long-term cardio-metabolic health of the offspring. Therefore, appropriate maternal physiological responses during pregnancy, in both humans and animals, are critical for ensuring the well-being of both the mother and her offspring.

During a normal pregnancy, maternal metabolic status changes dramatically, which associates with weight gain, elevated fasting blood glucose levels, insulin resistance, glucose intolerance, low grade inflammation and altered metabolic hormone levels. During early pregnancy, typically referred to as the anabolic phase, there is an increase in maternal lipid accumulation driven by increased food intake and lipogenesis [7]. Underlying the promotion of fat storage is an increase in the levels of hormones such as progesterone, leptin, prolactin and cortisol in addition to significant adipose tissue hypertrophy [7–9]. In contrast, glucose homeostasis is maintained by increasing insulin secretion through hypertrophy and hyperplasia of pancreatic β-cells [10]. This is critical as glucose is essential for supporting foetal development. In women, during the third trimester, metabolism switches into a net catabolic phase, due mainly to a decrease in insulin sensitivity and an increase in the levels of oestrogen. The result of these changes are significant increases in serum total cholesterol, triglycerides and low-density lipoprotein cholesterol [11–14]. Additionally, studies have observed significant rise in the levels of high-density lipoprotein cholesterol [11,15]. The relative gestational hypertriglyceridemia is driven by both an increased accumulation and production of triglyceride-rich lipoproteins in combination with a decreased clearance by the liver. The rise in the levels of these maternal lipids is critical for providing the necessary additional energy required to support foetal growth and placental function. Furthermore, in late pregnancy, the influence of increased oestrogen levels and insulin resistance decreases the activity of adipose lipoprotein lipase and hepatic lipase activity [16], resulting in increased levels of circulating lipids [17] and an increase in fatty acid synthesis. Typically, by late pregnancy, levels of plasma cholesterol are approximately 30% higher than in non-pregnant women, while levels of triglyceride are approximately 50% higher [18].

Paralleling the changes to the maternal metabolic status are modulations of the cardiovascular system. Following conception, the uterine spiral arteries are remodelled, promoted by vacuolation of endothelial cells and swelling of the vascular smooth muscle in response to the immune cells within the decidua [19]. The array of uterine immune cells, including T regulatory cells, dendritic cells and macrophages, as well as the factors they secrete such as vascular endothelial growth factor (VEGF), placental growth factors (PlGF), transforming growth factor-beta (TGF-β) and granulocyte macrophage colony stimulating factor (CSF2) [20,21], enable trophoblast invasion and the initiation of spiral artery remodelling. This remodelling forms the basis of subsequent changes in the maternal cardiovascular system. By the eighth week of gestation in women, changes in cardiac output are detected which increases by up to 50% by weeks 16–20 [22]. Such changes are enabled through reductions in vascular peripheral resistance [23] and allow for an increased uteroplacental blood flow while maintaining maternal blood pressure.

Previously, we have shown that a paternal low protein diet (LPD), with or without the addition of 1-carbon methyl donors and carriers such as folate, methionine and vitamin B12, impacts on a range of paternal, embryonic, foetal, placental and adult offspring parameters [24–28]. We have also observed differential profiles of maternal uterine cytokines and gene expression of immune cell markers following mating with LPD fed males [25]. These observations indicate that the quality of a father's diet at the time of mating impacts on both his developing offspring as well as the gestating mother. Paternally mediated changes in uterine vascular remodelling, embryo invasion and placental development and endocrine function could all influence maternal cardio-metabolic health and adaptations during pregnancy [29]. Such perturbations to maternal gestational physiology would also impact significantly on foetal well-being and long-term health. Therefore, the aim of this exploratory study was to explore the impact of paternal under (LPD) and over (a high fat/sugar Western diet; WD) nutrition on late gestation maternal cardio-metabolic status. Furthermore, we explore whether supplementation with a mix of methyl donors negates any detrimental influences of these poor-quality diets.

2. Materials and Methods

2.1. Animal Treatments and Tissue Collection

All experimental procedures were approved by the local ethics committee (AWERB) on the 06/10/2017 at the University of Nottingham. All procedures were conducted under the UK Home Office Animal (Scientific Procedures) Act 1986 Amendment Regulations 2012, which transposed Directive 2010/63/EU into UK law. Eight-week-old male C57BL/6J males and females (Harlan Ltd., Belton, Leicestershire, UK) were maintained within the Bio Support Unit at the University of Nottingham. Animals were housed in controlled 12/12 h light/dark conditions with a constant temperature (21 °C ± 3 °C) and access to food and water ad libitum. After a short period of acclimatization, males were allocated to one of five diets consisting of either a control diet (CD: 18% casein, 61% carbohydrate of which 21% are sugars, 10% fat), an isocaloric low protein diet (LPD: 9% casein, 69% carbohydrate of which 24% are sugars, 10% fat), the LPD supplemented with methyl donors and carriers (MD LPD; 5 g/kg diet choline chloride, 15 g/kg diet betaine, 7.5 g/kg diet methionine, 15 mg/kg diet folic acid, 1.5 mg/kg diet vitamin B12), a Western diet (WD: 19% casein, 44% carbohydrate of which 35% are sugars, 21% fat) or a Western diet supplemented with methyl donors and carriers (MD-WD; 5 g/kg diet choline chloride, 15 g/kg diet betaine, 7.5 g/kg diet methionine, 15 mg/kg diet folic acid, 1.5 mg/kg diet vitamin B12). All diets were commercially manufactured (Special Dietary Services Ltd., Witham, UK) and their full composition are detailed in Supplementary Materials (Table S1). Males (n = 8) were maintained on their respective diets for a minimum of 7 weeks prior to mating with virgin 8–12-week-old C57BL/6J females (n = 8). All females were mated with a separate male and maintained on standard chow (Rat and Mouse No.1 Maintenance chow diet, Special Dietary Services Ltd., Witham, UK). Pregnancy in females was confirmed by the presence of a vaginal plug (embryonic day 0.5; E0.5). Dams were culled via cervical dislocation at E17.5, and litter size was recorded. Samples of maternal liver and whole hearts were snap frozen in liquid nitrogen prior to storage at −80 °C. Maternal blood was collected via heart puncture, allowed to clot on ice and centrifuged at 10,000× g, 4 °C for 10 min prior to storage of the serum at −80 °C. Maternal faecal pellets were collected from the descending colon in nuclease-free tubes using sterile forceps and stored at −20 °C.

2.2. Assessment of Maternal Metabolic Status

Maternal liver cholesterol level was quantified using the Cholesterol Quantitation Kit (MAK043; Sigma-Aldrich, Gillingham, UK) in accordance with the manufacturer's instructions. Briefly, approximately 10 mg of frozen liver tissue was ground to a powder using a pestle and mortar placed on dry ice. Samples were disrupted in 200 μL of a chloroform–isopropanol–IGEPAL (7:11:0.1; Sigma-Aldrich, Gillingham, UK) solution using a TissueLyser II (Qiagen, Manchester, UK). Samples were centrifuged at 13,000× g for 10 min to remove cellular debris. The organic phase was transferred to a new tube and dried under vacuum to remove excess chloroform. Samples were reconstituted in an assay buffer, prior to incubation with a cholesterol probe and a cholesterol esterase enzyme mix, in accordance with the manufacturer's instructions. Samples were incubated at room temperature for 60 min and analysed in duplicate at 585 nm using an iMark Microplate Reader (Bio-Rad, Watford, UK).

Maternal liver free fatty acid levels were quantified using the Free Fatty Acid Quantitation Kit (MAK044; Sigma-Aldrich, Gillingham, UK) in accordance with the manufacturer's instructions. Briefly, approximately 10 mg of frozen liver tissue was ground to a powder using a pestle and mortar placed on dry ice prior to disruption in 200 μL of a 1% (w/v)

Triton X-100 in chloroform (Sigma-Aldrich, Gillingham, UK) solution using a TissueLyser II (Qiagen, Manchester, UK). The sample was centrifuged at 13,000× *g* for 10 min. The organic phase was removed and dried under vacuum before being reconstituted in assay buffer, prepared in accordance with the manufacturer's instructions. Samples were analysed in duplicate at 570 nm using an iMark Microplate Reader (Bio-Rad, Watford, UK).

Maternal liver triglyceride levels were measured using the Triglyceride Quantification Kit (MAK266; (Sigma-Aldrich, Gillingham, UK) in accordance with the manufacturer's instructions. Briefly, approximately 100 mg of frozen liver tissue was ground to a powder using a pestle and mortar placed on dry ice prior to disruption in 1 mL of 5% Nonidet P40 (Sigma-Aldrich, Gillingham, UK) using a TissueLyser II (Qiagen, Manchester, UK). The sample was centrifuged at 13,000× *g* for 10 min prior to analysis of the supernatant in accordance with the manufacturer's instructions. Samples were analysed in duplicate at 595 nm using an iMark Microplate Reader (Bio-Rad, Watford, UK).

Maternal Serum glucose levels were determined using a glucose colorimetric detection kit (EIAGLUC; Thermo Fisher Scientific, Loughborough, UK) in accordance with the manufacturer's instructions. Samples were measured in duplicate at 595 nm using an iMark Microplate Reader (Bio-Rad, Watford, UK). Serum insulin levels were determined using a rat/mouse insulin ELISA kit (EZRMI-13K; EMD Millipore Corporation, Burlington, MA, USA) in accordance with the manufacturer's instructions. Samples were measured in duplicate at 450 nm using an iMark Microplate Reader (Bio-Rad, Watford, UK).

2.3. Maternal Gut Microbiota Sequencing

DNA was isolated from maternal faecal pellets using the QIAamp DNA stool mini kit (Qiagen, Manchester, UK) in accordance with the manufacturer's instructions. Sequencing was conducted as described previously [28]. Briefly, sequencing was conducted on the V3-V4 region of the 16S rRNA gene in accordance with Illumina 16S Metagenomic Sequencing Library Preparation protocol. The 16S rRNA amplicons were generated using forward 5′ (TCGTCGGCAGCGTCAGATGTGTATAAGAGACAGCCTACGGGNGGCWGCAG) and Reverse 5′ (GTCTCGTGGGCTCGGAGATGTGTATAAGAGACAGGACTACHVGGGTATC-TAATCC) primers, flanked by Illumina adapter-overhang sequences. Illumina dual index barcodes (Illumina XT Index Kit v2, Set A: FC-131-2001; Illumina, Cambridge, UK) were attached to each amplicon. PCR clean-up was conducted using AMpure XP beads (Beckman; A63882; High Wycombe, UK). Library fragment-length distributions were analysed using the Agilent TapeStation 4200 and the Agilent D1000 ScreenTape Assay (Agilent; 5067-5582 and 5067-5583; Stockport, UK). Libraries were pooled in equimolar amounts, and the pool was size-selected using the Blue Pippin (Sage Science; Beverly, MA, USA) and a 1.5% Pippin Gel Cassette (Sage Science; BDF2010; Beverly, MA, USA). Sequencing was performed by Deep Seq at the University of Nottingham on an Illumina MiSeq using a MiSeq Reagent Kit v3 (600 cycle) (Illumina; MS-102-3003; Illumina, Cambridge, UK) to generate 300 bp paired-end reads. Raw reads were processed by the Qiime2 pipeline and trimmed. Greengenes version 13.8 was used in the classification [30].

2.4. Maternal Liver RNA-Seq

Total RNA was isolated from samples of maternal liver using the RNeasy Mini plus kit (Qiagen, Manchester, UK), in accordance with the manufacturer's instructions. RNA quantity and quality were initially assessed by Nanodrop. Stranded RNA-seq libraries were prepared from 500 ng of total RNA per sample, using the NEBNext Ultra II Directional RNA Library Preparation Kit for Illumina (NEB; E7760) and NEBNext Multiplex Oligos for Illumina (96 Unique Dual Index Pairs) (NEBNext; E6440). Prior to library preparation, total RNA was treated with QIAseq FastSelect -rRNA HMR (Cat: 334222, Qiagen, Manchester, UK) to prevent any rRNA present from being converted into the sequencing library. Libraries were quantified using the Qubit Fluorometer and the Qubit dsDNA HS Kit (ThermoFisher Scientific; Q32854; Loughborough, UK). Library fragment-length distributions were analysed using the Agilent 4200 TapeStation and the Agilent High Sensitivity

D1000 ScreenTape Assay (Agilent; 5067-5584 and 5067-5585; Stockport, UK). Libraries were pooled in equimolar amounts and final library quantification was performed using the KAPA Library Quantification Kit for Illumina (Roche; KK4824; Welwyn Garden City, UK). The library pool was sequenced on the Illlumina NextSeq500 over three NextSeq500 High Output 150 cycle kits (Illumina; 20024907; Cambridge, UK) to generate over 40 million pairs of 75 bp paired-end reads per sample. Raw reads were trimmed of Illumina adapters and low quality ($Q < 20$) nucleotides using TrimGalore (v 0.6.7). Reads shorter than 15 bp were discarded. Trimmed reads were aligned to the Mus musculus reference genome GRCm39 using HISAT2 (v 2.2.1). StringTie (v 2.2.1) was used to assemble genes and calculate gene abundance. Differential expression analysis was performed using DESeq2 (within version 3.0 of Bioconductor). Liver RNA-seq data have been submitted to the Gene Expression Omnibus (GEO) at NCBI under accession number: GSE265783.

2.5. Maternal Femur μ-CT Analysis

Maternal femurs were dissected free from muscle prior to fixation in 4% neutral buffered formalin (Sigma, UK) at room temperature overnight and subsequent storage in 70% ethanol prior to analysis. Whole femurs were scanned using a Skyscan 1174 μ-CT scanner (Bruker, Belgium). All scans were taken at 50 kVa and 800 μA with a 0.5 mm aluminium filter, 3600 ms exposure time, 180° tomographic rotation and a Voxel resolution of 17.84 μm^2. Individual two-dimensional cross-sectional images were reconstructed using Bruker NRecon software (version 1.7.4.6). Stacks of images for each individual femur were imported into BoneJ [31], and the total number of sections was defined. Identification of the trabecular and cortical regions for analysis was defined based on the first appearance of a bridging connection of the low-density growth plate chondrocyte seam. Using this as a standard anatomical set point, an offset of 3% of the total bone length towards the femoral head from the reference growth plate was used to define the start of the trabecular region of interest, and a series of trabecular sections representing 5% of the total bone length was analysed. For the analysis of the cortical bone, the mid bone position was defined based on the total length of the bone, and an offset of 2.5% of the total bone length towards the femoral head was used to define the start of the cortical region. A series of cortical sections representing 5% of the total bone length was analysed. Using BoneJ, measurements of trabecular and cortical bone volume (Bv), total volume (Tv), bone volume fraction (Bv/Tv), trabecular thickness (Tb.Th), medullary cavity volume (Mv) degree of anisotropy (Da), connectivity density (Con.D), maximum moment of inertia (Imax) and minimum moment of inertia (Imin) were defined. In addition, cortical bone volume (Cb-V), cross section area (Cb-Cx) and thickness (Cb-Th) and the volume of the medullary cavity (Cb-Mv) were also defined.

2.6. Maternal Cardiac Gene Expression

Total RNA was extracted from frozen maternal heart tissue using the RNeasy Plus Mini Kit (Qiagen, Manchester, UK). Whole hearts were powdered using a pestle and mortar over dry ice. No more than 25 mg of powdered tissue was disrupted in the kit's lysis buffer using the TissueLyser II (Qiagen, Manchester, UK) prior to RNA isolation in accordance with the manufacturer's instructions. cDNA synthesis was performed using the TaqMan™ Reverse Transcription Reagents kit (N8080234; ThermoFisher Scientific, Loughborough, UK) according to manufacturer's protocol with 1 μg input of RNA. Real-time quantitative PCR (RT-qPCR) was performed as previously outlined [27]. Briefly, reactions consisted of 5 ng cDNA and 175 nM forward and reverse primers (Eurofins Genomics, Ebersberg, Germany) with 1× Precision SYBR Green Mastermix (PrimerDesign, Southampton, UK), and RT-qPCR was performed using an Applied Biosystems 7500 Fast system. Gene expression was analysed using the delta–delta Ct method relative to CD expression. The GeNorm method used to normalise gene expression, as previously described [32] to two reference genes phosphoglycerate kinase 1 (Pgk1) and tubulin, alpha 1a (Tuba1a). All primer sequences are detailed in Supplementary Materials (Table S2).

2.7. Statistical Analyses

All maternal ($n = 8$) data were analysed using GraphPad Prism (version 10) or SPSS (version 28). Data were assessed initially for normality (Shapiro–Wilk and Kolmogorov–Smirnov tests) prior to analysis using one-way ANOVA for normally distributed data or a Kruskal–Wallis test for non-normally distributed data with an appropriate post hoc test for correction of multiple group comparisons. Where appropriate, data were corrected for multiple comparisons using the Benjamini–Hochberg False Discovery Rate (FDR) method. Correlations between parameters were conducted using Pearson correlation. Significance was taken at $p < 0.05$.

3. Results

3.1. Maternal Physiology and Metabolic Status

On the day immediately after mating (E0.5), there were no significant differences in mean body weight between any of the female groups. Similarly, there were no differences in mean maternal body weight when the females were culled at gestational day 17.5 (E17.5; Figure 1A), in the % weight gain from E0.5 to E17.5 (Figure 1B) or in the raw weight of the maternal heart (Figure 1C) or liver (Figure 1D). Additionally, there were no differences in mean litter size between groups (Figure 1E). While there were minimal differences between the groups in relation to mean body/organ weights, we identified group-specific correlative traits. In all groups, we identified significant positive correlations between maternal weight at day E17 and total weight gained over pregnancy. In females mated with CD fed males, we also observed a positive association between maternal E17.5 weight and the weight of the heart, liver and kidneys (Figure 1F, $p < 0.03$). However, in females mated with LPD and MD-LPD fed males, only a positive association between maternal weight and kidney weight existed (Figure 1G,H), while in in females mated with WD and MD-WD fed males, the association was between maternal weight and liver weight (Figure 1I,J). In females mated with CD fed males, we observed significant negative correlations between weight gained per foetus, maternal weight gain and litter size (Figure 1F, $p < 0.03$). However, such negative associations between increases in litter size (and hence maternal weight gain) and weight per foetus were not observed in any of the other groups (Figure 1G–J). Analysis of maternal hepatic and serum metabolite levels revealed no difference in mean hepatic triglyceride (Figure 1K), free fatty acids (Figure 1L) and cholesterol (Figure 1M), or in serum insulin (Figure 1N) or glucose (Figure 1O).

3.2. Impact of Paternal Diet on Late Gestation Maternal Gut Microbiota

To investigate whether paternal diet influenced late gestation maternal gut bacterial diversity, we sequenced the hypervariable V3–V4 region of the bacterial 16S rRNA gene (see Supplementary Materials for full sequencing data; Data S1). We observed no difference in bacterial diversity between groups (Figure 2A,B). Principle component analysis (PCA) showed all groups clustered similarly (Figure 2C). Analysis of bacterial profiles at the phylum (Figure 2D) and family (Supplementary Data S1) levels showed no significant differences between treatment groups. Furthermore, the assessment of the relationship between the abundance of Bacteroidetes and Firmicutes showed no differences between groups (Figure 2E). However, to understand whether poor paternal diet may influence the broader maternal bacterial profiles, we correlated each bacterial component at the family level with each other. All groups showed a significant positive correlation between the abundance of Actinobacteria and Bifidobacteriaceae ($p < 0.05$; Figure 2F–J), as well as a significant negative correlation between the abundance of Firmicutes and Bacteroidetes ($p < 0.05$; Figure 2F–J). In females mated with CD fed males, we observed a significant positive correlation between Tenericutes and Proteobacteria ($r = 0.99$, $p < 0.0001$; Figure 2F) which was not present in any of the other groups. In females mated with LPD fed males, we observed a significant positive association between the abundance of Proteobacteria and Actinobacteria ($r = 0.87$, $p = 0.024$; Figure 2G), an association also seen in females mated to WD fed males ($r = 0.95$, $p = 0.004$; Figure 2I) as well as a negative association between unclassified bacteria, termed 'Others' and Firmicutes (r = −0.82, p = 0.046). In females mated with MD-LPD fed males, we observed a positive correlation between 'Others' and Protobacteria (r = 0.92, p = 0.001), as well as significant associations between the abundance of Bifidobacteriaceae with 'Others' (r = 0.90, p = 0.014) and with Tenericutes (r = 0.091; p = 0.013) (Figure 2H). In females mated with MD-LPD fed males, we also observed significant positive correlations between Deferribacteres and Firmicutes (r = 0.89, p = 0.017; Figure 2H) and a negative association between Deferribacteres and Bacteroidetes (r = −0.95, p = 0.004; Figure 2H), associations which were not present in the other groups. In females mated with WD fed males, we observed a significant positive correlation between Proteobacteria and Bifidobacteriaceae (r = 0.91, p = 0.01; Figure 2I) and a negative association between Proteobacteria and Firmicutes (r = −0.87, p = 0.024; Figure 2J). Finally, in females mated with MD-WD fed males, we observed a positive association between 'Others' and Protobacteria (r = 0.92, p = 0.008; Figure 2J).

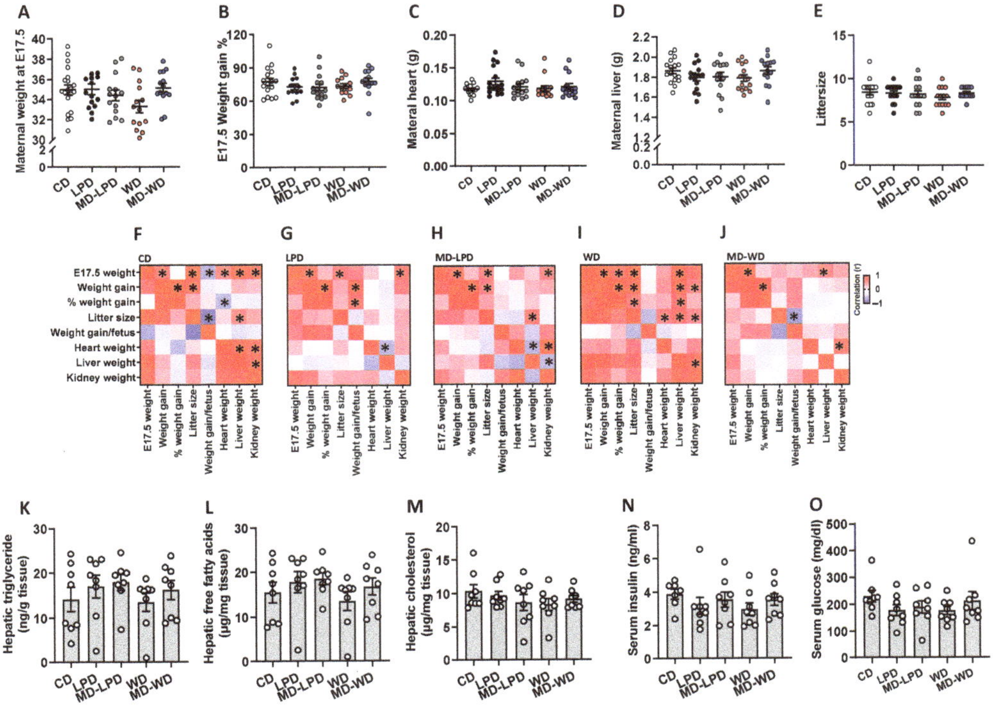

Figure 1. Impact of paternal diet on late gestation maternal physiology and metabolic status. Maternal weight immediately following mating on embryonic day (E)0.5 (**A**) and at the time of cull on E17.5 (**B**) in females mated to males fed either a control diet (CD), low protein diet (LPD), methyl donor-supplemented LPD (MD-LPD), Western diet (WD) or methyl donor-supplemented WD (MD-WD). Maternal heart (**C**) and liver (**D**) weight and litter size (**E**) at E17.5. Correlations between maternal weight at E17, gestational weight gain (E0.5–17.5), weight gain as a % of body weight, litter size, weight gain/foetus, heart, liver, and kidney weight in females mated to CD (**F**), LPD (**G**), MD-LPD (**H**), WD (**I**) or MD-WD (**J**) fed males (red boxes denote positive correlations, blue denotes a negative correlation, boxes with an asterisk denote statistically significant correlations). Late gestation maternal hepatic triglyceride (**K**), free fatty acids (**L**) and cholesterol (**M**) and serum insulin (**N**) and glucose (**O**) levels. n = 8 females per treatment group, each mated with a separate male. Data are expressed as mean ± SEM (**A–E**,**K–O**) or Spearman's correlation (**F–J**). Statistical differences were determined using a one-way ANOVA or Kruskal–Wallis test with post hoc correction (**A–E**,**K–O**), or by Spearman's correlation (**F–J**). Statistical significance was taken when $p < 0.05$.

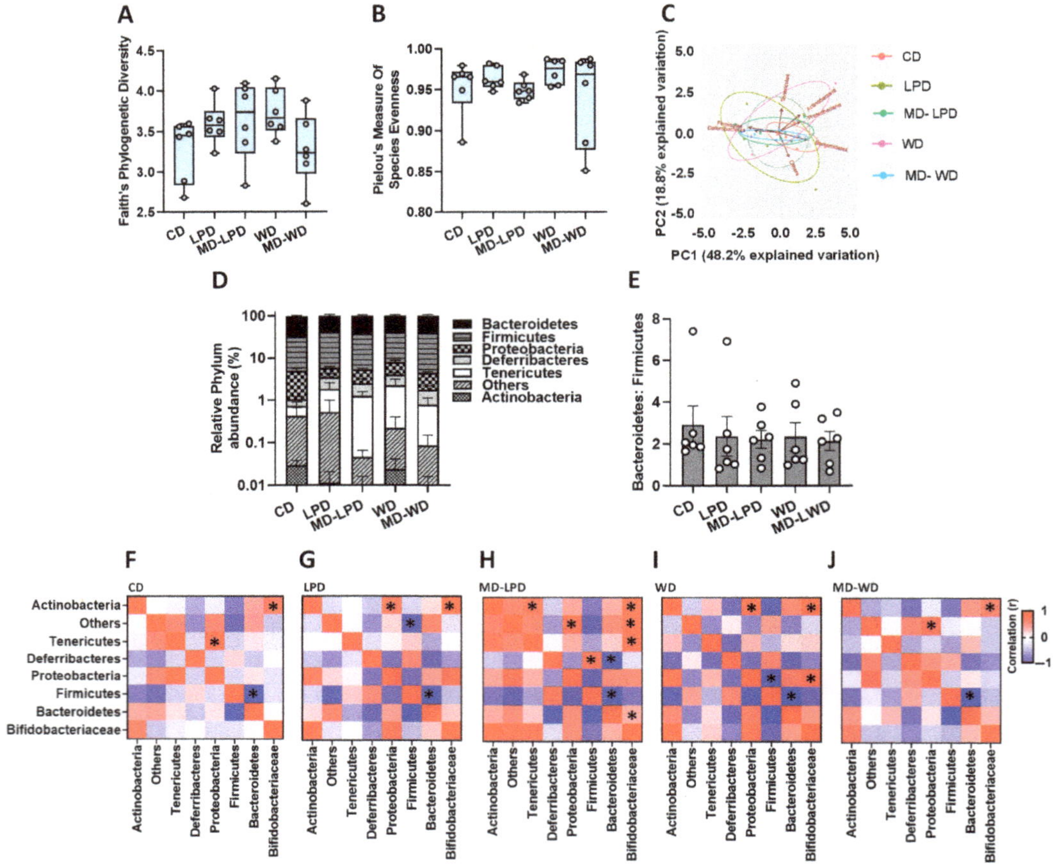

Figure 2. Impact of paternal diet on late gestation maternal gut microbiota. Assessment of maternal faecal bacterial profiles using Faith's phylogenetic diversity (**A**) and Pielou's measure of species evenness (**B**) in females mated to males fed either a control diet (CD), low protein diet (LPD), methyl donor-supplemented LPD (MD-LPD), Western diet (WD) or methyl donor-supplemented WD (MD-WD). Principal component analysis (PCA) based on bacterial abundance at the phylum level (**C**) and relative abundance (**D**). Ratio of gut Bacteroidetes to Firmicutes (**E**). Correlations between bacterial populations in females mated to males fed either CD (**F**), LPD (**G**), MD-LPD (**H**), WD (**I**) or MD-WD (**J**) (boxes with an asterisk denote statistically significant correlations). $n = 6$ females per treatment group, each mated with a separate male. Data are expressed as mean ± minimum to maximum value (**A,B**), mean ± SEM (**D,E**) or Spearman's correlation (**F–J**). Statistical differences were determined using a one-way ANOVA or Kruskal–Wallis test with post hoc correction (**A,B,D,E**), or by Spearman's correlation (**F–J**). Statistical significance was taken when $p < 0.05$.

3.3. Comparison of Maternal Late Gestation Liver Transcriptome

To explore further the potential influences of paternal diet on maternal metabolic status, we conducted RNA-Seq on samples of maternal liver (RNA-Seq data are currently being deposited in the Gene Expression Omnibus). Using a p-value of < 0.05 and no log2 fold-change cut-off, we detected 743, 934, 540 and 508 differentially expressed genes between females mated with CD and LPD, MD-LPD, WD and MD-WD males, respectively (Figure 3A–D). We conducted subsequent pathway and ontology analysis using differentially expressed genes with a p-value of < 0.05 and no log2 fold-change cut-off in order to obtain a broader understanding of potentially subtle changes in maternal metabolic status and the interplay between them. Interestingly, we observed minimal overlap in the number of shared up- and down-regulated genes between each group (Figure 3E,F). Between females mated with LPD and MD-LPD males, there were only 33 uniquely shared down-regulated genes (Figure 3E). Similarly, between females mated with WD and MD-WD males, we identified only 15 unique genes in common between the two groups. When comparing the differentially expressed down-regulated gene profiles across all four experimental diet groups, 17 genes were common to all four diets (Figure 3E). A similar pattern was observed for common up-regulated genes. Here, 56 and 28 genes were common between females mated with LPD and MD-LPD fed males and between females mated with WD and MD-WD fed males, respectively (Figure 3F). Only one single gene, Src-like-adapter 2 (Sla2), was upregulated in all four groups (Figure 3F). In total, 23 genes were identified as being significantly differentially expressed in all groups (Figure 3G). Cluster analysis of these common genes indicated that females mated with WD and MD-WD males clustered together, while females mated with LPD fed males showed the most variability in levels of expression. These data indicate that each of the paternal diets has a largely unique impact on the transcriptomic profile in the maternal liver.

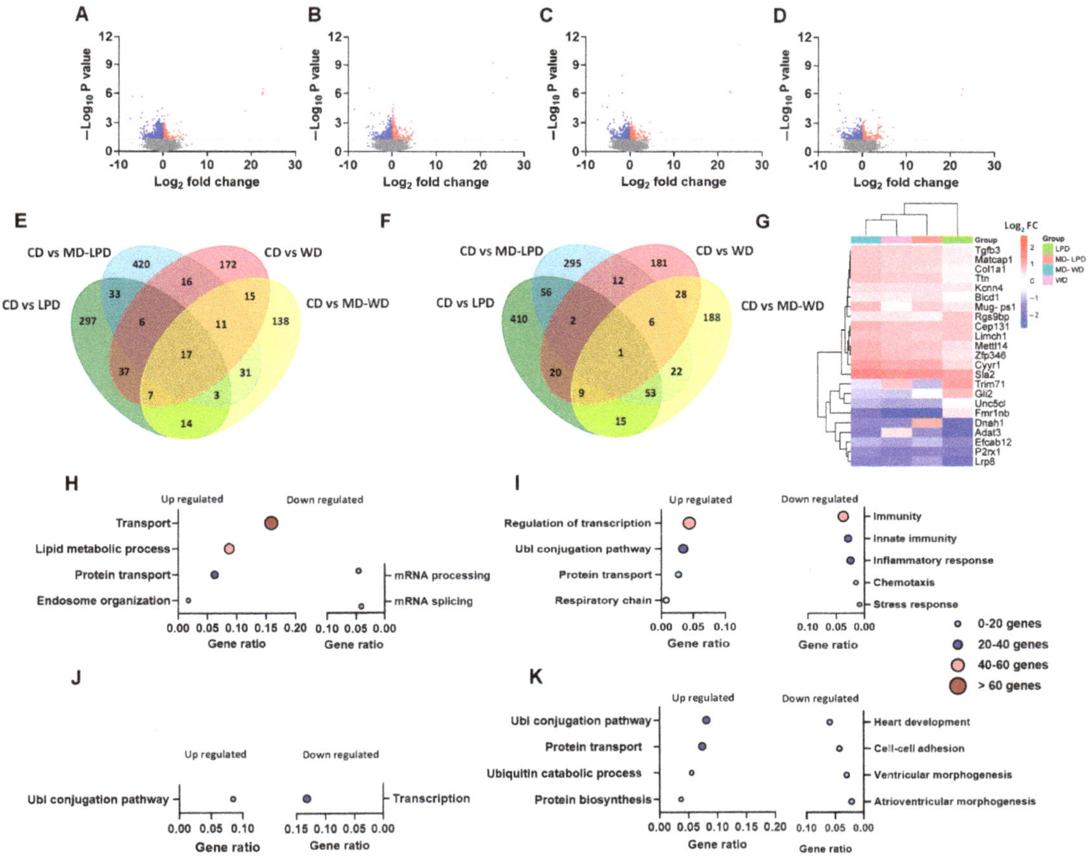

Figure 3. Comparison of maternal late gestation liver transcriptome in response to paternal diet. Volcano plot of differentially expressed hepatic genes between females mated to males fed either a control diet (CD) and (**A**) a low protein diet (LPD), (**B**) methyl donor-supplemented LPD (MD-LPD), (**C**) Western diet (WD) or, (**D**) methyl donor-supplemented WD (MD-WD). Significantly ($p < 0.05$)

down- and up-regulated genes are shown in blue and red, respectively while non-differentially expressed genes are shown in grey. Venn diagrams showing the number of unique and common, significantly downregulated (**E**) and upregulated (**F**) genes in each group when compared to females mated with CD fed males. Heat map displaying all common genes (23 in total) in each group when compared to females mated with CD fed males (**G**). Gene ontology and pathway analysis of significantly up- and downregulated genes between females mated to males fed either a control diet (CD) and (**H**) LPD, (**I**) MD-LPD, (**J**) WD or, (**K**) or MD-WD. Colour and size of circles represent the number of genes differentially expressed in each pathway. Sequencing data are from 6 females per group, each mated with a separate male.

Pathway and gene ontology analysis of the differentially expressed maternal genes identified significant (Padj < 0.01) upregulation of genes associated with protein transport (GO:0015031), lipid metabolism (GO:0006629) and endosome organisation (GO:0007032) in females mated with LPD fed males when compared to female mated with CD fed males (Figure 3H). In contrast, pathways associated with mRNA splicing (GO:0008380) and processing (GO:0006397) were significantly (Padj < 0.01) down regulated. In females mated with MD-LPD fed males, genes associated with protein transport were also upregulated (Figure 3I), regulation of transcription and ubiquitin-dependent protein catabolic process (GO:0006511) and the respiratory chain (GO:0042773) were also upregulated (Padj < 0.01). In contrast, multiple genes associated with innate immunity, chemotaxis and stress responses were down regulated when compared to females mated with CD fed males. In females mated with WD fed males, only ubiquitin-dependent protein catabolic (GO:0006511) and transcription (KW-0804) processes (Figure 3J) were significantly up- and down-regulated, respectively (Padj < 0.05). Finally, in females mated with MD-WD fed males, we observed up-regulation of genes involved in ubiquitin-dependent protein catabolic process (GO:0006511) and protein transport (GO:0015031), while genes associated with heart development (GO:0007507) and cell–cell adhesion (GO:0098609) were decreased (Figure 3K) when compared to females mated with CD fed males (Padj < 0.001).

3.4. *Maternal Late Gestation Cardiac Gene Expression*

To determine if maternal cardiac health may be altered in response to poor paternal diet, we assessed the expression of genes central in the regulation of cardiac function. We observed no difference in the Ct values of the reference genes phosphoglycerate kinase 1 (Pgk1) or tubulin, alpha 1A (Tuba1a) (Figure 4A,B). Additionally, we observed no difference in the relative expression of angiotensin converting enzyme 2 (Ace2, Figure 4C), adrenergic receptor, beta 1 (Adrb1, Figure 4D), angiotensin II receptor, type 1a (Agtr1a, Figure 4E) or the ATPase, Ca2+ transporting, plasma membrane 1 (Atp2b1, Figure 4F) between groups.

3.5. *Maternal Late Gestation Femur Trabecular Bone Morphology*

Finally, to determine whether paternal diet at the time of conception could also affect other maternal systems, we analysed maternal femur morphology by μCT. Initially, we observed that femurs from females mated with MD-LPD fed males were significantly longer than femurs from CD mated females ($p = 0.023$; Figure 5A). However, there were no differences in mean femur volume between any of the groups (Figure 5B). Morphological analysis of the trabecular bone (Figure 5C) identified significant differences in total trabecular volume (Tv) between females mated with MD-LPD fed males and females mated with MD-WD fed males ($p = 0.047$; Figure 5D). However, there were no differences in trabecular bone volume (BV; Figure 5E), in Bv:Tv ratio (Figure 5F) or in trabecular thickness (Tb:Th; Figure 5G). Finally, analysis of the number of connected trabeculae in the image (ConD) revealed a significant reduction in females mated with MD-LPD fed males when compared to female mated to LPD fed males ($p = 0.049$; Figure 5H).

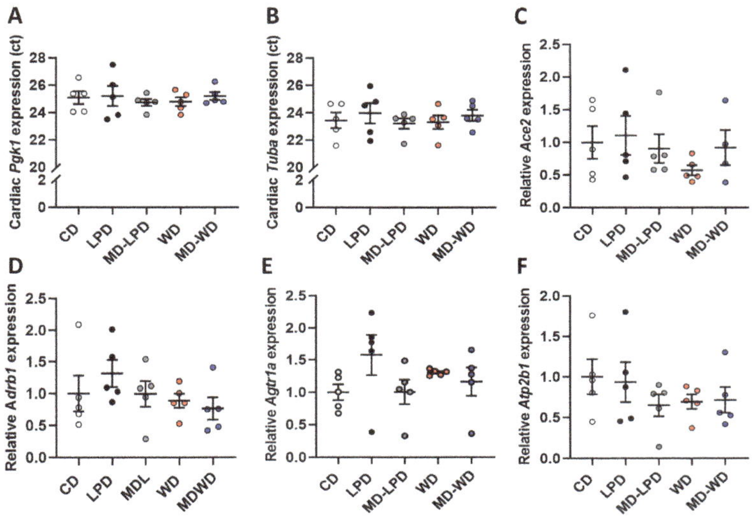

Figure 4. Analysis of maternal late gestation cardiac gene expression. The cycle threshold (Ct) values for the reference genes phosphoglycerate kinase 1 (Pgk1) (**A**) and tubulin, alpha 1a (Tuba1a) (**B**) in females mated to males fed either a control diet (CD), low protein diet (LPD), methyl donor-supplemented LPD (MD-LPD), Western diet (WD) or methyl donor-supplemented WD (MD-WD). Relative expression of angiotensin converting enzyme 2 (Ace2) (**C**), adrenergic receptor, beta 1 (Adrb1) (**D**), angiotensin II receptor, type 1a (Agtr1a) (**E**) and ATPase, Ca2+ transporting, plasma membrane 1 (Atp2b1) (**F**). $n = 5$ females per treatment group, each mated with a separate male. Data are mean ± SEM. Statistical differences were determined using a one-way ANOVA or Kruskal–Wallis test with post hoc correction.

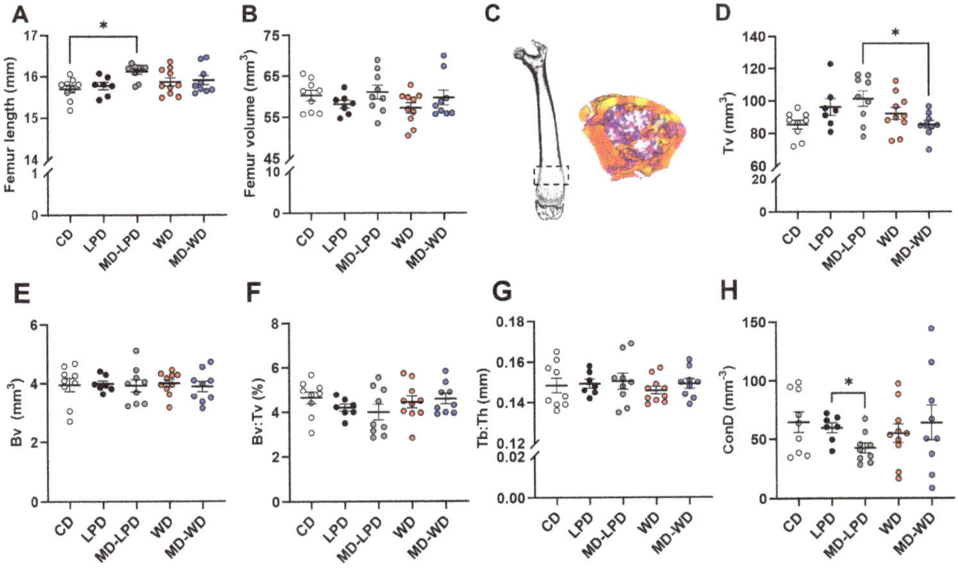

Figure 5. Assessment of maternal late gestation femur trabecular bone morphology. Whole femur length (**A**) and volume (**B**) in females mated to males fed either a control diet (CD), low protein diet (LPD), methyl donor-supplemented LPD (MD-LPD), Western diet (WD) or methyl donor-supplemented WD (MD-WD). Representative image of a whole scanned femur highlighting

the trabecular bone region (dashed box) and the re-composed trabecular bone (**C**). Trabecular volume (Tv) (**D**), bone volume (Bv) (**E**), bone volume–trabecular volume ratio (Bv:Tv) (**F**), trabecular thickness (Tb:Th) (**G**) and connectivity (ConD) (**H**). n = 7–10 females per treatment group. Data are mean ± SEM. Statistical differences were determined using a one-way ANOVA or Kruskal–Wallis test with post hoc correction. * $p < 0.05$.

3.6. Maternal Late Gestation Femur Cortical Bone Morphology

Analysis of cortical bone morphology (Figure 6A) indicated all females had a similar volume of cortical bone (C-Bv; Figure 6B). However, females mated with males fed a MD-WD had a reduced volume of the cortical medullary cavity when compared to female mated with CD. LPD and WD fed males (Cb-Mv, $p < 0.05$; Figure 6D). Similarly, females mated with males fed a MD-WD displayed an increase in mean cortical bone thickness (Cb-Th) when compared to all other groups ($p < 0.05$; Figure 6E). However, there were no differences in the mean cortical bone cross-section areas between groups (Figure 6F). Finally, analysis of the second moment of area around major (Imax) and minor (Imin) axis showed no differences between groups (Figure 6G,H).

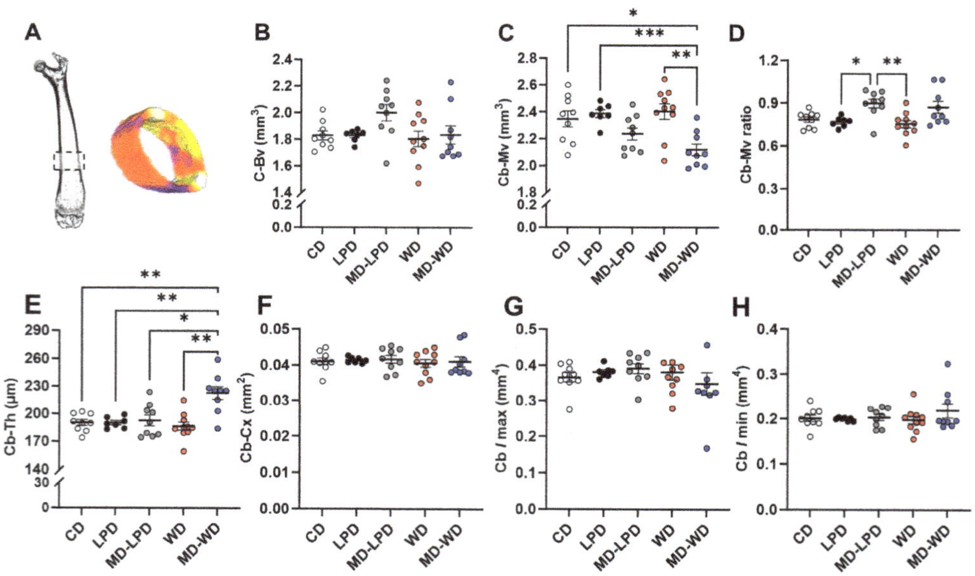

Figure 6. Assessment of maternal late gestation femur cortical bone morphology. Representative image of a whole scanned femur, highlighting the cortical bone region (dashed box) and the re-composed cortical bone (**A**). Cortical bone volume (C-Bv) (**B**), cortical bone medullary cavity volume (Cb-Mv) (**C**), cortical bone: medullary cavity ratio (**D**), cortical bone thickness (Cb-Th) (**E**), cortical bone cross section area (**F**), cortical bone maximum moment of inertia (Imax) (**G**) and minimum moment of inertia (Imin) (**H**) in females mated to males fed either a control diet (CD), low protein diet (LPD), methyl donor-supplemented LPD (MD-LPD), Western diet (WD) or methyl donor-supplemented WD (MD-WD). n = 7–10 females per treatment group. Data are mean ± SEM. Statistical differences were determined using a one-way ANOVA or Kruskal–Wallis test with post hoc correction. * $p < 0.05$, ** $p < 0.01$, *** $p < 0.001$.

4. Discussion

The impacts of poor maternal diet in pregnancy on the health of her offspring, as well as her own health, are well-established [33]. However, the influence that a father has on this fundamental period is less well-defined. In the current study, we explored the impact of paternal nutrition on maternal cardio-metabolic and bone health in late gestation. We observed subtle changes in patterns of association between maternal weight gain, organ weight and litter size despite no change in metabolic status. In contrast, we observed differential hepatic expression of genes associated with protein transport, lipid metabolism, transcription, innate immunity, chemotaxis and ubiquitin process in females mated with males fed sub-optimal diets. Finally, we observed differential remodelling of cortical and trabecular regions of the maternal femur in response to paternal diet. These data highlight the broad influences that paternal nutrition can have on maternal health during pregnancy, irrespective of her own dietary status.

We observed minimal differences in maternal weight gain or organ sizing during pregnancy between groups. However, we observed subtle differences in the way maternal weight gain and organ weight interacted. Additionally, we observed a negative association between litter size and the amount of weight gained by the dam in females mated with CD fed males. This observation suggests that as litter size increases, so does the metabolic demand on the dam, and the female's resources are utilised to a greater extent, resulting in a diminishing body weight. Interestingly, only females mated with MD-WD fed males also showed this association. In late gestation, females have entered a catabolic state of metabolism, characterised by an increased breakdown of adipose deposits and liberation of non-esterified fatty acids and glycerol. The non-esterified fatty acids can be oxidised to acetyl-CoA, while glycerol can be utilised in glucose synthesis [33]. While the changes in patterns of gestational body size might suggest differential nutritional partitioning between groups, we observed no significant differences in circulating or hepatic metabolites.

To establish how paternal diet might influence maternal metabolic status in more detail, we next analysed the maternal microbiota. The significance of the gut microbiota in the metabolism and synthesis of vitamins [34], regulation of lipid metabolism [35], inflammatory [36] and metabolic diseases [37] as well as cardiovascular disease [38] is now accepted. In this study, we sampled from the lower region (ileum and post ileocecal) of the gut which is predominated by coliforms and anaerobic species including Bacteroides, Bifidobateria, Clostridia and Lactobacilli [39]. In all females, a significant positive correlation between the abundance of Actinobacteria and Bifidobacteriaceae was observed. Actinobacteria are a diverse phylum of Gram-positive bacteria which includes the anaerobe families (Bifidobacteria, Propionibacteria and Corynebacteria) and an aerobe family (Streptomyces) [40]. Levels of Bifidobacteria have been inversely linked with BMI and insulin levels in women [41]. In contrast, reduced levels of Bifidobacteria have been linked to enhanced gut permeability [42] and immune system activation [42]. Therefore, the lack of any difference in the abundance of these bacterial groups is in line with no significant differences in the central metabolic status. Similarly, all females showed a negative association between Firmicutes and Bacteroidetes. The Firmicutes–Bacteroidetes ratio is widely cited as a marker of obesity with a higher ratio seen in obese mice and women when compared to females of a healthy weight [43]. However, as all our females displayed the same overall weight gain and body morphometry then a change in the Firmicutes–Bacteroidetes ratio might not be anticipated. However, some subtle differences in the associations between different bacterial groups were observed. Females mated with LPD fed males showed a positive association between the abundance of Actinobacteria and Protobacteria, while females mated with WD fed males showed a positive association between the abundance of Protobacteria and Bifidobacteriaceae. As the gut microbiota has a wide range of influences on physiology, subtle shifts in patterns of relative abundance could still have influences on maternal health.

To define the wider effects of poor paternal diet on maternal health in pregnancy, we conducted RNA-Seq analysis on the maternal liver. Females mated with LPD and MD-LPD fed males displayed the largest number of differentially expressed genes, 743 and 934 genes, respectively. However, only 89 genes were common to both groups. Here, pathway analysis identified changes in lipid metabolism, mRNA processing, protein transport and immunity. Increases in the expression of genes such as phosphatidate cytidylyltransferase 2 (*Cds2*), adiponectin receptors 1 and 2 (*Adipor 1/2*), long-chain fatty acid transport protein 2 (*Slc27a2*), sterol O-acyltransferase 1 (*Soat1*), cardiolipin synthase 1 (*Crls1*) and several genes involved in acyl-CoA metabolism (*Acaca, Acsf2, Acss3, Acadl, Acads*) in females mated to LPD fed males suggest differential patterns of lipid uptake and metabolism within the mitochondria. Beta-oxidation is a source of significant metabolic energy during periods of high energy demand [44]. Impairments in β-oxidation have been linked to the accumulation of hepatic lipid-species, mitochondrial dysfunction and the development of metabolic diseases such as non-alcoholic fatty liver disease [45]. Therefore, a more detailed lipidomic profiling, in combination with mitochondrial function, metabolism and lipid composition, in our females is warranted.

In females mated with MD-LPD males, we observed differential regulation of genes associated with transcription and immunity. The immune system modulates throughout pregnancy, shifting from a pro-inflammatory state in early gestation to a more anti-inflammatory state in later gestation [46]. This concept is supported by observations that the severity of autoimmune disorders appears diminished during pregnancy but then returns postpartum [47]. Other conditions such as multiple sclerosis also show decreased progression during pregnancy [48]. We observed a decreased expression of multiple complement genes (*C8a, C8b, C9*), histocompatibility genes (*H2-D1, H2-Ab1, H2-Eb1*), antigen presentation (*Fcgr1, Tapbpl*), and inflammatory mediators (*Card9, Cx3cr1, Mefv, Tlr12*). Changes in maternal immunological status are directly linked to hormonal levels. Progesterone has been shown to inhibit the proliferation and cytokine secretion of CD8+ T cells [49], while in T-helper cells, progesterone promotes the expression of LIF [50], both of which would support the maintenance of pregnancy. Whether the potential immunological changes observed in our current study are due to paternal influences on maternal hormonal status in pregnancy are currently unknown as maternal endocrine homeostasis is still to be determined in this model. It is also unclear whether any potential immunological perturbations in our females might put the dams, and their offspring, at greater risk of infection. Finally, we are unable to ascertain whether the reduction in maternal immunological status is a paternally mediated adaptive mechanism to increase maternal tolerance of his offspring, thus enhancing their survival.

In females mated with WD and MD-WD fed males, we observed 540 and 508 differentially expressed genes, respectively, with only 43 genes shared exclusively between them. Gene ontology and pathway analysis identified few significant changes in females mated with WD fed males. Genes associated with the ubiquitin pathway were up-regulated while genes involved in transcription were down-regulated. In females mated with MD-WD fed males, we also observed an up-regulation of ubiquitin pathway genes suggesting an increase in protein turnover in both groups. In contrast, a down regulation of cardiac and cardiovascular genes was also observed in these females. Closer analysis of these genes identified regulators of transcription (*Kdm6b*), vessel formation and angiogenesis (*Apln, Tgfb2*), cell and muscle structure (*Flrt3, Myh10, Ttn*). Whether any potential disruptions in hepatic immune status and vascular function in our females are also reflective of wider cardiovascular and immune impairments remains to be determined. However, such impairments in pregnancy have been associated with gestational conditions such as preeclampsia [51]. However, we observed no differences in central cardiac regulatory gene expression between groups.

Our final observation was that poor paternal diet influenced maternal femur structure in late gestation. We observed modest changes in femur length, volume of trabecular bone (Tv), trabecular connectivity (ConD), volume of the cortical medullar cavity (Cb-Mv) and thickness of the cortical bone (Cb-Th) between groups. Similar to many of the other maternal systems, the skeleton undergoes dramatic adaptation throughout pregnancy so as to provide sufficient calcium and other minerals to support foetal development [52]. Underlying the modulation of maternal bone physiology is parathyroid hormone-related protein (PTHrP). During pregnancy, PTHrP production by the placenta and breasts increases due to the action of oestradiol, placental lactogen and prolactin [52]. During early pregnancy, an increase in intestinal calcium absorption is sufficient to meet the foetal demands [53];

however, in later gestation and during lactation, the decline in oestradiol levels in combination with PTHrP results in greater maternal bone resorption [54,55]. During lactation, there is a continued (3–10%) loss in bone mineral density in women which occurs predominantly in the trabecular bone (i.e., lumbar spine) and some cortical regions such as the hip [56]. However, much of this is reversed by 6–12 months after breast feeding stops [57,58]. While studies indicate that maternal bone physiology can recover post-lactation, the short- and long-term effects of inappropriate bone remodelling during pregnancy are unknown.

5. Conclusions

While our study indicates that poor paternal diet at the time of conception can influence maternal physiological and metabolic status in late gestation, the underlying mechanisms remain unanswered. One central mechanism is a change in sperm epigenetic status. Previously, we have shown that sperm from LPD fed males display genome-wide DNA hypomethylation and the expression of several epigenetic regulators within the testis of LPD fed males was perturbed [25]. Other studies have linked paternal diet [59], ageing [60], and environmental pollutants [61] to perturbed patterns of offspring health and placental development. As many aspects of maternal health in pregnancy are regulated via placental function, metabolism and the range of endocrine factors it produces [62], paternal epigenetic modulation of the placenta provides one link between paternal diet and maternal health.

Our study widens our understanding of the range of influences that poor paternal diet has on the environment in which his offspring development. There are several limitations within our findings. First, the use of a methyl-donor-supplemented control diet (MD-CD) would have provided additional insight into the role that the methyl donors on their own might play. However, comparing the size of effect between the respective diets with (MD-LPD, MD-WD) and without (LPD, WD) methyl-donor supplementation suggests that they are having a minimal impact on maternal health. Similarly, our WD is deficient in methionine when compared to the other diets. However, as minimal differences were seen in the effects of WD vs. MD-WD, we do not believe this to be a significant factor for maternal well-being. Also, as already discussed, our understanding of the maternal metabolic changes is currently limited. Future studies will adopt a wider analysis of maternal serum lipid and hormone profiles, as well as microbiota-associated metabolites such as butyrate and other short-chain fatty acids. A final limitation of this study was that we were unable to perform a full characterisation of maternal cardiovascular health. Analysis of blood pressure, vascular reactivity or cardiac morphology would have provided a more insight into how paternal diet affects maternal cardiovascular health in pregnancy.

Supplementary Materials: The following supporting information can be downloaded at: https://www.mdpi.com/article/10.3390/nu16121879/s1; Data S1: Maternal faecal microbiome sequencing data; Table S1: Control and experimental dietary compositions; Table S2: RT-qPCR primer sequences and details.

Author Contributions: Conceptualization, A.J.W.; Methodology, A.K., N.H., S.H., F.S., H.L.M. and A.J.W.; Investigation, A.K., N.E., N.H., S.H., F.S. and H.L.M.; Data curation, A.K., N.E., N.H., S.H., F.S. and H.L.M.; Formal analysis, A.K., N.B., N.E., N.H., S.H., F.S., V.W. and H.L.M.; Funding acquisition, A.J.W.; Writing—original draft, A.J.W.; Writing—review and editing, A.K., N.E., V.B., N.H., S.H., F.S., V.W., J.M., K.S., K.J.W., H.L.M. and A.J.W. All authors have read and agreed to the published version of the manuscript.

Funding: This work was funded by a Biotechnology and Biological Sciences Research Council (BBSRC) grant (BB/R003556/1) to A.J.W.

Institutional Review Board Statement: All animal experiments and procedures were conducted in accordance with UK Home Office Animal (Scientific Procedures) Act 1986 Amendment Regulations 2012, which transposed Directive 2010/63/EU into UK law and with approval of the Animal Welfare and Ethical Review Board (AWERB) on the 6 October 2017 at the University of Nottingham.

Informed Consent Statement: Not applicable.

Data Availability Statement: All data are either contained within the Supplementary Materials or are available upon reasonable request sent to the corresponding author. Liver RNA-seq data have been submitted to the Gene Expression Omnibus (GEO) at NCBI under accession number: GSE265783 https://www.ncbi.nlm.nih.gov/geo/query/acc.cgi (accessed on 10 June 2024).

Acknowledgments: The authors would like to thank the staff within the Bio Support Unit at the University of Nottingham for animal provision and maintenance.

Conflicts of Interest: The authors declare no conflict of interest.

References

1. Vinnars, M.T.; Bixo, M.; Damdimopoulou, P. Pregnancy-related maternal physiological adaptations and fetal chemical exposure. *Mol. Cell. Endocrinol.* **2023**, *578*, 112064. [CrossRef]
2. Casagrande, S.S.; Linder, B.; Cowie, C.C. Prevalence of gestational diabetes and subsequent Type 2 diabetes among U.S. women. *Diabetes Res. Clin. Pract.* **2018**, *141*, 200–208. [CrossRef]
3. Say, L.; Chou, D.; Gemmill, A.; Tunçalp, Ö.; Moller, A.-B.; Daniels, J.; Gülmezoglu, A.M.; Temmerman, M.; Alkema, L. Global causes of maternal death: A WHO systematic analysis. *Lancet Glob. Health* **2014**, *2*, e323–e333. [CrossRef]
4. Brett, K.E.; Ferraro, Z.M.; Yockell-Lelievre, J.; Gruslin, A.; Adamo, K.B. Maternal-fetal nutrient transport in pregnancy pathologies: The role of the placenta. *Int. J. Mol. Sci.* **2014**, *15*, 16153–16185. [CrossRef]
5. Dimasuay, K.G.; Boeuf, P.; Powell, T.L.; Jansson, T. Placental Responses to Changes in the Maternal Environment Determine Fetal Growth. *Front. Physiol.* **2016**, *7*, 12. [CrossRef]
6. Kingdom, J.C.; Audette, M.C.; Hobson, S.R.; Windrim, R.C.; Morgen, E. A placenta clinic approach to the diagnosis and management of fetal growth restriction. *Am. J. Obstet. Gynecol.* **2018**, *218*, S803–S817. [CrossRef]
7. Mulder, J.; Kusters, D.M.; Roeters van Lennep, J.E.; Hutten, B.A. Lipid metabolism during pregnancy: Consequences for mother and child. *Curr. Opin. Lipidol.* **2024**, *35*, 133–140. [CrossRef]
8. Napso, T.; Yong, H.E.J.; Lopez-Tello, J.; Sferruzzi-Perri, A.N. The Role of Placental Hormones in Mediating Maternal Adaptations to Support Pregnancy and Lactation. *Front. Physiol.* **2018**, *9*, 1091. [CrossRef]
9. Wiznitzer, A.; Mayer, A.; Novack, V.; Sheiner, E.; Gilutz, H.; Malhotra, A.; Novack, L. Association of lipid levels during gestation with preeclampsia and gestational diabetes mellitus: A population-based study. *Am. J. Obstet. Gynecol.* **2009**, *201*, 482.e1–482.e8. [CrossRef] [PubMed]
10. Ernst, S.; Demirci, C.; Valle, S.; Velazquez-Garcia, S.; Garcia-Ocana, A. Mechanisms in the adaptation of maternal beta-cells during pregnancy. *Diabetes Manag.* **2011**, *1*, 239–248. [CrossRef] [PubMed]
11. Dathan-Stumpf, A.; Vogel, M.; Jank, A.; Thiery, J.; Kiess, W.; Stepan, H. Reference intervals of serum lipids in the second and third trimesters of pregnancy in a Caucasian cohort: The LIFE Child study. *Arch. Gynecol. Obstet.* **2019**, *300*, 1531–1539. [CrossRef]
12. Emet, T.; Üstüner, I.; Güven, S.G.; Balık, G.; Ural, M.; Tekin, Y.B.; Şentürk, Ş.; Şahin, F.K.; Avşar, A.F. Plasma lipids and lipoproteins during pregnancy and related pregnancy outcomes. *Arch. Gynecol. Obstet.* **2013**, *288*, 49–55. [CrossRef]
13. Friis Petersen, J.; Friis-Hansen, L.J.; Jensen, A.K.; Nyboe Andersen, A.; Lokkegaard, E.C.L. Early pregnancy reference intervals; 29 serum analytes from 4 to 12 weeks' gestation in naturally conceived and uncomplicated pregnancies resulting in live births. *Clin. Chem. Lab. Med.* **2019**, *57*, 1956–1967. [CrossRef]
14. Zheng, W.; Huang, W.; Zhang, L.; Tian, Z.; Wang, T.; Zhang, T.; Zhang, Z.; Zhang, W.; Li, G. Changes in Serum Lipid Levels During Pregnancy and Association with Neonatal Outcomes: A Large Cohort Study. *Reprod. Sci.* **2018**, *25*, 1406–1412. [CrossRef]
15. Jo, Y.S.; Cheon, J.Y.; Ahn, J.W.; Kim, Y.; Lee, G.S.R. Pentraxin 3 and lipid profile status in pregnancy. *J. Obstet. Gynaecol.* **2017**, *37*, 727–730. [CrossRef]
16. Alvarez, J.J.; Montelongo, A.; Iglesias, A.; Lasuncion, M.A.; Herrera, E. Longitudinal study on lipoprotein profile, high density lipoprotein subclass, and postheparin lipases during gestation in women. *J. Lipid Res.* **1996**, *37*, 299–308. [CrossRef]
17. Wang, J.; Li, Z.; Lin, L. Maternal lipid profiles in women with and without gestational diabetes mellitus. *Medicine* **2019**, *98*, e15320. [CrossRef]
18. Roeters van Lennep, J.E.; Tokgözoğlu, L.S.; Badimon, L.; Dumanski, S.M.; Gulati, M.; Hess, C.N.; Holven, K.B.; Kavousi, M.; Kayıkçıoğlu, M.; Lutgens, E.; et al. Women, lipids, and atherosclerotic cardiovascular disease: A call to action from the European Atherosclerosis Society. *Eur. Heart J.* **2023**, *44*, 4157–4173. [CrossRef]
19. Pijnenborg, R.; Vercruysse, L.; Hanssens, M. The uterine spiral arteries in human pregnancy: Facts and controversies. *Placenta* **2006**, *27*, 939–958.
20. Lash, G.E.; Robson, S.C.; Bulmer, J.N. Review: Functional role of uterine natural killer (uNK) cells in human early pregnancy decidua. *Placenta* **2010**, *31*, S87–S92. [CrossRef]
21. Sferruzzi-Perri, A.N.; Macpherson, A.M.; Roberts, C.T.; Robertson, S.A. Csf2 null mutation alters placental gene expression and trophoblast glycogen cell and giant cell abundance in mice. *Biol. Reprod.* **2009**, *81*, 207–221. [CrossRef]
22. Hunter, S.; Robson, S.C. Adaptation of the maternal heart in pregnancy. *Br. Heart J.* **1992**, *68*, 540–543. [CrossRef]
23. Duvekot, J.J.; Peeters, L.L. Maternal cardiovascular hemodynamic adaptation to pregnancy. *Obstet. Gynecol. Surv.* **1994**, *49* (Suppl. S12), S1–S14. [CrossRef]

24. Watkins, A.J.; Sirovica, S.; Stokes, B.; Isaacs, M.; Addison, O.; Martin, R.A. Paternal low protein diet programs preimplantation embryo gene expression, fetal growth and skeletal development in mice. *Biochim. Biophys. Acta* **2017**, *1863*, 1371–1381. [CrossRef]
25. Watkins, A.J.; Dias, I.; Tsuro, H.; Allen, D.; Emes, R.D.; Moreton, J.; Wilson, R.; Ingram, R.J.; Sinclair, K.D. Paternal diet programs offspring health through sperm- and seminal plasma-specific pathways in mice. *Proc. Natl. Acad. Sci. USA* **2018**, *115*, 10064–10069. [CrossRef]
26. Morgan, H.L.; Paganopoulou, P.; Akhtar, S.; Urquhart, N.; Philomin, R.; Dickinson, Y.; Watkins, A.J. Paternal diet impairs F1 and F2 offspring vascular function through sperm and seminal plasma specific mechanisms in mice. *J. Physiol.* **2020**, *598*, 699–715. [CrossRef] [PubMed]
27. Morgan, H.L.; Aljumah, A.; Rouillon, C.; Watkins, A.J. Paternal low protein diet and the supplementation of methyl-donors impact fetal growth and placental development in mice. *Placenta* **2021**, *103*, 124–133. [CrossRef]
28. Morgan, H.L.; Furse, S.; Dias, I.H.K.; Shabir, K.; Castellanos, M.; Khan, I.; May, S.T.; Holmes, N.; Carlile, M.; Sang, F.; et al. Paternal low protein diet perturbs inter-generational metabolic homeostasis in a tissue-specific manner in mice. *Commun. Biol.* **2022**, *5*, 929. [CrossRef]
29. Morgan, H.L.; Watkins, A.J. The influence of seminal plasma on offspring development and health. *Semin. Cell Dev. Biol.* **2020**, *97*, 131–137. [CrossRef]
30. Caporaso, J.G.; Kuczynski, J.; Stombaugh, J.; Bittinger, K.; Bushman, F.D.; Costello, E.K.; Fierer, N.; Gonzalez Peña, A.; Goodrich, J.K.; Gordon, J.I.; et al. QIIME allows analysis of high-throughput community sequencing data. *Nat. Methods* **2010**, *7*, 335–336. [CrossRef]
31. Doube, M.; Kłosowski, M.M.; Arganda-Carreras, I.; Cordelières, F.P.; Dougherty, R.P.; Jackson, J.S.; Schmid, B.; Hutchinson, J.R.; Shefelbine, S.J. BoneJ: Free and extensible bone image analysis in ImageJ. *Bone* **2010**, *47*, 1076–1079. [CrossRef]
32. Lucas, E.S.; Watkins, A.J.; Cox, A.L.; Marfy-Smith, S.J.; Smyth, N.; Fleming, T.P. Tissue-specific selection of reference genes is required for expression studies in the mouse model of maternal protein undernutrition. *Theriogenology* **2011**, *76*, 558–569. [CrossRef]
33. Sferruzzi-Perri, A.N.; Lopez-Tello, J.; Napso, T.; Yong, H.E.J. Exploring the causes and consequences of maternal metabolic maladaptations during pregnancy: Lessons from animal models. *Placenta* **2020**, *98*, 43–51. [CrossRef]
34. Pham, V.T.; Dold, S.; Rehman, A.; Bird, J.K.; Steinert, R.E. Vitamins, the gut microbiome and gastrointestinal health in humans. *Nutr. Res.* **2021**, *95*, 35–53. [CrossRef]
35. Brown, E.M.; Clardy, J.; Xavier, R.J. Gut microbiome lipid metabolism and its impact on host physiology. *Cell Host Microbe* **2023**, *31*, 173–186. [CrossRef] [PubMed]
36. Clemente, J.C.; Manasson, J.; Scher, J.U. The role of the gut microbiome in systemic inflammatory disease. *BMJ* **2018**, *360*, j5145. [CrossRef]
37. Aron-Wisnewsky, J.; Vigliotti, C.; Witjes, J.; Le, P.; Holleboom, A.G.; Verheij, J.; Nieuwdorp, M.; Clément, K. Gut microbiota and human NAFLD: Disentangling microbial signatures from metabolic disorders. *Nat. Rev. Gastroenterol. Hepatol.* **2020**, *17*, 279–297. [CrossRef]
38. Masenga, S.K.; Hamooya, B.; Hangoma, J.; Hayumbu, V.; Ertuglu, L.A.; Ishimwe, J.; Rahman, S.; Saleem, M.; Laffer, C.L.; Elijovich, F.; et al. Recent advances in modulation of cardiovascular diseases by the gut microbiota. *J. Hum. Hypertens.* **2022**, *36*, 952–959. [CrossRef]
39. Arumugam, M.; Raes, J.; Pelletier, E.; Le Paslier, D.; Yamada, T.; Mende, D.R.; Fernandes, G.R.; Tap, J.; Bruls, T.; Batto, J.M.; et al. Enterotypes of the human gut microbiome. *Nature* **2011**, *473*, 174–180. [CrossRef]
40. Binda, C.; Lopetuso, L.R.; Rizzatti, G.; Gibiino, G.; Cennamo, V.; Gasbarrini, A. Actinobacteria: A relevant minority for the maintenance of gut homeostasis. *Dig. Liver Dis.* **2018**, *50*, 421–428. [CrossRef]
41. Teixeira, T.F.S.; Grześkowiak, Ł.M.; Salminen, S.; Laitinen, K.; Bressan, J.; Peluzio, M.D.C.G. Faecal levels of Bifidobacterium and Clostridium coccoides but not plasma lipopolysaccharide are inversely related to insulin and HOMA index in women. *Clin. Nutr.* **2013**, *32*, 1017–1022. [CrossRef] [PubMed]
42. Duca, F.A.; Sakar, Y.; Covasa, M. The modulatory role of high fat feeding on gastrointestinal signals in obesity. *J. Nutr. Biochem.* **2013**, *24*, 1663–1677. [CrossRef] [PubMed]
43. Magne, F.; Gotteland, M.; Gauthier, L.; Zazueta, A.; Pesoa, S.; Navarrete, P.; Balamurugan, R. The Firmicutes/Bacteroidetes Ratio: A Relevant Marker of Gut Dysbiosis in Obese Patients? *Nutrients* **2020**, *12*, 1474. [CrossRef] [PubMed]
44. Houten, S.M.; Violante, S.; Ventura, F.V.; Wanders, R.J. The Biochemistry and Physiology of Mitochondrial Fatty Acid Beta-Oxidation and Its Genetic Disorders. *Annu. Rev. Physiol.* **2016**, *78*, 23–44. [CrossRef] [PubMed]
45. Peng, K.-Y.; Watt, M.J.; Rensen, S.S.; Greve, J.W.; Huynh, K.; Jayawardana, K.S.; Meikle, P.J.; Meex, R.C.R. Mitochondrial dysfunction-related lipid changes occur in nonalcoholic fatty liver disease progression. *J. Lipid Res.* **2018**, *59*, 1977–1986. [CrossRef]
46. Mor, G.; Cardenas, I.; Abrahams, V.; Guller, S. Inflammation and pregnancy: The role of the immune system at the implantation site. *Ann. N. Y. Acad. Sci.* **2011**, *1221*, 80–87. [CrossRef] [PubMed]
47. Elenkov, I.J.; Wilder, R.L.; Bakalov, V.K.; Link, A.A.; Dimitrov, M.A.; Fisher, S.; Crane, M.; Kanik, K.S.; Chrousos, G.P. IL-12, TNF-alpha, and hormonal changes during late pregnancy and early postpartum: Implications for autoimmune disease activity during these times. *J. Clin. Endocrinol. Metab.* **2001**, *86*, 4933–4938. [PubMed]

48. Confavreux, C.; Hutchinson, M.; Hours, M.M.; Cortinovis-Tourniaire, P.; Moreau, T.; Pregnancy in Multiple Sclerosis Group. Rate of pregnancy-related relapse in multiple sclerosis. *N. Engl. J. Med.* **1998**, *339*, 285–291. [CrossRef] [PubMed]
49. Shen, Z.; Rodriguez-Garcia, M.; Patel, M.V.; Wira, C.R. Direct and Indirect endocrine-mediated suppression of human endometrial CD8+T cell cytotoxicity. *Sci. Rep.* **2021**, *11*, 1773. [CrossRef]
50. Hellberg, S.; Raffetseder, J.; Rundquist, O.; Magnusson, R.; Papapavlou, G.; Jenmalm, M.C.; Ernerudh, J.; Gustafsson, M. Progesterone Dampens Immune Responses in In Vitro Activated CD4+ T Cells and Affects Genes Associated with Autoimmune Diseases That Improve During Pregnancy. *Front. Immunol.* **2021**, *12*, 672168. [CrossRef]
51. Flint, E.J.; Cerdeira, A.S.; Redman, C.W.; Vatish, M. The role of angiogenic factors in the management of preeclampsia. *Acta Obstet. Gynecol. Scand.* **2019**, *98*, 700–707. [CrossRef] [PubMed]
52. Dimas, A.; Politi, A.; Bargiota, A.; Panoskaltsis, T.; Vlahos, N.F.; Valsamakis, G. The Gestational Effects of Maternal Bone Marker Molecules on Fetal Growth, Metabolism and Long-Term Metabolic Health: A Systematic Review. *Int. J. Mol. Sci.* **2022**, *23*, 8328. [CrossRef] [PubMed]
53. Kovacs, C.S. Maternal Mineral and Bone Metabolism During Pregnancy, Lactation, and Post-Weaning Recovery. *Physiol. Rev.* **2016**, *96*, 449–547. [CrossRef] [PubMed]
54. Kovacs, C.S. Osteoporosis presenting in pregnancy, puerperium, and lactation. *Curr. Opin. Endocrinol. Diabetes Obes.* **2014**, *21*, 468–475. [CrossRef] [PubMed]
55. Masiukiewicz, U.S.; Mitnick, M.; Gulanski, B.I.; Insogna, K.L. Evidence that the IL-6/IL-6 soluble receptor cytokine system plays a role in the increased skeletal sensitivity to PTH in estrogen-deficient women. *J. Clin. Endocrinol. Metab.* **2002**, *87*, 2892–2898. [CrossRef] [PubMed]
56. Moller, U.K.; Vieth Streym, S.; Mosekilde, L.; Rejnmark, L. Changes in bone mineral density and body composition during pregnancy and postpartum. A controlled cohort study. *Osteoporos. Int.* **2012**, *23*, 1213–1223. [CrossRef] [PubMed]
57. Chan, S.M.; Nelson, E.A.; Leung, S.S.; Cheng, J.C. Bone mineral density and calcium metabolism of Hong Kong Chinese postpartum women—A 1-y longitudinal study. *Eur. J. Clin. Nutr.* **2005**, *59*, 868–876. [CrossRef]
58. Sowers, M. Pregnancy and lactation as risk factors for subsequent bone loss and osteoporosis. *J. Bone Miner. Res.* **1996**, *11*, 1052–1060. [CrossRef]
59. Lambrot, R.; Xu, C.; Saint-Phar, S.; Chountalos, G.; Cohen, T.; Paquet, M.; Suderman, M.; Hallett, M.; Kimmins, S. Low paternal dietary folate alters the mouse sperm epigenome and is associated with negative pregnancy outcomes. *Nat. Commun.* **2013**, *4*, 2889. [CrossRef] [PubMed]
60. Alio, A.P.; Salihu, H.M.; McIntosh, C.; August, E.M.; Weldeselasse, H.; Sanchez, E.; Mbah, A.K. The effect of paternal age on fetal birth outcomes. *Am. J. Men's Health* **2012**, *6*, 427–435. [CrossRef]
61. Bellinger, D.C. Teratogen update: Lead and pregnancy. *Birth Defects Res. Part A Clin. Mol. Teratol.* **2005**, *73*, 409–420. [CrossRef] [PubMed]
62. Burton, G.J.; Fowden, A.L. The placenta: A multifaceted, transient organ. *Philos. Trans. R. Soc. Lond. B Biol. Sci.* **2015**, *370*, 20140066. [CrossRef] [PubMed]

nutrients

MDPI

Article

Exogenous Ketone Supplement Administration Abrogated Isoflurane-Anesthesia-Induced Increase in Blood Glucose Level in Female WAG/Rij Rats

Enikő Rauch [1,2], Csilla Ari [3,4,*], Dominic P. D'Agostino [3,5,6] and Zsolt Kovács [1]

1 Department of Biology, Berzsenyi Dániel Teacher Training Centre, ELTE Eötvös Loránd University, Károlyi Gáspár tér 4, 9700 Szombathely, Hungary; raucheniko9810@gmail.com (E.R.); kovacs.zsolt@sek.elte.hu or zskovacsneuro@gmail.com (Z.K.)
2 Institute of Biology, University of Pécs, Ifjúság Str. 6, 7624 Pécs, Hungary
3 Ketone Technologies LLC, Tampa, FL 33612, USA; ddagosti@usf.edu
4 Behavioral Neuroscience Research Laboratory, Department of Psychology, University of South Florida, Tampa, FL 33620, USA
5 Laboratory of Metabolic Medicine, Department of Molecular Pharmacology and Physiology, Morsani College of Medicine, University of South Florida, Tampa, FL 33612, USA
6 Institute for Human and Machine Cognition, Ocala, FL 34471, USA
* Correspondence: csari2000@yahoo.com

Abstract: It has been demonstrated that isoflurane-induced anesthesia can increase the blood glucose level, leading to hyperglycemia and several adverse effects. The administration of a mix of ketone diester (KE) and medium-chain triglyceride (MCT) oil, named KEMCT, abolished the isoflurane-anesthesia-induced increase in blood glucose level and prolonged the recovery time from isoflurane anesthesia in a male preclinical rodent model, Wistar Albino Glaxo/Rijswijk (WAG/Rij) rats. While most preclinical studies use exclusively male animals, our previous study on blood glucose changes in response to KEMCT administration showed that the results can be sex-dependent. Thus, in this study, we investigated female WAG/Rij rats, whether KEMCT gavage (3 g/kg/day for 7 days) can change the isoflurane (3%)-anesthesia-induced increase in blood glucose level and the recovery time from isoflurane-evoked anesthesia using the righting reflex. Moreover, KEMCT-induced ketosis may enhance both the extracellular level of adenosine and the activity of adenosine A1 receptors (A1Rs). To obtain information on the putative A1R mechanism of action, the effects of an A1R antagonist, DPCPX (1,3-dipropyl-8-cyclopentylxanthine; intraperitoneal/i.p. 0.2 mg/kg), on KEMCT-generated influences were also investigated. Our results show that KEMCT supplementation abolished the isoflurane-anesthesia-induced increase in blood glucose level, and this was abrogated by the co-administration of DPCPX. Nevertheless, KEMCT gavage did not change the recovery time from isoflurane-induced anesthesia. We can conclude that intragastric gavage of exogenous ketone supplements (EKSs), such as KEMCT, can abolish the isoflurane-anesthesia-induced increase in blood glucose level in both sexes likely through A1Rs in WAG/Rij rats, while recovery time was not affected in females, unlike in males. These results suggest that the administration of EKSs as an adjuvant therapy may be effective in mitigating metabolic side effects of isoflurane, such as hyperglycemia, in both sexes.

Keywords: isoflurane anesthesia; ketone supplement; glucose level; recovery time; adenosine receptor; female WAG/Rij rat

1. Introduction

Inhalational anesthetic isoflurane (1-chloro-2,2,2-trifluoroethyl difluoromethyl ether) crosses the blood–brain barrier (BBB) and neuronal membranes, leading to not only general anesthesia, but also neuroprotective influences [1–3]. For example, it has been demonstrated that isoflurane can inhibit apoptosis, reduce excitotoxicity, prevent mitochondrial dysfunction and have anti-inflammatory influence [4–6]. Consequently, isoflurane may generate alleviating effects in the treatment of several central nervous system diseases, such as cerebral ischemia, epilepsy, depression and Alzheimer's disease [3,7,8]. Moreover, isoflurane can modulate metabolic processes in the brain, including cortical metabolism [9]. It has also been demonstrated that the administration of isoflurane [2,10] can increase the blood glucose level through enhanced glucose production and reduced glucose clearance [11,12], leading to hyperglycemia as a consequence [13]. Moreover, hyperglycemic effects can decrease the cardioprotective effect of isoflurane [14,15]. Consequently, theoretically, a decrease in isoflurane-induced hyperglycemic effects may allow us to not only avoid adverse side effects but also preserve the potential protective effects of isoflurane.

It was also previously demonstrated that different exogenous ketone supplements (EKSs) and their combinations, such as the mix of ketone diester (KE) and medium-chain triglyceride (MCT) oil (named KEMCT), may enhance and maintain the blood ketone body level (e.g., R-β-hydroxybutyrate, R-βHB) and reduce the level of blood glucose not only in animals but also in humans [16–18]. These effects of EKSs were generated not only under physiological but also in pathological conditions, such as epilepsy and anxiety [16,19,20]. Furthermore, it was suggested that EKSs can modulate both sleep-like influences and isoflurane anesthesia [21,22]. Consequently, it was hypothesized that EKSs would reduce isoflurane-anesthesia-evoked changes in blood glucose level and recovery time. Indeed, the administration of KEMCT by intragastric gavage enhanced the blood R-βHB level and reduced the blood glucose level, as well as decreased the isoflurane-anesthesia-evoked increase in blood glucose level and prolonged the time required for recovery from isoflurane-generated anesthesia in an animal model of human absence epilepsy, Wistar Albino Glaxo Rijswijk (WAG/Rij) rats [23]. Moreover, KEMCT treatment delayed the onset of isoflurane-generated anesthesia in WAG/Rij rats [24]. These results suggest that, theoretically, the administration of EKSs as an adjuvant therapy may not only alter isoflurane-induced anesthesia but also attenuate isoflurane-anesthesia-generated side effects, such as hyperglycemia, not only in animal models but also in humans.

It has also been demonstrated that an EKS-generated increase in ketone body levels can enhance the level of adenosine in the brain [25,26], leading to enhanced activity of adenosine A1 receptors (A1Rs) and neuronal hyperpolarization [27,28]. These last effects may modulate the influences evoked by EKSs, such as KEMCT, on the isoflurane-generated onset of anesthesia and, theoretically, on the isoflurane-evoked increase in both glucose level and recovery time from isoflurane anesthesia [23,24,29]. Although the KEMCT-induced reduction in blood glucose level and increase in R-βHB level is sex-dependent in WAG/Rij rats [30], the influence of KEMCT administration on isoflurane-anesthesia-induced changes in blood R-βHB and glucose levels and recovery time were investigated only in males so far, but not in female WAG/Rij rats [23].

While most preclinical studies use exclusively male animals, our previous study on blood glucose changes in response to KEMCT administration shows that the results can be sex-dependent. It appears that KEMCT affects female and male WAG/Rij rats differently and that these differences are also influenced by age [30]. In that earlier study, KEMCT gavage induced significantly lower glucose levels at the 4th, 7th, 9th, 10th, 12th, and 13th months in females, whereas it induced significantly lower glucose levels between the 4th and 6th months and between the 9th and 13th months in males, compared with the results at the 1st month. In addition, KEMCT treatment induced lower blood glucose levels in female rats than in male rats between the 1st and 8th months, but significantly higher glucose levels were measured in female rats at the 17th month than in males. These results suggest that blood glucose level changes may be significantly different between female and male rats in response to KEMCT treatment during isoflurane anesthesia. Thus, in this study, we extended our previous experiments on ten-month-old males [23] to similar-age female WAG/Rij rats.

Thus, as a continuation of our previous study on male WAG/Rij rats [23], we investigated whether (i) KEMCT administration can generate changes in isoflurane-evoked alterations in blood glucose level and recovery time from anesthesia in female WAG/Rij rats (putative sex-dependent difference) and (ii) A1Rs can modulate EKS- and isoflurane-anesthesia-evoked effects in female WAG/Rij rats. In accordance with the main goals of this study, we investigated (i) the KEMCT administration (gavage, 3 g/kg once a day for 7 days)-generated effects on isoflurane (3%)-anesthesia-evoked influences on blood glucose and R-βHB levels, as well as recovery time from anesthesia, and (ii) the influence of an A1R antagonist, DPCPX (1,3-dipropyl-8-cyclopentylxanthine; intraperitoneal/i.p. administration of 0.2 mg/kg), on the above-mentioned KEMCT-evoked putative effects by assessing the righting reflex in female WAG/Rij rats. We hypothesized that KEMCT administration can abolish the isoflurane-induced increase in blood glucose level and alter recovery time from anesthesia through A1Rs in female rats.

2. Methods

2.1. Experimental Animals

Experiments on animals were performed according to (i) the Hungarian Act of Animal Care and Experimentation (1998, XXVIII, section 243), European Communities Council Directive (86/609; EEC) and EU Directive (2010/63; EU), as well as (ii) approval of the Animal Care and Experimentation Committee (Savaria University Centre, Eötvös Loránd University) and National Scientific Ethical Committee on Animal Experimentation (license: VA/ÉBÁF-ÁO/00279-4/2021; Hungary).

Ten-month-old female WAG/Rij rats (n = 56; 182–212 g) were housed in groups containing 4 animals in each group (breeding colony in Savaria University Centre, Eötvös Loránd University) under standard laboratory conditions (light–dark cycle 12:12 h; with light on between 08.00 A.M. and 08.00 P.M.; ad libitum access to food/SSNIFF RM-Z+H rat breeding and maintenance diet, by TOXI-COOP Ltd., Budapest, Hungary, and water; room temperature kept at 22 ± 2 °C). On the last day of experiments, the rats were humanely euthanized using isoflurane. All efforts were made to minimize pain, suffering and the number of animals used in the study.

2.2. Experimental Design

It has been demonstrated in our previous studies that gavage of 3 g/kg KEMCT for 7 days not only increased the blood R-βHB level, but also reduced the blood glucose level effectively without side effects, such as diarrhea [23]. Thus, similarly to our previous studies on WAG/Rij rats [23,24], in the present study, we fed all rats with a standard rodent chow diet, supplemented with KEMCT (3 g/kg/day) through intragastric needle (gavage) for 7 days. KEMCT contained KE (1,3-butanediol-acetoacetate diester), which was mixed with MCT oil (≈60% caprylic triglyceride and 40% capric triglyceride) in a 1:1 ratio (purchased from University of South Florida, Tampa, FL, USA; Savind, Inc. Urbana, IL, USA; Now Foods, Bloomingdale, IL, USA).

To adapt the animals for treatment methods, i.p. injection of saline (1 mL/kg) and, half an hour later, water gavage (3 g/kg) were administered for 5 days (adaptation period) to female WAG/Rij rats (Figure 1). After the adaptation period, rats were assigned into 4 groups (14 animals/group). Animals of the first and second groups (group 1 and group 2) were i.p. injected with saline (1 mL/kg) and, after 30 min, treated with water gavage between the 6th and 11th day of experiments. On these same days, rats of the third and fourth groups (group 3 and group 4) were treated with KEMCT gavage 30 min after the i.p. saline (1 mL/kg) (Figure 1). On the 12th day of experiments, animals of group 1 received 1 mL/kg 10% dimethyl sulfoxide (DMSO) solution i.p. and, after 30 min, water gavage. Animals of group 2 received similar treatments to animals in group 1, but the DMSO solution contained 0.2 mg/kg DPCPX (group 2). Animals in group 3 were treated with i.p. 1 mL/kg 10% DMSO solution, and 30 min later, KEMCT gavage was administered, whereas animals in group 4 received KEMCT 30 min after the i.p. injection of 0.2 mg/kg DPCPX in 10% DMSO solution.

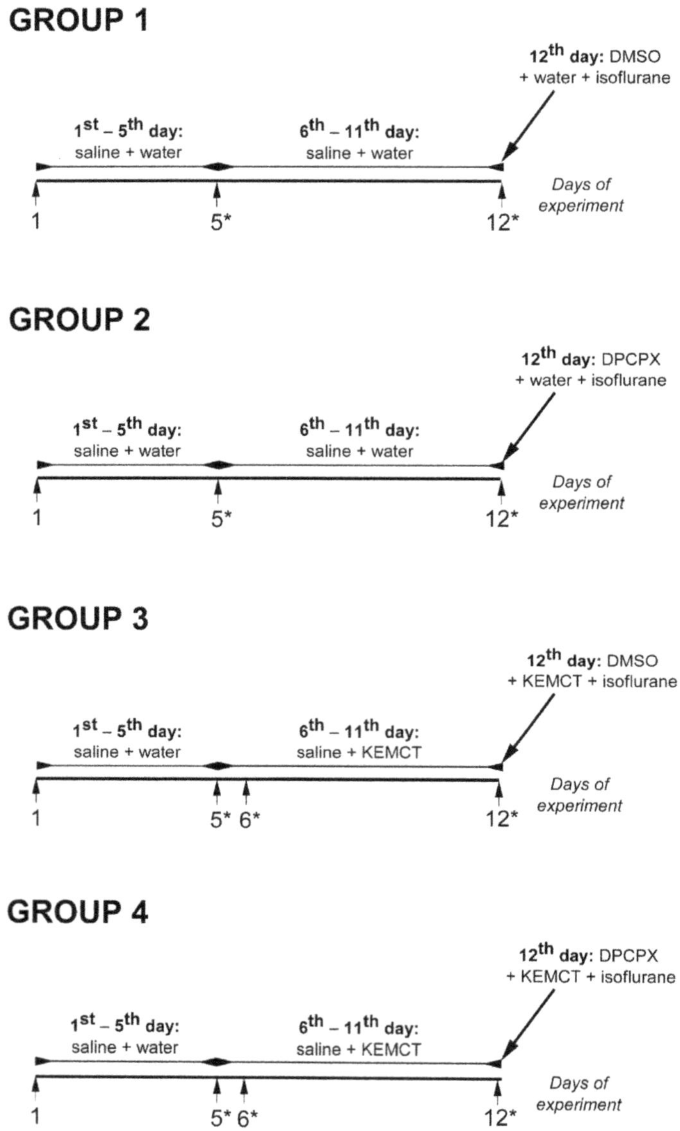

Figure 1. Details of the experimental design. Abbreviations: *, days of blood R-βHB and glucose level measurement; DMSO, dimethyl sulfoxide; DPCPX, 1,3-dipropyl-8-cyclopentylxanthine; KEMCT, mix of KE (1,3-butanediol-acetoacetate diester) and MCT oil (≈60% caprylic triglyceride + 40% capric triglyceride) in 1:1 ratio.

In relation to all animal groups (group 1–group 4), 60 min after the last gavage, an isoflurane (3%)–air mixture was applied to evoke general anesthesia in an airtight anesthesia chamber for 20 min, as previously described [23]. The righting reflex was used to measure the time required for recovery from isoflurane-induced anesthesia immediately after 20 min of anesthesia. The righting reflex is a well-known postural response of animals by which animals are able to reorient themselves when placed on their side or back. As a result, after recovery of the righting reflex from (isoflurane) anesthesia, the animals' paws will be reoriented to the ground again, which has been considered as an indicator of emergence from anesthesia in previous studies [1,31,32]. In this study, after 20 min anesthesia, the rats were placed on their backs in a Plexiglas box, and the recovery time (or emergence time/period: the time between termination of isoflurane anesthesia and recovery of the righting reflex) was video recorded and determined by a blinded observer [23].

2.3. Blood R-βHB and Glucose Levels, as Well as Body Weight

The blood levels of glucose and R-βHB were measured by a ketone and glucose monitoring system (Precision Xtra™, Abbott Laboratories, Green Oaks, IL, USA) from the blood taken from the tail vein of rats [23]. The measuring of blood R-βHB and glucose levels was carried out 90 min after the gavage on the 5th (control) and 12th (last) days (group 1 and group 2) or the 5th (control), 6th and 12th (last) days (group 3 and group 4) of experiments (Figure 1). On day 12, the blood levels of R-βHB and glucose were measured several minutes after recovery from isoflurane-induced anesthesia.

Body weight was measured on the 5th day (control) and on the 12th day of experiments (group 3 and group 4).

2.4. Statistics

The mean and the standard error of the mean (S.E.M.) were used for data presentation. The levels of blood glucose and R-βHB and body weight were compared to control values (measured on the 5th day of experiments). In relation to recovery time from isoflurane anesthesia, results from group 1 (1 mL/kg 10% DMSO solution i.p. + 3 g/kg water gavage) were compared to group 3 (1 mL/kg 10% DMSO solution i.p. + 3 g/kg KEMCT gavage), whereas results from group 2 (0.2 mg/kg DPCPX in 1 mL/kg 10% DMSO solution i.p. + 3 g/kg water gavage) were compared with group 4 (0.2 mg/kg DPCPX in 1 mL/kg 10% DMSO solution i.p. + 3 g/kg KEMCT gavage). Two-way ANOVA, Tukey's multiple comparisons test, Šídák's multiple comparisons test and *t*-test were performed using GraphPad version 9.2.0. [23]. Statistical significance was accepted when $p < 0.05$.

3. Results

3.1. Influence of DPCPX and KEMCT on Isoflurane-Anesthesia-Evoked Alterations in Level of Blood Glucose and R-βHB and on Body Weight

Isoflurane-generated anesthesia significantly enhanced the blood level of both glucose and R-βHB compared to the control (left part of Figure 2A,B; Table 1: group 1). We also demonstrated that the administration of i.p. 0.2 mg/kg DPCPX alone (without KEMCT) was not able to change the isoflurane-generated effects on blood glucose and R-βHB levels (right parts of Figure 2A,B; Table 1: group 2).

On the 6th experimental day, the first KEMCT gavage significantly decreased the blood level of glucose and increased the level of R-βHB compared to the control (Figure 2C,D; Table 1: group 3 and group 4). Nevertheless, the seventh gavage of KEMCT (on the 12th day of experiments) alone (without DPCPX) was able not only to increase the blood R-βHB level more efficiently (left part of Figure 2D; Table 1: group 3), but also to maintain the normal (control) level of blood glucose under isoflurane-generated anesthesia (left part of Figure 2C). Namely, the isoflurane-anesthesia-induced increase in blood glucose level was abolished by administration of KEMCT compared to the control in female WAG/Rij rats (left part of Figure 2C; Table 1: group 3), which is similar to our previous results for male WAG/Rij rats [23]. Moreover, i.p. 0.2 mg/kg DPCPX abolished the seventh KEMCT-gavage-evoked alleviating effect on the isoflurane-induced increase in blood glucose level. Thus, under these circumstances, the level of blood glucose significantly increased compared to the control (right part of Figure 2C; Table 1: group 4). However, DPCPX did not alter the KEMCT-generated increase in blood R-βHB level (Figure 2D; Table 1: group 4).

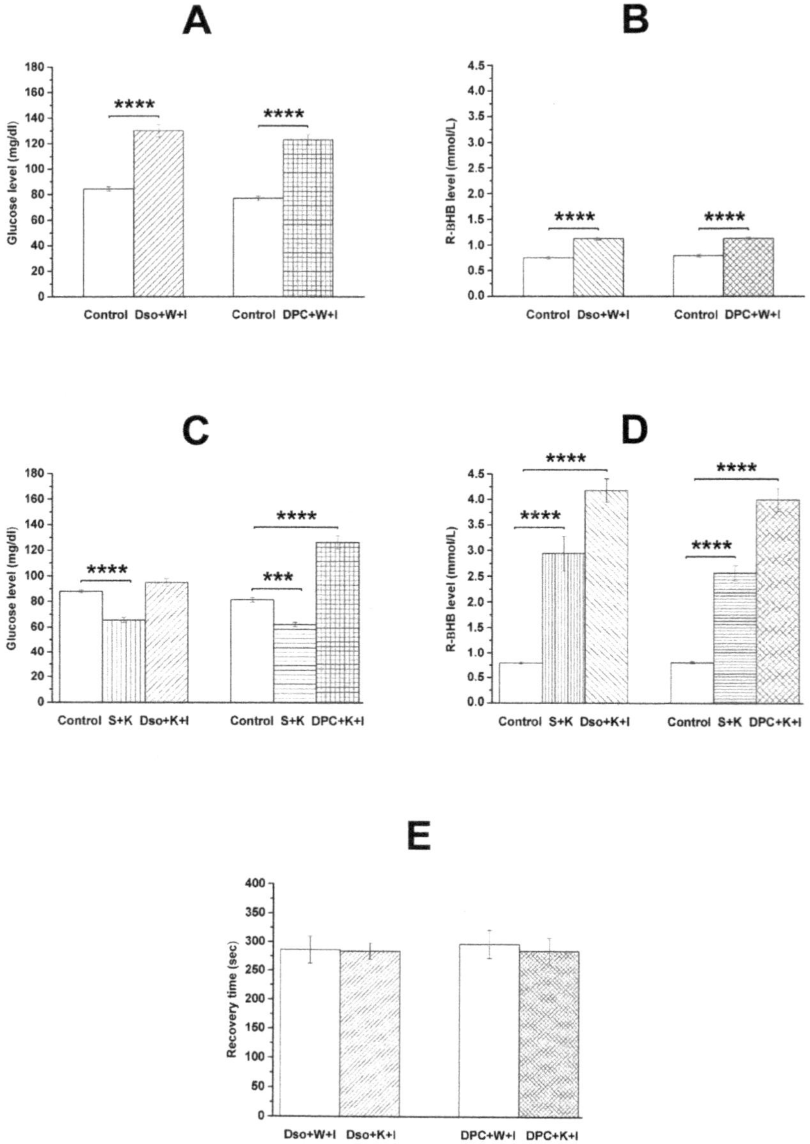

Figure 2. KEMCT-, DPCPX- and their combination-generated effects on isoflurane-anesthesia-evoked changes in blood glucose (**A**,**C**) and R-βHB (**B**,**D**) levels, as well as recovery time (**E**), compared to control. Abbreviations: DPC+K+I, group 4 (0.2 mg/kg DPCPX in 1 mL/kg 10% DMSO solution i.p. + 3 g/kg KEMCT gavage + isoflurane anesthesia); DPC + W + I, group 2 (0.2 mg/kg DPCPX in 1 mL/kg 10% DMSO solution i.p. + 3 g/kg water gavage + isoflurane anesthesia); Dso + K + I, group 3 (1 mL/kg 10% DMSO solution i.p. + 3 g/kg KEMCT gavage + isoflurane anesthesia); Dso + W + I, group 1 (1 mL/kg 10% DMSO solution i.p. + 3 g/kg water gavage + isoflurane anesthesia); R-βHB, R-beta-hydroxybutyrate; S + K, group 3 and group 4 (1 mL/kg saline i.p. + 3 g/kg KEMCT gavage); *** $p < 0.001$; **** $p < 0.0001$.

Table 1. Effects of KEMCT gavage and i.p. DPCPX on isoflurane-anesthesia-generated changes in blood R-βHB and glucose levels, compared to control. Abbreviations: ns, non-significant; R-βHB, R-beta-hydroxybutyrate; *** $p < 0.001$; **** $p < 0.0001$.

Treatments	Glucose (mg/dL)	R-βHB (mmol/L)
	Group 1	
Control	84.43 ± 1.775	0.75 ± 0.027
12th day of experiments	130.07 ± 4.748 ****/<0.0001	1.13 ± 0.027 ****/<0.0001
	Group 2	
Control	77.00 ± 1.816	0.79 ± 0.027
12th day of experiments	123.00 ± 4.235 ****/<0.0001	1.13 ± 0.022 ****/<0.0001
	Group 3	
Control	88.00 ± 1.371	0.79 ± 0.019
6th day of experiments	65.71 ± 1.962 ****/<0.0001	2.94 ± 0.338 ****/<0.0001
12th day of experiments	94.79 ± 3.015 ns/0.0913	4.18 ± 0.228 ****/<0.0001
	Group 4	
Control	81.36 ± 1.833	0.80 ± 0.019
6th day of experiments	62.29 ± 1.588 ***/0.0005	2.56 ± 0.146 ****/<0.0001
12th day of experiments	126.14 ± 5.122 ****/<0.0001	4.00 ± 0.228 ****/<0.0001

Similar to our previous study on male WAG/Rij rats [23], the body weight of female WAG/Rij rats was not changed compared to the control (control/KEMCT alone, group 3: 200.5 ± 1.82 g/199.1 ± 2.50 g, $p = 0.5778$; control/DPCPX + KEMCT, group 4: 191.6 ± 1.57 g/190.1 ± 2.01 g, $p = 0.5472$).

3.2. *Effects of DPCPX and KEMCT on Recovery Time*

In contrast to our previous study on male WAG/Rij rats [23], recovery time from isoflurane-evoked anesthesia did not change after KEMCT gavage alone or combined administration of KEMCT gavage with DPCPX (0.2 mg/kg i.p.) in female WAG/Rij rats (DMSO + water + isoflurane/group 1 v. DMSO + KEMCT + isoflurane/group 3: $p = 0.9312$; DPCPX + water + isoflurane/group 2 v. DPCPX + KEMCT + isoflurane/group 4: $p = 0.7225$) (Figure 2E).

4. Discussion

As putative sex-dependent effects of EKSs can have important implications during anesthesia, we studied and compared females to our previous study on male rats. In this study, we investigated the KEMCT-administration-generated effects on isoflurane-induced changes in blood glucose level and recovery time using female WAG/Rij rats. We demonstrated that like in male WAG/Rij rats [23], isoflurane anesthesia (i) can increase blood glucose and R-βHB levels; KEMCT administration (ii) was able to abolish the isoflurane-anesthesia-induced augmentation in blood glucose level, whereas (iii) it did not alter the body weight of female rats. These results strengthened our previous results on male WAG/Rij rats and suggested sex-independent effects of EKSs. Moreover, we also demonstrated a sex-dependent effect of KEMCT administration: in contrast to our previous results on male WAG/Rij rats, KEMCT gavage did not alter recovery time from isoflurane-generated anesthesia. We extended our previous results, as we also suggested that (i) KEMCT administration may evoke its beneficial effect on the isoflurane-anesthesia-induced increase in blood glucose level likely through a ketosis-induced increase in adenosine level and A1R activity in the brain and (ii) inhibition of A1Rs alone did not change the isoflurane-anesthesia-induced effects on blood glucose and R-βHB levels.

Earlier studies suggest that insulin resistance and hyperglycemia may be related to, among others, immunosuppression, cardiovascular problems and ischemic brain damage [13,33,34]. Isoflurane anesthesia may impair glucose clearance and produce hyperglycemia likely through processes evoked by ATP-sensitive potassium channels in pancreatic β-cells leading to a decrease in insulin release and glucose utilization [35,36]. Indeed, it was evidenced that isoflurane anesthesia increased the blood glucose level in both male [23] and female WAG/Rij rats (left part of Figure 2A), and this influence was abolished by KEMCT administration [23] (left part of Figure 2C). In relation to the putative mechanism of action of the KEMCT-evoked influence on isoflurane-anesthesia-generated effects on blood glucose levels, it was demonstrated that EKSs are able to increase the blood ketone body level as well as mitigate the blood glucose level [23,37], suggesting that EKS-evoked ketosis is able to increase insulin sensitivity [38]. It has also been demonstrated that insulin sensitivity and glucose tolerance were impaired in A1R KO mice [39], whereas A1R overexpression can protect mice from insulin resistance [40]. Furthermore, A1R activation can increase insulin sensitivity [41,42]. Thus, as KEMCT-induced ketosis can increase the adenosine level in the brain [25,26], it is possible that KEMCT could modulate the effect of isoflurane anesthesia on the blood glucose level through A1Rs. Indeed, we demonstrated that DPCPX alone was not effective on the isoflurane-anesthesia-induced elevation in blood glucose level (right part of Figure 2A), but i.p. injection of DPCPX in combination with KEMCT gavage abrogated the beneficial influence of KEMCT treatment on isoflurane-anesthesia-generated changes in blood glucose levels (Figure 2C).

We demonstrated previously that isoflurane anesthesia alone, KEMCT alone and the combination of KEMCT gavage with isoflurane anesthesia increased the blood R-βHB level in male WAG/Rij rats [23]. These results were strengthened by the present study in female WAG/Rij rats (left parts of Figure 2B,D), suggesting that the isoflurane anesthesia increases the blood R-βHB level sex-independently. Moreover, KEMCT administration can enhance the R-βHB level of blood under isoflurane-induced anesthesia, also independently of sex. These effects on the blood R-βHB level evoked by the administration of isoflurane alone and isoflurane in combination with KEMCT were not modulated by DPCPX (right parts of Figure 2B,D), revealing that A1Rs likely do not have a role in these processes, at least in female WAG/Rij rats.

Previous studies suggest that adenosine and A1Rs may modulate the recovery time from isoflurane-evoked anesthesia. For example, not only non-selective antagonists of adenosine receptors, such as caffeine and theophylline in both rats and mice [43–45], but also DPCPX in mice decreased the time spent to recover from isoflurane anesthesia [44]. Nevertheless, an A1R agonist, N-p-sulfophenyl adenosine, augmented the recovery time from isoflurane-evoked anesthesia in mice, likely through A1Rs [29]. Although KEMCT-administration-evoked ketosis can exert its effects on isoflurane-anesthesia-evoked influences through increased adenosine levels and A1R activity [25,26,43–45] and KEMCT administration for 7 days prolonged the recovery time from isoflurane anesthesia in male WAG/Rij rats, perhaps through A1Rs [23], in this study, the administration of KEMCT alone (group 3) did not alter the recovery time in female WAG/Rij rats (left part of Figure 2E). It was demonstrated previously that metabolic enzymes and processes modulating the adenosine level and, consequently, the extracellular level of adenosine, as well as the expression of nucleoside transporters and activity of A1Rs in the brain, are sex-dependent [46–48]. Thus, these results suggest that the effect of KEMCT administration on the time required to recover from isoflurane anesthesia may be sex-dependent, likely through the adenosinergic system, at least in WAG/Rij rats. Nonetheless, it was demonstrated that the recovery time from isoflurane anesthesia was similar in male and female rats [49] and was not modulated by the estrus cycle [50].

We demonstrated previously that the administration of different i.p. doses of DPCPX (0.15–0.25 mg/kg) alone was not effective on absence epileptic activity, anxiety level and the onset of the isoflurane-evoked light phase of anesthesia [19,20,24]. Nevertheless, the administration of the same doses of DPCPX in combination with EKSs, such as KEMCT, KSMCT (mix of ketone salt/KS and MCT oil) and/or KEKS (mix of KE and KS), was effective in the inhibition of EKS-generated effects on absence epileptic activity (0.2 mg/kg DPCPX i.p.), anxiety (0.15 and 0.25 mg/kg DPCPX i.p.) and the onset of the isoflurane-generated light phase of anesthesia (0.2 mg/kg DPCPX i.p.) in male WAG/Rij rats [19,20,24]. Similarly, in this study, DPCPX alone (group 2) was ineffective in changing the recovery time from isoflurane anesthesia in female WAG/Rij rats (right part of Figure 2E). Nevertheless, based on (i) the above-mentioned results suggesting that the applied dose of DPCPX could be effective only in combination with EKSs if the administration of KEMCT alone was effective, for example, in the modulation of isoflurane-evoked effects [19,20,24] and (ii) the lack of efficacy of KEMCT gavage alone (group 3) on recovery time (left part of Figure 2E), it is not astonishing that combined administration of DPCPX and KEMCT (group 4) was also not effective in changing the recovery time (right part of Figure 2E). However, higher doses of KEMCT may change the time required to recover from isoflurane anesthesia. Moreover, there may be a need to test whether different doses and types of EKSs evoke significant effects on different physiological and pathophysiological processes under different circumstances in male and female rats [30] through the interaction of several neurotransmitter systems (e.g., adenosinergic, serotonergic and GABAergic system) and receptors [51]. Thus, further studies are needed to investigate the effect of different doses and types of EKSs, as well as their exact mechanism of action, on not only the isoflurane-evoked increase in blood glucose level but also the recovery time from isoflurane-induced anesthesia through different neurotransmitter systems, such as the adenosinergic system and A1Rs.

5. Conclusions

Based on our earlier study on male WAG/Rij rats, together with the present study performed on female WAG/Rij rats, we can conclude that isoflurane-induced anesthesia can increase the blood level of both glucose and R-βHB independently of sex. Moreover, KEMCT administration was able to abolish the isoflurane-anesthesia-evoked increase in blood glucose level, also independently of sex, likely through A1Rs. Nevertheless, unlike the results obtained on male WAG/Rij rats, KEMCT gavage did not change the recovery time from isoflurane-evoked anesthesia in female WAG/Rij rats, suggesting a sex-dependent influence. These results further strengthened our previous suggestion that the administration of EKSs (e.g., KEMCT) as an adjuvant therapy may be effective in mitigating the metabolic side effects of isoflurane, such as hyperglycemia, in both sexes. However, further studies are needed to reveal the putative sex-dependent effects, as well as the exact mechanism of action, of EKSs on influences evoked by isoflurane and other types of anesthesia in both males and females.

Author Contributions: Conceptualization, E.R. and Z.K.; data curation, Z.K.; formal analysis, C.A.; investigation, E.R.; methodology, E.R. and Z.K.; resources, Z.K.; validation, Z.K.; writing—original draft, E.R., C.A. and Z.K.; writing—review and editing, D.P.D. All authors have read and agreed to the published version of the manuscript.

Funding: This work was supported by the ELTE BDPK Excellence Program (12/2021, 12/2022 and 12/2023) and Ketone Technologies LLC (12/2021). The funding body had no influence on the design of the study, collection and analysis of data, interpretation of results, and writing of the manuscript.

Institutional Review Board Statement: The study was conducted according to the guidelines of the Declaration of Helsinki and approved by the Animal Care and Experimentation Committee of the Eötvös Loránd University (Savaria University Centre) and National Scientific Ethical Committee on Animal Experimentation (Hungary) under license number VA/ÉBÁF-ÁO/00279-4/2021 (1 April 2021).

Informed Consent Statement: Not applicable.

Data Availability Statement: The data presented in this study are available on request from the corresponding author. The data are not publicly available due to ethical reasons.

Conflicts of Interest: Patent US10980764B1 C.A., D.P.D., "Exogenous ketone supplements for reducing anxiety-related behavior". C.A. and D.P.D. are co-owners of Ketone Technologies LLC, and C.A. is owner of Fortis World LLC. The authors declare that this study received funding from Ketone Technologies LLC. The funders had the following involvement with the study: formal analysis; writing—original draft C.A., writing—review and editing, D.P.D. These interests have been reviewed and managed by the University in accordance with its Institutional and Individual Conflict of Interest policies. All authors declare that there are no additional conflicts of interest.

Abbreviations

A1R: adenosine A1 receptor; DMSO, dimethyl sulfoxide; DPCPX, 1,3-dipropyl-8-cyclopentylxanthine; EKS, exogenous ketone supplement; i.p., intraperitoneal; KE (ketone diester), 1,3 butanediol-acetoacetate diester; KEMCT, mix of KE and MCT oil; MCT, medium-chain triglyceride; R-βHB, R-beta-hydroxybutyrate; WAG/Rij, Wistar Albino Glaxo/Rijswijk.

References

1. Franks, N.P. General anaesthesia: From molecular targets to neuronal pathways of sleep and arousal. *Nat. Rev. Neurosci.* **2008**, *9*, 370–386. [CrossRef]
2. Hao, X.; Ou, M.; Zhang, D.; Zhao, W.; Yang, Y.; Liu, J.; Yang, H.; Zhu, T.; Li, Y.; Zhou, C. The Effects of General Anesthetics on Synaptic Transmission. *Curr. Neuropharmacol.* **2020**, *18*, 936–965. [CrossRef] [PubMed]
3. Deng, J.; Lei, C.; Chen, Y.; Fang, Z.; Yang, Q.; Zhang, H.; Cai, M.; Shi, L.; Dong, H.; Xiong, L. Neuroprotective gases—Fantasy or reality for clinical use? *Prog. Neurobiol.* **2014**, *115*, 210–245. [CrossRef] [PubMed]
4. Altay, O.; Suzuki, H.; Hasegawa, Y.; Ostrowski, R.P.; Tang, J.; Zhang, J.H. Isoflurane on brain inflammation. *Neurobiol. Dis.* **2014**, *62*, 365–371. [CrossRef] [PubMed]
5. Bickler, P.E.; Warner, D.S.; Stratmann, G.; Schuyler, J.A. Gamma-Aminobutyric acid-A receptors contribute to isoflurane neuroprotection in organotypic hippocampal cultures. *Anesth. Analg.* **2003**, *97*, 564–571. [CrossRef] [PubMed]
6. Wang, L.; Traystman, R.J.; Murphy, S.J. Inhalational anesthetics as preconditioning agents in ischemic brain. *Curr. Opin. Pharmacol.* **2008**, *8*, 104–110. [CrossRef]
7. Jiang, J.; Jiang, H. Effect of the inhaled anesthetics isoflurane, sevoflurane and desflurane on the neuropathogenesis of Alzheimer's disease (review). *Mol. Med. Rep.* **2015**, *12*, 3–12. [CrossRef] [PubMed]
8. Tadler, S.C.; Mickey, B.J. Emerging evidence for antidepressant actions of anesthetic agents. *Curr. Opin. Anaesthesiol.* **2018**, *31*, 439–445. [CrossRef] [PubMed]
9. Newberg, L.A.; Milde, J.H.; Michenfelder, J.D. The cerebral metabolic effects of isoflurane at and above concentrations that suppress cortical electrical activity. *Anesthesiology* **1983**, *59*, 23–28. [CrossRef]
10. Sakai, E.M.; Connolly, L.A.; Klauck, J.A. Inhalation anesthesiology and volatile liquid anesthetics: Focus on isoflurane, desflurane, and sevoflurane. *Pharmacotherapy* **2005**, *25*, 1773–1788. [CrossRef]
11. Lattermann, R.; Schricker, T.; Wachter, U.; Georgieff, M.; Goertz, A. Understanding the mechanisms by which isoflurane modifies the hyperglycemic response to surgery. *Anesth. Analg.* **2001**, *93*, 121–127. [CrossRef] [PubMed]
12. Saha, J.K.; Xia, J.; Grondin, J.M.; Engle, S.K.; Jakubowski, J.A. Acute hyperglycemia induced by ketamine/xylazine anesthesia in rats: Mechanisms and implications for preclinical models. *Exp. Biol. Med. Maywood.* **2005**, *230*, 777–7784. [CrossRef] [PubMed]
13. Ljungqvist, O.; Jonathan, E. Rhoads lecture 2011: Insulin resistance and enhanced recovery after surgery. *J. Parenter. Enteral. Nutr.* **2012**, *36*, 389–398. [CrossRef] [PubMed]
14. Tanaka, K.; Kehl, F.; Gu, W.; Krolikowski, J.G.; Pagel, P.S.; Warltier, D.C.; Kersten, J.R. Isoflurane-induced preconditioning is attenuated by diabetes. *Am. J. Physiol. Heart Circ. Physiol.* **2002**, *282*, 2018–2023. [CrossRef]
15. Tanaka, K.; Weihrauch, D.; Ludwig, L.M.; Kersten, J.R.; Pagel, P.S.; Warltier, D.C. Mitochondrial adenosine triphosphate-regulated potassium channel opening acts as a trigger for isoflurane-induced preconditioning by generating reactive oxygen species. *Anesthesiology* **2003**, *98*, 935–943. [CrossRef]
16. Ari, C.; Kovács, Z.; Juhasz, G.; Murdun, C.; Goldhagen, C.R.; Koutnik, A.M.; Poff, A.M.; Kesl, S.L.; D'Agostino, D. Exogenous ketone supplements reduce anxiety-related behavior in Sprague-Dawley and Wistar Albino Glaxo/Rijswijk rats. *Front. Mol. Neurosci.* **2016**, *9*, 137. [CrossRef] [PubMed]
17. Hashim, S.A.; VanItallie, T.B. Ketone body therapy: From the ketogenic diet to the oral administration of ketone ester. *J. Lipid Res.* **2014**, *55*, 1818–1826. [CrossRef]

18. Stubbs, B.J.; Cox, P.J.; Evans, R.D.; Santer, P.; Miller, J.J.; Faull, O.K.; Magor-Elliott, S.; Hiyama, S.; Stirling, M.; Clarke, K. On the metabolism of exogenous ketones in humans. *Front. Physiol.* **2017**, *8*, 848. [CrossRef]
19. Kovács, Z.; D'Agostino, D.P.; Dobolyi, A.; Ari, C. Adenosine A1 receptor antagonism abolished the anti-seizure effects of exogenous ketone supplementation in Wistar Albino Glaxo Rijswijk rats. *Front. Mol. Neurosci.* **2017**, *10*, 235. [CrossRef]
20. Kovács, Z.; D'Agostino, D.P.; Ari, C. Anxiolytic effect of exogenous ketone supplementation is abolished by adenosine A1 receptor inhibition in Wistar Albino Glaxo/Rijswijk rats. *Front. Behav. Neurosci.* **2018**, *12*, 29. [CrossRef]
21. Constantinides, C.; Murphy, K. Molecular and integrative physiological effects of isoflurane anesthesia: The paradigm of cardiovascular studies in rodents using magnetic resonance imaging. *Front. Cardiovasc. Med.* **2016**, *3*, 23. [CrossRef] [PubMed]
22. Hallböök, T.; Lundgren, J.; Rosén, I. Ketogenic diet improves sleep quality in children with therapy-resistant epilepsy. *Epilepsia* **2007**, *48*, 59–65. [CrossRef] [PubMed]
23. Kovács, Z.; D'Agostino, D.P.; Ari, C. Ketone supplementation abolished isoflurane anesthesia-induced elevation in blood glucose level and increased recovery time from anesthesia in Wistar Albino Glaxo Rijswijk rats. *BMC Anesthesiol.* **2023**, *23*, 43. [CrossRef] [PubMed]
24. Kovács, Z.; Brunner, B.; D'Agostino, D.P.; Ari, C. Inhibition of adenosine A1 receptors abolished the nutritional ketosis-evoked delay in the onset of isoflurane-induced anesthesia in Wistar Albino Glaxo Rijswijk rats. *BMC Anesthesiol.* **2020**, *20*, 30. [CrossRef] [PubMed]
25. Sharma, A.K.; Rani, E.; Waheed, A.; Rajput, S.K. Pharmacoresistant epilepsy: A current update on non-conventional pharmacological and non-pharmacological interventions. *J. Epilepsy Res.* **2015**, *5*, 1–8. [CrossRef] [PubMed]
26. Yang, Q.; Guo, M.; Wang, X.; Zhao, Y.; Zhao, Q.; Ding, H.; Dong, Q.; Cui, M. Ischemic preconditioning with a ketogenic diet improves brain ischemic tolerance through increased extracellular adenosine levels and hypoxia-inducible factors. *Brain Res.* **2017**, *1667*, 11–18. [CrossRef] [PubMed]
27. Kawamura, M., Jr.; Ruskin, D.N.; Masino, S.A. Metabolic autocrine regulation of neurons involves cooperation among pannexin hemichannels, adenosine receptors, and KATP channels. *J. Neurosci.* **2010**, *30*, 3886–3895. [CrossRef] [PubMed]
28. Ribeiro, J.A. Purinergic inhibition of neurotransmitter release in the central nervous system. *Pharmacol. Toxicol.* **1995**, *77*, 299–305. [CrossRef] [PubMed]
29. Gettys, G.C.; Liu, F.; Kimlin, E.; Baghdoyan, H.A.; Lydic, R. Adenosine A(1) receptors in mouse pontine reticular formation depress breathing, increase anesthesia recovery time, and decrease acetylcholine release. *Anesthesiology* **2013**, *118*, 327–336. [CrossRef]
30. Kovács, Z.; Brunner, B.; D'Agostino, D.P.; Ari, C. Age- and Sex-Dependent Modulation of Exogenous Ketone Supplement-Evoked Effects on Blood Glucose and Ketone Body Levels in Wistar Albino Glaxo Rijswijk Rats. *Front. Neurosci.* **2021**, *14*, 618422. [CrossRef]
31. Leung, L.S.; Luo, T.; Ma, J.; Herrick, I. Brain areas that influence general anesthesia. *Prog. Neurobiol.* **2014**, *122*, 24–44. [CrossRef]
32. Wang, L.; Holland, L.; Fong, R.; Khokhar, S.; Fox, A.P.; Xie, Z. A pilot study showing that repeated exposure to stress produces alterations in subsequent responses to anesthetics in rats. *PLoS ONE* **2019**, *14*, e0214093. [CrossRef] [PubMed]
33. Turina, M.; Miller, F.N.; Tucker, C.F.; Polk, H.C. Short-term hyperglycemia in surgical patients and a study of related cellular mechanisms. *Ann. Surg.* **2006**, *243*, 845–851. [CrossRef] [PubMed]
34. Warner, D.S.; Gionet, T.X.; Todd, M.M.; McAllister, A.M. Insulin-induced normoglycemia improves ischemic outcome in hyperglycemic rats. *Stroke* **1992**, *23*, 1775–1780. [CrossRef] [PubMed]
35. Tanaka, T.; Nabatame, H.; Tanifuji, Y. Insulin secretion and glucose utilization are impaired under general anesthesia with sevoflurane as well as isoflurane in a concentration-independent manner. *J. Anesth.* **2005**, *19*, 277–281. [CrossRef] [PubMed]
36. Tanaka, K.; Kawano, T.; Tomino, T.; Kawano, H.; Okada, T.; Oshita, S.; Takahashi, A.; Nakaya, Y. Mechanisms of impaired glucose tolerance and insulin secretion during isoflurane anesthesia. *Anesthesiology* **2009**, *111*, 1044–1051. [CrossRef]
37. Kashiwaya, Y.; Pawlosky, R.; Markis, W.; King, M.T.; Bergman, C.; Srivastava, S.; Murray, A.; Clarke, K.; Veech, R.L. A ketone ester diet increases brain malonyl-CoA and Uncoupling proteins 4 and 5 while decreasing food intake in the normal Wistar Rat. *J. Biol. Chem.* **2010**, *285*, 25950–25956. [CrossRef] [PubMed]
38. Sato, K.; Kashiwaya, Y.; Keon, C.; Tsuchiya, N.; King, M.T.; Radda, G.K.; Chance, B.; Clarke, K.; Veech, R.L. Insulin, ketone bodies, and mitochondrial energy transduction. *FASEB J.* **1995**, *9*, 651–658. [CrossRef] [PubMed]
39. Faulhaber-Walter, R.; Jou, W.; Mizel, D.; Li, L.; Zhang, J.; Kim, S.M.; Huang, Y.; Chen, M.; Briggs, J.P.; Gavrilova, O.; et al. Impaired glucose tolerance in the absence of adenosine A1 receptor signaling. *Diabetes* **2011**, *60*, 2578–2587. [CrossRef]
40. Dong, Q.; Ginsberg, H.N.; Erlanger, B.F. Overexpression of the A1 adenosine receptor in adipose tissue protects mice from obesity-related insulin resistance. *Diabetes Obes. Metab.* **2001**, *3*, 360–366. [CrossRef]
41. Dhalla, A.K.; Wong, M.Y.; Voshol, P.J.; Belardinelli, L.; Reaven, G.M. A1 adenosine receptor partial agonist lowers plasma FFA and improves insulin resistance induced by high-fat diet in rodents. *Am. J. Physiol. Endocrinol. Metab.* **2007**, *292*, 1358–1363. [CrossRef] [PubMed]
42. Töpfer, M.; Burbiel, C.E.; Müller, C.E.; Knittel, J.; Verspohl, E.J. Modulation of insulin release by adenosine A1 receptor agonists and antagonists in INS-1 cells: The possible contribution of 86Rb+ efflux and 45Ca2+ uptake. *Cell Biochem. Funct.* **2008**, *26*, 833–843. [CrossRef] [PubMed]
43. Fong, R.; Wang, L.; Zacny, J.P.; Khokhar, S.; Apfelbaum, J.L.; Fox, A.P.; Xie, Z. Caffeine accelerates emergence from isoflurane anesthesia in humans: A randomized, double-blind, crossover study. *Anesthesiology* **2018**, *129*, 912–920. [CrossRef] [PubMed]

44. Van Dort, C.J.; Baghdoyan, H.A.; Lydic, R. Adenosine A(1) and A(2A) receptors in mouse prefrontal cortex modulate acetylcholine release and behavioral arousal. *J. Neurosci.* **2009**, *29*, 871–881. [CrossRef]
45. Wang, Q.; Fong, R.; Mason, P.; Fox, A.P.; Xie, Z. Caffeine accelerates recovery from general anesthesia. *J. Neurophysiol.* **2014**, *111*, 1331–1340. [CrossRef] [PubMed]
46. Kovács, Z.; Juhász, G.; Palkovits, M.; Dobolyi, A.; Kékesi, K.A. Area, age and gender dependence of the nucleoside system in the brain: A review of current literature. *Curr. Top. Med. Chem.* **2011**, *11*, 1012–1033. [CrossRef]
47. Lu, H.; Chen, C.; Klaassen, C. Tissue distribution of concentrative and equilibrative nucleoside transporters in male and female rats and mice. *Drug Metab. Dispos.* **2004**, *32*, 1455–1461. [CrossRef] [PubMed]
48. Rucker, B.; Pereira, G.S.; Furstenau, C.R.; Izquierdo, I.; Bonan, C.D.; Sarkis, J.J. Inhibitory avoidance task reveals differences in ectonucleotidase activities between male and female rats. *Neurochem. Res.* **2004**, *29*, 2231–2237. [CrossRef] [PubMed]
49. Mansouri, M.T.; Fidler, J.A.; Meng, Q.C.; Eckenhoff, R.G.; García, P.S. Sex effects on behavioral markers of emergence from propofol and isoflurane anesthesia in rats. *Behav. Brain Res.* **2019**, *367*, 59–67. [CrossRef]
50. Vincent, K.F.; Mallari, O.G.; Dillon, E.J.; Stewart, V.G.; Cho, A.J.; Dong, Y.; Edlow, A.G.; Ichinose, F.; Xie, Z.; Solt, K. Oestrous cycle affects emergence from anaesthesia with dexmedetomidine, but not propofol, isoflurane, or sevoflurane, in female rats. *Br. J. Anaesth.* **2023**, *131*, 67–78. [CrossRef]
51. Newman, J.C.; Verdin, E. Ketone bodies as signaling metabolites. *Trends Endocrinol. Metab.* **2014**, *25*, 42–52. [CrossRef] [PubMed]

nutrients

MDPI

Article

Carbohydrate Ingestion before Exercise for Individuals with McArdle Disease: Survey Evidence of Implementation and Perception in Real-World Settings

Sam L. Torrens [1,*], Evelyn B. Parr [2], Craig McNulty [1], Lynda Ross [1], Helen MacLaughlin [1] and Robert A. Robergs [1]

1 School of Exercise and Nutrition Sciences, Faculty of Health, Queensland University of Technology, Victoria Park Road, Kelvin Grove, QLD 4058, Australia; c.mcnulty@qut.edu.au (C.M.); l20.ross@qut.edu.au (L.R.); h.maclaughlin@qut.edu.au (H.M.); rob.robergs@qut.edu.au (R.A.R.)

2 Mary Mackillop Institute for Health Research, Australian Catholic University, Level 5, 215 Spring Street, Melbourne, VIC 3000, Australia; evelyn.parr@acu.edu.au

* Correspondence: sam.torrens@qut.edu.au; Tel.: +61-3138-0339

Abstract: In individuals with McArdle disease (IWMD), the ingestion of carbohydrates before exercise has previously been shown in laboratory studies to significantly decrease the exercising symptoms of the condition and increase exercise tolerance during the early stages of exercise. As a result, carbohydrate ingestion pre-exercise is currently included in management guidelines, and often advised by medical professionals treating the condition. The aim of the current study was to determine whether positive lab-based results for the ingestion of carbohydrate before exercise in laboratory studies are being effectively translated into practice and produce perceptions of the same positive outcomes in real-world settings (RWS). An online survey method was used to collect responses from 108 IWMD. Data collected on the amount and type of carbohydrate consumed prior to exercise found that most surveyed participants (69.6%) who supplied qualitative data (n = 45) consumed less than the 37 g currently recommended in management guidelines. Survey data also revealed a large variation in the type and amount of carbohydrate ingested when IWMDs are applying carbohydrate ingestion before exercise in RWS. Consistent with these findings, only 17.5% of participants stated that they found carbohydrate ingestion before exercise relieved or minimised their MD symptoms. Results suggest that positive lab-based findings (increased exercise tolerance) of carbohydrate ingestion before exercise are not being effectively translated to RWS for many IWMD. There is a need for improved patient education of IWMD on the application of carbohydrate ingestion before exercise in RWS.

Keywords: McArdle disease; patient experience; carbohydrate; sucrose; exercise; clinical practice guidelines

1. Introduction

McArdle disease (MD) is a rare (1:100,000) genetic myopathy whereby individuals are unable to produce the enzyme glycogen phosphorylase in skeletal muscle tissue. Muscle glycogen phosphorylase catalyses the first step of glycogenolysis through phosphorylation from muscle glycogen to glucose-1-phosphate. As a result, stored muscle glycogen cannot be accessed to fuel glycolysis and subsequent oxidative phosphorylation within the mitochondria. The inaccessibility to muscle glycogen causes a reduction in the capacity for energy release from carbohydrate oxidation to support muscle contraction and other chemical-driven systems of cellular work. Therefore, individuals with MD (IWMD) heavily rely on alternative substrates such as fatty acids, liver glycogen, and blood glucose to fuel the cellular energy demands of skeletal muscle during activities of daily living, as well as exercise, for health promotion and disease prevention [1,2]. As muscle glycogen plays an integral role in mediation between anaerobic and aerobic cellular ATP resynthesis, deleterious symptoms (increased HR, increased RPE, increased muscle stiffness, cramping, contractures, decreased exercise tolerance) are most notable in IWMD during rapid increases in cellular ATP demand, such as during the early stages of exercise and abrupt increases in exercise intensity.

For non-MD individuals, ATP resynthesis during the first few minutes of exercise is predominantly met by the breakdown of creatine phosphate and muscle glycolysis fuelled by intramuscular glycogenolysis [3,4]. As IWMD do not have muscle phosphorylase (hence no access to muscle glycogen stores) and creatine phosphate is limited in its supply, times of rapidly increased metabolic demand result in a cellular energy crisis due to the lack of available substrate for rapid ATP resynthesis. As a result, IWMD must wait for alternative fuel sources to arrive at the working muscle for subsequent breakdown and ATP resynthesis to meet the increased metabolic demand. The alternative fuels (free fatty acids and liver-derived blood glucose) must first be mobilised from their site of origin and then transported to the active muscle, which results in the delay of their utilisation for ATP resynthesis. Once these alternative fuels reach the active muscle, IWMD can replace the substrate demand that would normally (in non-MD individuals) be met by muscle glycogen, though the capacity and rate response of their muscle metabolism remains relatively constrained compared to non-MD individuals. Hence, IWMD see a dramatic decrease in symptoms after ~6–8 min of activity. This phenomenon, termed the MD 'second wind' [5] is a distinct feature of the condition. However, the term 'second wind' is somewhat of a misnomer. This is because the response is unrelated to ventilation and/or external respiration and results from a systemic metabolic adjustment whereby the additional substrate becomes available to the working muscle.

Early research studies [6,7] discovered that the MD metabolic adjustment (second wind) could be prematurely induced during the early stages of exercise when participants were infused via cannula with glucose for 30 min before the initiation of exercise. By infusing glucose 30 min before exercise, researchers were able to prime the individual with free plasma glucose that could be immediately utilized by the exercising muscle at the onset of exercise (hyperglycaemia). The infused carbohydrate could then enter the working muscle via its conversion into glucose-6-phosphate (hexokinase reaction, thereby bypassing glycogenolysis), increasing the supply of available substrate for ATP production.

While direct glucose infusion was an effective measure to induce hyperglycaemia and elicit a MD metabolic adjustment during the early stages of exercise [6,8], its application to real-world settings (RWS) was unrealistic due to the need for venous cannulation and a constant infusion of carbohydrate. In order to provide a more practical alternative, subsequent research focused on the oral ingestion of simple carbohydrates (sucrose) to elicit hyperglycaemia prior to exercise and induce an MD metabolic adjustment [9,10], producing results similar to that of the previous infusion studies (i.e., improved exercise tolerance).

The ingestion of 37 g of sucrose 5–10 min before exercise has been advised in the most recent MD clinical management guidelines [11] and often advised as an initial management method for the condition by medical professionals. However, despite research [9,10] demonstrating consistent and significant benefits when employing carbohydrate ingestion prior to exercise in research settings, the translation of these positive results into RWS has never been assessed. Consequently, the purpose of this research was to determine whether the management technique of carbohydrate ingestion before exercise, advised in the most recent clinical management guidelines, is of benefit to those IWMD employing it in RWS to relieve and minimise their MD symptoms during exercise.

2. Materials and Methods

With custodian consent, an online survey (conducted via Qualtrics, Seattle, WA, USA) was advertised on the International Association for Muscle Glycogen Storage Disease's (IamGSD) community Facebook page, an MD patient advocacy group. The survey was available for completion for a 7-week period (from 17 March 2022) and interested participants were invited to take part in capturing individual responses. The survey was approved by the Queensland University of Technology Ethics Committee (approval number 4583) and participants were required to be 18 years of age or older and have a genetic diagnosis of McArdle disease.

The survey comprised twenty-seven questions including a broad range of topics including participant characteristics, exercise habits, the use of carbohydrates before exercise as a MD management technique, daily activity, medical history, disease impact, disease education, and access to disease-related educational resources (see Supplementary File S1). Participants were not required to complete all questions available on the survey for their responses to be recorded. Seventeen questions were multiple choice where only a single option could be selected; two questions were multiple choice where one or more options could be selected for the same question; seven were short response questions; and one was a multiple choice question, where more than one option could be selected with the option of a short response. Open text boxes were available to capture short responses.

Data Analysis

For the purpose of this manuscript in which we specifically investigated the implementation and perceptions of carbohydrate ingestion before exercise of IWMD in RWS, questions referring to participant characteristics, exercise habits, and the use of carbohydrates before exercise as a MD management technique were utilized for analysis (nine in total). Specific questions utilized are labelled in Supplementary File S1. Where participants were asked to report the type and amount of carbohydrate ingested before attempting exercise, participants could enter multiple responses and sufficient carbohydrate ingestion was determined as a total amount of carbohydrate intake equal to or exceeding the 37 g of carbohydrate currently advised in the most recent clinical practice guidelines. Results are presented as totals.

3. Results

A total of 108 participants provided survey responses. Because participants were not required to complete all questions available on the survey for their responses to be recorded, the total number of responses for each question varied as not all participants elected to answer all questions provided.

A summary of participant details is presented in Table 1.

Table 1. A summary of participant details who provided data on nationality (n = 108) and their gender, age, and weight (n = 104). Data on participant height were also collected for the subsequent calculation of BMI.

n = 104 (Males 27, Females 77)	Mean ± SD	Range
Age (y)	46 ± 15	18–82
Weight (kg)	86.7 ± 34.1	46–260
BMI (kg·m^2)	30.4 ± 11.2	16.3–80.0
Nationality	**Number (n = 108)**	
Argentina	1	
Australia	15	
Belgium	1	
Canada	9	
Colombia	1	
Czech Republic	1	
Germany	3	
Hong Kong	2	
Ireland	2	
Italy	1	
México	1	
The Netherlands	3	
New Zealand	2	
Portugal	1	
Spain	4	
Sweden	1	
United States of America	44	
United Kingdom	16	

Questions 10 and 11 asked participants about their continuous physical activity participation. The responses of 94 participants are detailed in Table 2. Among 104 survey responses to question 10 asking the number of days they participated in continuous physical activity, 10 participants selected that they do not undertake any continuous physical activity and therefore recorded no response for question 11.

Table 2. Collated responses from the number of days where continuous physical activity is performed (Questions 10) and the duration of that physical activity (Question 11) from participants self-reporting as IWMDs (n = 94) *.

	Question 11				
Question 10	**15 min or Less**	**15–30 min**	**30–60 min**	**More than 60 min**	**Total**
Once a week	5	5	2	1	13
2–3 times a week	4	14	12	4	34
More than 3 times a week	0	11	23	13	47
Total	9	30	37	18	94

* Of 104 survey responses in Q10, 10 participants selected that they do not undertake any continuous physical activity and therefore recorded no response for Q11.

In response to survey question 15. "Do you find the current McArdle disease management guidelines of consuming sugary drinks or foods before you are physically active relieves/minimises your McArdle symptoms", of 103 responses, only 17.5% of participants confirmed that the management technique of pre-exercise carbohydrate ingestion relieved or minimised their MD symptoms (Figure 1). The majority (39.8%) of participants confirmed that pre-exercise carbohydrate ingestion did not relieve or minimize their MD symptoms. When those who answered 'Sometimes' were combined with those who answered 'Yes', this represented less than half the participants surveyed (46.6%), experiencing at least one positive event where carbohydrate ingestion before exercise was of benefit (Figure 1). The remaining 13.6% of participants had never tried carbohydrate ingestion before exercise.

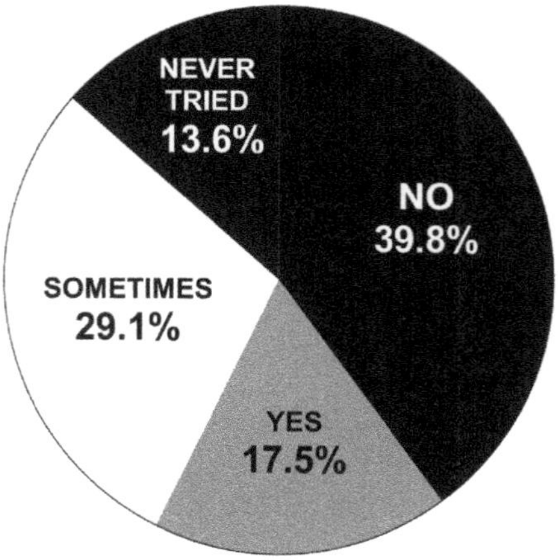

Figure 1. Results from 103 individuals with McArdle disease (IWMDs) when asked "Do you find the current McArdle disease management guidelines of consuming sugary drinks or foods before you are physically active relieves/minimises your McArdle symptoms?"

The reported sources of carbohydrate utilised (Figure 2) and the amounts ingested (Figure 3) when employing the technique of carbohydrate ingestion before exercise displayed large variation. Fruit or fruit juice and sports drinks were the most reported carbohydrate sources ingested prior to exercise (Figure 2). Where participants were asked to report the type and amount of carbohydrate ingested before attempting exercise, 45 participants reported a total of 79 exercise attempts (participants could record more than one exercise attempt), where most attempts (69.6%) occurred with the participant consuming less than the 37 g of carbohydrate currently advised in the management guidelines (Figure 3).

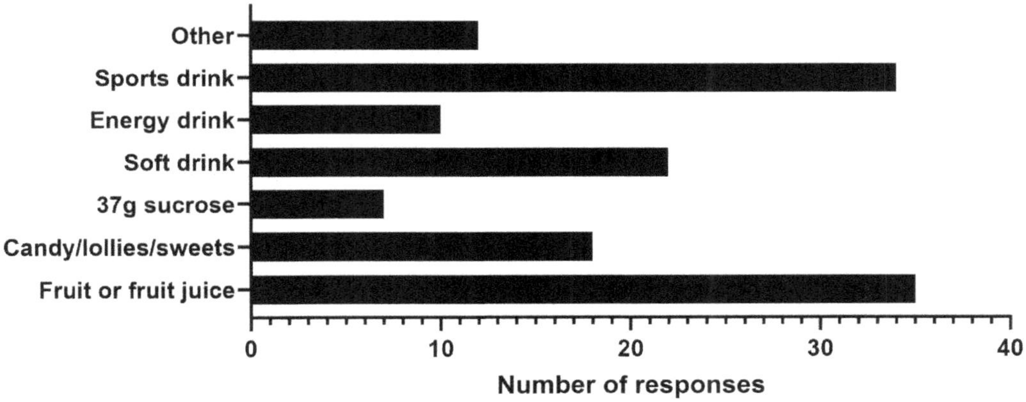

Figure 2. Results from 72 individuals with McArdle disease (IWMD) when asked "With regard to the current McArdle disease management guidelines of consuming sugary drinks or foods before you are physically active, what sugary drinks or foods have you tried before exercise?" Individual participants could enter multiple responses to this question (138 total responses were recorded).

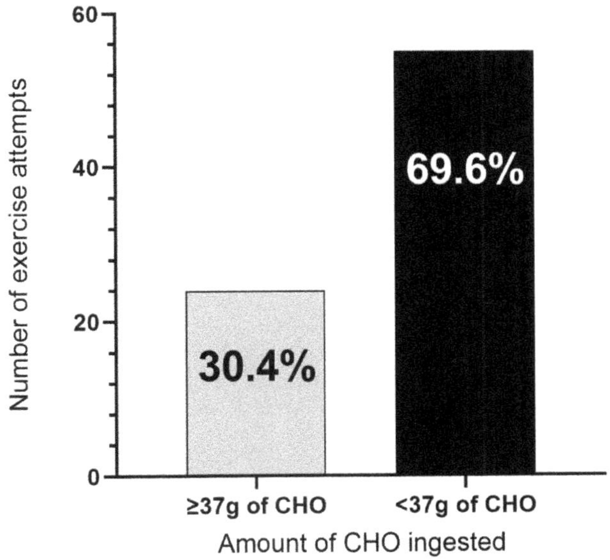

Figure 3. Comparison of the relative reporting of exercise attempts where participants consumed < vs. ≥37 g of carbohydrate pre-exercise. CHO = carbohydrate.

4. Discussion

The current investigation sought to provide further understanding and detail into the perceived effectiveness of the currently advised MD management technique of carbohydrate ingestion before exercise in RWS. Participants provided survey responses to questions of their exercise habits and their use of carbohydrate before exercise to decrease their MD symptoms during exercise.

The major findings of this study were that, for many surveyed IWMD, the details of pre-exercise carbohydrate ingestion were: (1) not well understood and/or not applied in RWS as per the current guidelines; and (2) the positive lab-based results (increased exercise tolerance/reduction in MD symptoms) occurring after carbohydrate ingestion were not being experienced. This lack of translation should be of concern to researchers, policy makers, and medical professionals working with IWMD as the advice currently being given is not benefiting a large portion of those it is intended to support. Regarding item 1 above, there was a large variation in the type and amount of carbohydrate ingested when surveyed IWMD applied the technique of carbohydrate ingestion before exercise in RWS (Figures 2 and 3). The large variation in the type and amount of carbohydrate ingested when applying the technique may explain the poor outcomes in RWS. Consequently, there is a need for improved patient education about the practice of pre-exercise carbohydrate ingestion for IWMD as well as further research investigation into the most effective strategies for its implementation.

There are likely multiple factors contributing to the poor translation of positive lab-based research results to perceptions of the effectiveness of carbohydrate ingestion before exercise in RWS; however, it is acknowledged that carbohydrate ingestion before exercise in IWMD has received little investigation to date. It is often stated in MD management advice that the ingestion of carbohydrate before exercise to improve exercise tolerance is 'well established'. However, to our knowledge, there are only six published studies to date investigating the effects of orally ingested carbohydrates on exercise tolerance in MD populations [9,10,12–15], and only two of the six studies are utilised in the current clinical management guidelines. MD is a rare condition and the difficulties in completing research on such rare populations are obvious; however, this does not decrease the need for multiple studies across multiple cohorts to confirm findings before they are considered to be well established.

While there have been nine studies investigating carbohydrate infusion and exercise tolerance in IWMD [2,6,8,12,16–20], to group research of infusion and ingestion as one and the same would be a profound error, as both methods elicit vastly different metabolic, central nervous system, and hormonal responses [21,22]. Thus, the topic of carbohydrate ingestion before exercise in IWMD remains sparsely researched.

The large variation in how IWMD may utilise carbohydrate ingestion before exercise is illustrated by large disparities across both the amount and type of carbohydrate ingested (Figures 2 and 3). While it is possible that some IWMD have found their own unique ways to ingest carbohydrate that specifically works for them based on their own body size, exercise regime, and basal blood glucose levels (BGLs), the large discrepancies suggest a lack of clear instruction/understanding for the vast majority surveyed on how to optimally ingest carbohydrate before exercise to improve exercise tolerance. This may explain the varying reports of success for carbohydrate ingestion before exercise improving exercise tolerance not only between participants, but also within individuals' attempts at employing it (Figure 1).

To optimally utilise carbohydrate ingestion before exercise to improve exercise tolerance, an understanding of the basic mechanistic reasoning for its application should be expressed. While metabolic signalling and benefits to athletic performance (in non-MD populations) have been shown to occur at the mouth, prior to the arrival of carbohydrates in the stomach or blood stream [23,24], the mechanistic principle of the technique in regard to IWMD, is to rapidly elicit an increase in BGLs (hyperglycaemia) before exercise. The increase in BGLs allows for the exogenous carbohydrate to partially replace the metabolism of endogenous carbohydrate (muscle and liver glycogen) during the early stages of exercise, partially circumventing the blockage in glycogenolysis. This needs to be clearly understood by those employing (and advising) the technique. Therefore, to achieve this rapid rise in BGLs, the amount of carbohydrate, the type of carbohydrate, and the duration over which the carbohydrate is ingested is important. Survey results suggest that these combinations of factors are not well understood by those employing the technique, as the results from 45 participants who provided data on the type and amount of carbohydrate they consumed before exercise showed that only 30.4% attempted exercise following a carbohydrate consumption equal to or greater than 37 g of carbohydrate, as currently advised in clinical management guidelines (Figure 3). To date, this is the minimum amount of carbohydrate that has been shown in research to decrease the exercise-induced symptoms of MD. A further concern is the small proportion of individuals that reported ingesting 'sugar-free' beverages when trying to improve exercise tolerance, when such beverages do not contain carbohydrates that would facilitate such an outcome.

While no studies have been carried out to determine an accurate dose response (i.e., amount of carbohydrate relative to body mass), it is highly likely that the amount of carbohydrate required for successful improvement in exercise tolerance will vary between individuals. As a secondary investigation of the original oral carbohydrate ingestion study [10] where participants ingested 75 g 30–40 min before exercise, researchers had the patient with the lowest body weight (weight not specified in the research paper) undertake an additional trial, ingesting only half the original amount of carbohydrate (37.5 g of sucrose). Results found that the lower dose of carbohydrate only partially induced an MD metabolic adjustment (i.e., second wind) in comparison to fully inducing the response following the ingestion of 75 g. This indicates that there is a dose–response relationship and suggests that ingestion levels below the currently advised 37 g of carbohydrate may result in even further reductions to the induction of a MD metabolic adjustment (i.e., second wind), or completely removed any benefits to employing the technique.

Considering the research underpinning the current MD management guidelines involving carbohydrate ingestion prior to exercise suggests consuming 37 g of sucrose, it is interesting that, of the 72 participant responses, only 7 participants had actually tried the technique with 37 g of sucrose (Figure 2). If IWMD in RWS are not undertaking the same measures that showed positive results in lab studies, it is to be expected that results are not likely to be replicated. While there may be more practical substitutes to apply the technique in RWS, there is currently no research to support this. As such, currently, it would be advised that 37 g of sucrose diluted in water (method of positive lab results) is the first application applied before alternative carbohydrate sources are trialled in RWS.

The most utilized source of carbohydrate ingested when undertaking the technique was fruit or fruit juice, closely followed by sports drinks (Figure 2). Among the 72 participant responses, 35 had undertaken the technique with fruit or fruit juice, representing 48.6% of participants. This high percentage should be of interest as the ingestion of solid foods (as opposed to liquids or gels) such as whole fibrous or starchy fruit (apples, oranges, grapefruit) would delay and slow the rise in BGLs due to the additional time required for digestion and therefore would not be considered optimal options to produce a rapid spike in BGLs. Liquid or gel forms of carbohydrate would be advisable for speed of digestion, but considering the high percentage of IWMD surveyed that consumed fruit and fruit juice as the source of carbohydrate, it is important to understand that not all liquid (or solid) forms of carbohydrate are digested, absorbed, and metabolized in the same way.

Carbohydrate content within a fruit juice like apple juice primarily comprises the monosaccharide fructose (~70%), with glucose making up the remaining ~30% [25,26]. Sucrose (which is the carbohydrate advised in management guidelines) is a disaccharide and consists of one molecule of glucose (50%) and one molecule of fructose (50%) bound together. The difference in ratios of monosaccharides is important to understand as fructose and glucose have different metabolic fates once ingested. Glucose is a unique monosaccharide, because it can be directly absorbed into the blood stream from the small intestine, transported into the working muscle (insulin or contraction mediated uptake), and immediately utilized in the glycolysis pathway to provide ATP and other metabolites that can provide added ATP from complete oxidation in mitochondria [27,28]. As skeletal muscle does not possess the enzyme fructokinase which allows fructose to enter the glycolysis pathway, fructose must first be absorbed into the blood stream, transported via the portal vein to the liver (which contains the enzyme fructokinase), and then be converted to either glucose (~50%), lactate (~25%), or liver glycogen [29–31]. The liver-derived glucose can then be transported around the body for cellular uptake, oxidation, and ATP production. If apple juice were ingested as a substitution for sucrose diluted in water, the individual would need to drink nearly twice as much apple juice to consume and absorb the same amount of free glucose within the same time-period. This is just one example of why it should be strongly advised that IWMD undertake the technique (at least the first few attempts) with known types and quantities of monosaccharides within the beverage.

The large variation in application for the ingestion of carbohydrate before exercise is perhaps to be expected considering the guidelines around the techniques' utilisation were not developed and published until as recently as 2021 [11]. This is nearly two decades after the original research on which the technique is founded was completed [10]. Without specific and detailed guidelines as to the appropriate steps to be taken when ingesting carbohydrate before exercise, it is likely that the large variation in its application plays a role in the lack of translation from positive lab results to RWS. The development of published guidelines for IWMD (and medical professionals) to follow and understand when implementing the technique is an important step in the management of the condition. However, current guidelines are limited in their description and instruction for optimal application, and this may be a contributing factor to poor perceived outcomes in RWS.

There is a clear need for more detailed resources explaining the principles underlying the technique of carbohydrate ingestion before exercise for IWMD and practitioners, as well as more descriptive instructions on how to optimally employ the technique. Based on the comments above, the following recommendations are founded on current scientific research and principles of exercise nutrition and physiology. In being research evidence-based, each recommendation is based on the best available evidence for the optimal benefit to IWMD in RWS.

(1) Type of carbohydrate consumed—The carbohydrate consumed when first attempting the technique should be sucrose. Sucrose is regular table sugar and can easily be diluted in water and consumed.

Current MD clinical management guidelines suggest that 37 g of sucrose approximately corresponds to one can of soda (330 mL). However, soda can have vast differences in monosaccharide ratios as well as total carbohydrate content [25]. As such, sucrose diluted in water is advised when initially utilising this technique.

The ingestion of pure glucose is also a viable and potentially more effective option than sucrose. While not directly attempting to induce a MD metabolic adjustment (second wind), Coakley and colleagues provided direct evidence of an increase in exercise tolerance following the ingestion of 75 g of glucose 20–30 min before exercise in IWMD [12,13]. Based on the research by Coakley, maltodextrin (glucose molecules connected together) should also result in substantial free glucose availability in the blood stream. However, research on maltodextrin consumption prior to exercise has yet to be completed.

Previous research in MD populations has shown increases in exercise tolerance following the infusion of sodium lactate during exercise in IWMD [6]. As discussed prior, it is also important to understand that the increased hepatic production of lactate via the ingestion, liver uptake, and subsequent metabolism of fructose derived from the sucrose disaccharide may provide a valuable substrate to the working muscle during the early stages of exercise. However, this is yet to be clearly established for ingestion protocols.

With the guidance of a medical professional, alternatives should be trialled by the individual if they have no success with sucrose or seek to improve upon the technique once success has been established with sucrose.

(2) Amount of carbohydrate consumed—The minimum amount of carbohydrate consumed should be 37 g, as any amount less than this has not yet been investigated.

It is important to note that this is not the total amount of substance to be consumed, but the total amount of pure carbohydrate to be consumed. This is why sucrose is a practical method of application as sucrose (table sugar) is pure carbohydrate. As no participant data were provided on body weight for either of the successful research trials utilising carbohydrate ingestion before exercise to induce a MD metabolic adjustment (second wind) [9,10], it can only be assumed that bigger individuals may require higher doses of carbohydrate to achieve the same benefits as smaller individuals.

(3) The consumption period—The ingestion of the carbohydrate needs to be consumed in one single bolus.

The ingestion of the carbohydrate over an extended period of time in small quantities (i.e., sipping sucrose diluted in water over the course of an exercise session) will not produce the same blood glucose response as if it were consumed all in one quantity before exercise. It is likely that the longer the time taken to ingest the total amount of carbohydrate before exercise, the less effect it will have on increasing BGLs, and as a result, likely decrease the effectiveness of the technique.

(4) The waiting/priming period prior to exercise—The optimal waiting or 'priming' period between the ingestion of carbohydrates and the initiation of exercise is yet to be clearly established. Nevertheless, based on past research, it would be advised that a 25–30 min waiting period is used between ingestion and the initiation of exercise.

The current recommendation in management guidelines advises a 5–10 min waiting period between ingestion and the initiation of exercise. This is based on the research by Andersen [9] in which they suggest a 5 min wait period between ingestion and exercise. This recommendation requires further research enquiry as the results of the study have never been verified by a second source, and the published data of the study suggest critical variables may have been unaccounted for during the experiments. Andersen [9] failed to acknowledge in the findings of their study that the average overnight fasted BGLs of their participants before each trial were ~7.00 mmol/L. This presents a major limitation to the results of the study as it suggests that participants were either insulin-compromised or had not followed the dietary conditions specified in testing protocols (overnight fasted). The overnight fasted BGLs reported in Andersen [9] are significantly higher than would be expected under overnight fasting conditions (normal overnight fasted BGL 4.0–5.5 mmol/L). Due to this variable, the validity of the results is severely compromised as the probability of superior results (increased exercise tolerance) after ingestion 5 min before exercise is at a significantly advantage when compared to a placebo or the ingestion of carbohydrate with a longer wait/priming period before exercise.

BGLs following the ingestion of a bolus of carbohydrate typically peak 30 min post ingestion [32]. Based on research by Kowalski, Moore, Hamley, Selathurai, and Bruce [32] and in line with previous carbohydrate ingestion research on IWMD [10,12,13], a 25–30 min wait/priming period between ingestion and the initiation of exercise is recommended until the research by Andersen [9] can be validated by a second source.

Additional Considerations

Exercise mode and exercise intensity—The current clinical management guidelines are based upon research that was completed on stationary exercise bikes at intensities that elicited a near-maximal heart rate and level of perceived exertion before the MD metabolic adjustment (second wind) occurred. As such, the mode of exercise (i.e., cycling vs. walking vs. activities of daily living) and intensity of exercise may have a significant impact on the effectiveness of carbohydrate ingestion prior to exercise for IWMD. There is a need for continued research, education and understanding on pre-exercise carbohydrate ingestion for different types of exercise or activities of daily living.

Fasted vs. fed—The ingestion of carbohydrates in both studies showing positive effects on exercise tolerance for IWMD was completed following an overnight fast. Nutritional status prior to the ingestion of carbohydrates is also likely to influence its effectiveness as changes to nutritional status will impact liver glycogen levels [33], blood metabolite levels including blood glucose, ketone, and free fatty acids [34,35], as well as the function of the central nervous system [22].

Rebound and metabolic deficiency hypoglycaemia—It is encouraged for IWMD and medical professionals assisting them to understand the potential for the development of hypoglycaemia to occur under situations of both carbohydrate ingestion before exercise (rebound hypoglycaemia), as well as no carbohydrate intake during exercise (metabolic deficiency). While it may be argued that, because IWMD exercise at very low absolute exercise intensities, absolute carbohydrate utilisation would also be very low. However, because the relative intensity of exercise for this population is high and the only source of carbohydrate available is derived from blood glucose supplemented by the liver (due to the blockage in muscle glycogenolysis), the increased utilisation of blood glucose for muscular contraction has the potential to negatively impact BGLs and euglycemia.

As with the prescription of any management method, to prevent adverse outcomes and develop the best management method for each individual, the manipulation of BGLs should be performed under guidance and consultation with an appropriately qualified medical professional that is aware of the individual's full medical history. The above recommendations should be thoroughly considered and manipulated for each individual and not prescribed on a one-size-fits-all basis.

5. Conclusions

Survey results suggest that the positive outcomes (increase in exercise tolerance/decrease in exercising MD symptoms) achieved in lab-based settings regarding carbohydrate ingestion before exercise are not being effectively translated to RWS for the vast majority of surveyed IWMD. Further research is required to determine the reasoning for this lack of transfer with potential factors including a poor understanding of the underlying principles of the management technique for optimal application by those employing and advising it, limited detail and clarity of guidelines around the utilisation of the technique, and the accuracy of prior research supporting the utilisation of the technique.

A current appraisal of this prior scientific research, in combination with the results of this survey research, revealed the need to provide clear recommendations to IWMD on how to best support the energy demands of their muscles during exercise and activities of daily living. Such recommendations should be updated as more research evidence becomes available.

Supplementary Materials: The following supporting information can be downloaded at: https://www.mdpi.com/article/10.3390/nu16101423/s1, File S1. Survey—The Effectiveness of Current McArdle Disease Management Guidelines of Nutrition and Exercise.

Author Contributions: Conceptualization, S.L.T., E.B.P., L.R. and R.A.R.; Methodology, S.L.T., E.B.P., C.M., L.R. and H.M.; Validation, S.L.T. and R.A.R.; Formal analysis, S.L.T. and R.A.R.; Investigation, S.L.T.; Resources, S.L.T.; Writing—original draft, S.L.T. and R.A.R.; Writing—review & editing, S.L.T., E.B.P., C.M., L.R. and R.A.R.; Visualization, S.L.T. and R.A.R.; Supervision, C.M. and R.A.R.; Project administration, S.L.T. All authors have read and agreed to the published version of the manuscript.

Funding: This research received no external funding.

Institutional Review Board Statement: The study was conducted in accordance with the Australian Code for the Responsible Conduct of Research, and approved by the Queensland University of Technology Ethics Committee (Approval number 4583; 12 January 2022).

Informed Consent Statement: Informed consent was obtained from all subjects involved in the study.

Data Availability Statement: The original contributions presented in the study are included in the article/supplementary material, further inquiries can be directed to the corresponding author.

Conflicts of Interest: The authors declare no conflicts of interest.

Abbreviations

ATP	Adenosine triphosphate
BGL	Blood glucose level
IamGSD	International Association for Muscle Glycogen Storage Disease
IWMDs	Individuals with McArdle disease
MD	McArdle disease
RWSs	Real-world settings

References

1. Ørngreen, M.C.; Jeppesen, T.D.; Andersen, S.T.; Taivassalo, T.; Hauerslev, S.; Preisler, N.; Haller, R.G.; van Hall, G.; Vissing, J. Fat metabolism during exercise in patients with McArdle disease. *Neurology* **2009**, *72*, 718–724. [CrossRef] [PubMed]
2. Vissing, J.; Lewis, S.F.; Galbo, H.; Haller, R.G.; Gonzalez, J.T.; Fuchs, C.J.; Betts, J.A.; van Loon, L.J.C.; Krag, T.O.; Pinós, T.; et al. Effect of deficient muscular glycogenolysis on extramuscular fuel production in exercise. *J. Appl. Physiol.* **1992**, *72*, 1773–1779. [CrossRef]
3. Romijn, J.A.; Coyle, E.F.; Sidossis, L.S.; Gastaldelli, A.; Horowitz, J.F.; Endert, E.; Wolfe, R.R. Regulation of endogenous fat and carbohydrate metabolism in relation to exercise intensity and duration. *Am. J. Physiol. Endocrinol. Metab.* **1993**, *265*, E380–E391. [CrossRef] [PubMed]
4. Gastin, P.B. Energy System Interaction and Relative Contribution during Maximal Exercise. *Sports Med.* **2001**, *31*, 725–741. [CrossRef]
5. Braakhekke, J.; De Bruin, M.; Stegeman, D.; Wevers, R.; Binkhorst, R.; Joosten, E. The second wind phenomenon in McArdle's disease. *Brain* **1986**, *109*, 1087–1101. [CrossRef] [PubMed]
6. Pearson, C.M.; Rimer, D.G.; Mommaerts, W.F. A metabolic myopathy due to absence of muscle phosphorylase. *Am. J. Med.* **1961**, *30*, 502–517. [CrossRef]
7. Schmid, R.; Mahler, R. Chronic progressive myopathy with myoglobinuria: Demonstration of a glycogenolytic defect in the muscle. *J. Clin. Investig.* **1959**, *38*, 2044–2058. [CrossRef] [PubMed]
8. Porte, D., Jr.; Crawford, D.W.; Jennings, D.B.; Aber, C.; McIlroy, M.B. Cardiovascular and metabolic responses to exercise in a patient with McArdle's syndrome. *N. Engl. J. Med.* **1966**, *275*, 406–412. [CrossRef]
9. Andersen, S.T.; Haller, R.G.; Vissing, J. Effect of Oral Sucrose Shortly Before Exercise on Work Capacity in McArdle Disease. *Arch. Neurol.* **2008**, *65*, 786–789. [CrossRef]
10. Vissing, J.; Haller, R.G. The Effect of Oral Sucrose on Exercise Tolerance in Patients with McArdle's Disease. *N. Engl. J. Med.* **2003**, *349*, 2503–2509. [CrossRef]
11. Lucia, A.; Martinuzzi, A.; Nogales-Gadea, G.; Quinlivan, R.; Reason, S.; Bali, D.; Godfrey, R.; Haller, R.; Kishnani, P.; Laforêt, P.; et al. Clinical practice guidelines for glycogen storage disease V & VII (McArdle disease and Tarui disease) from an international study group. *Neuromuscul. Disord.* **2021**, *31*, 1296–1310. [CrossRef]
12. Coakley, J.; Wagenmakers, A.; Edwards, R. Relationship between ammonia, heart rate, and exertion in McArdle's disease. *Am. J. Physiol. Endocrinol. Metab.* **1992**, *262*, E167–E172. [CrossRef]
13. Coakley, J.; Wagenmakers, A.; Edwards, R. Plasma ammonia is reduced by glucose and branched chain keto-acids during exercise in McArdle's disease. *Clin. Sci.* **1989**, *76*, 49P. [CrossRef]
14. Løkken, N.; Khawajazada, T.; Slipsager, A.; Voermans, N.C.; Vissing, J. Repeated oral sucrose dosing after the second wind is unnecessary in patients with McArdle disease: Results from a randomized, placebo-controlled, double-blind, cross-over study. *J. Inherit. Metab. Dis.* **2023**, *46*, 1139–1146. [CrossRef]
15. Valenzuela, P.L.; Santalla, A.; Alejo, L.B.; Merlo, A.; Bustos, A.; Castellote-Bellés, L.; Ferrer-Costa, R.; Maffiuletti, N.A.; Barranco-Gil, D.; Pinós, T.; et al. Dose-response effect of pre-exercise carbohydrates under muscle glycogen unavailability: Insights from McArdle disease. *J. Sport Health Sci.* **2023**, *13*, 398–408. [CrossRef]
16. Vissing, J.; Haller, R.G. A diagnostic cycle test for McArdle's disease. *Ann. Neurol. Off. J. Am. Neurol. Assoc. Child Neurol. Soc.* **2003**, *54*, 539–542. [CrossRef]
17. Lewis, S.F.; Haller, R.G.; Cook, J.; Nunnally, R. Muscle fatigue in McArdle's disease studied by 31P-NMR: Effect of glucose infusion. *J. Appl. Physiol.* **1985**, *59*, 1991–1994. [CrossRef]
18. Haller, R.G.; Vissing, J. Spontaneous "Second Wind" and Glucose-Induced Second "Second Wind" in McArdle Disease: Oxidative Mechanisms. *Arch. Neurol.* **2002**, *59*, 1395–1402. [CrossRef]
19. Haller, R.G.; Lewis, S.F.; Cook, J.; Blomqvist, C. Myophosphorylase deficiency impairs muscle oxidative metabolism. *Ann. Neurol. Off. J. Am. Neurol. Assoc. Child Neurol. Soc.* **1985**, *17*, 196–199. [CrossRef] [PubMed]
20. Andersen, S.T.; Jeppesen, T.D.; Taivassalo, T.; Sveen, M.-L.; Heinicke, K.; Haller, R.G.; Vissing, J. Effect of Changes in Fat Availability on Exercise Capacity in McArdle Disease. *Arch. Neurol.* **2009**, *66*, 762–766. [CrossRef] [PubMed]
21. Leelayuwat, N.; Tsintzas, K.; Patel, K.; Macdonald, I.A. Metabolic Responses to Exercise after Carbohydrate Loads in Healthy Men and Women. *Med. Sci. Sports Exerc.* **2005**, *37*, 1721–1727. [CrossRef] [PubMed]
22. Jeukendrup, A.E.; Rollo, I.; Carter, J.M. Carbohydrate mouth rinse: Performance effects and mechanisms. *Sports Sci. Exch.* **2013**, *26*, 1–8.

23. Rollo, I.; Homewood, G.; Williams, C.; Carter, J.; Goosey-Tolfrey, V.L. The Influence of Carbohydrate Mouth Rinse on Self-Selected Intermittent Running Performance. *Int. J. Sport Nutr. Exerc. Metab.* **2015**, *25*, 550–558. [CrossRef]
24. Carter, J.M.; Jeukendrup, A.E.; Jones, D.A. The Effect of Carbohydrate Mouth Rinse on 1-h Cycle Time Trial Performance. *Med. Sci. Sports Exerc.* **2004**, *36*, 2107–2111. [CrossRef]
25. Walker, R.W.; Dumke, K.A.; Goran, M.I. Fructose content in popular beverages made with and without high-fructose corn syrup. *Nutrition* **2014**, *30*, 928–935. [CrossRef] [PubMed]
26. Wosiacki, G.; Nogueira, A.; Frederico, F.; Vieira, R.G. Sugar composition of depectinized apple juices. *Semin. Ciências Agrárias* **2007**, *28*, 645–652. [CrossRef]
27. Ferraris, R.P. Dietary and developmental regulation of intestinal sugar transport. *Biochem. J.* **2001**, *360*, 265–276. [CrossRef] [PubMed]
28. Semenza, G.; Kessler, M.; Hosang, M.; Weber, J.; Schmidt, U. Biochemistry of the Na+, d-glucose cotransporter of the small-intestinal brush-border membrane: The state of the art in 1984. *Biochim. Biophys. Acta (BBA)—Rev. Biomembr.* **1984**, *779*, 343–379. [CrossRef]
29. Lecoultre, V.; Benoit, R.; Carrel, G.; Schutz, Y.; Millet, G.P.; Tappy, L.; Schneiter, P. Fructose and glucose co-ingestion during prolonged exercise increases lactate and glucose fluxes and oxidation compared with an equimolar intake of glucose. *Am. J. Clin. Nutr.* **2010**, *92*, 1071–1079. [CrossRef]
30. Sun, S.Z.; Empie, M.W. Fructose metabolism in humans—What isotopic tracer studies tell us. *Nutr. Metab.* **2012**, *9*, 89. [CrossRef]
31. Rowlands, D.S.; Thorburn, M.S.; Thorp, R.M.; Broadbent, S.; Shi, X. Effect of graded fructose coingestion with maltodextrin on exogenous ^{14}C-fructose and ^{13}C-glucose oxidation efficiency and high-intensity cycling performance. *J. Appl. Physiol.* **2008**, *104*, 1709–1719. [CrossRef] [PubMed]
32. Kowalski, G.M.; Moore, S.M.; Hamley, S.; Selathurai, A.; Bruce, C.R. The Effect of Ingested Glucose Dose on the Suppression of Endogenous Glucose Production in Humans. *Diabetes* **2017**, *66*, 2400–2406. [CrossRef]
33. Awad, S.; Constantin-Teodosiu, D.; Constantin, D.; Rowlands, B.J.; Fearon, K.C.; Macdonald, I.A.; Lobo, D.N. Cellular mechanisms underlying the protective effects of preoperative feeding: A randomized study investigating muscle and liver glycogen content, mitochondrial function, gene and protein expression. *Ann. Surg.* **2010**, *252*, 247–253. [CrossRef] [PubMed]
34. Garber, A.J.; Menzel, P.H.; Boden, G.; Owen, O.E. Hepatic Ketogenesis and Gluconeogenesis in Humans. *J. Clin. Investig.* **1974**, *54*, 981–989. [CrossRef]
35. Wolever, T.M.; Bolognesi, C. Prediction of glucose and insulin responses of normal subjects after consuming mixed meals varying in energy, protein, fat, carbohydrate and glycemic index. *J. Nutr.* **1996**, *126*, 2807–2812. [PubMed]

nutrients

MDPI

Article

Impact of Brewers' Spent Grain-Containing Biscuit on Postprandial Glycaemic Response in Individuals with Metabolic Syndrome: A Crossover Randomised Controlled Trial

Yujing Xu [1], Zi Ning Leong [1], Weijia Zhang [1], Xinrui Jin [1], Jia Wen Kong [1], Gregory Chung Tsing Chan [2] and Jung Eun Kim [1,*]

[1] Department of Food Science and Technology, Faculty of Science, National University of Singapore, 2 Science Drive 2, Singapore 117543, Singapore; yujingxu@u.nus.edu (Y.X.); leong.zining@nus.edu.sg (Z.N.L.); zhangweijia0817@163.com (W.Z.); e0950404@u.nus.edu (X.J.); kongjw@u.nus.edu (J.W.K.)

[2] Saw Swee Hock School of Public Health, National University of Singapore, Singapore 117549, Singapore; ephgchan@nus.edu.sg

* Correspondence: fstkje@nus.edu.sg; Tel.: +65-6516-1136; Fax: +65-6775-7895

Abstract: Brewers' spent grain (BSG) is a fibre and protein-rich by-product of beer-brewing. Fermenting BSG with *Rhizopus oligosporus* can further increase its content of soluble fibre, protein and certain antioxidants. Since nutrients rich in BSG can improve postprandial glycaemic response, this study assessed the postprandial glucose response (PPGR) and postprandial insulin response (PPIR) controlling effect of consuming 30% wheat flour substituted biscuits with autoclaved BSG (ABSG) or fermented BSG (FBSG) in individuals with metabolic syndrome (MetS). The effect on postprandial lipid panel, breath hydrogen (H_2) and methane (CH_4) concentration and subjective appetite response was also examined. Fifteen subjects with MetS participated in this crossover randomised controlled trial, and blood was collected at 9 time-points for 4 h after consumption of control biscuits (Control), ABSG and FBSG. A significant interaction effect was observed ($P_{interaction} = 0.013$) for the glucose time-points concentration. At 180 min, the glucose concentration was lowered after the consumption of ABSG ($p = 0.010$) and FBSG ($p = 0.012$) compared to the Control. Moreover, the FBSG resulted in a significantly lower glucose incremental area under curve (iAUC) compared to the Control ($p = 0.028$). Insulin level was also lowered at 180 min after the ABSG ($p = 0.010$) and FBSG ($p = 0.051$) consumption compared to the Control. However, no difference was noted for postprandial lipid panel, breath H_2 and CH_4 concentration and subjective appetite response. In conclusion, the consumption of BSG-incorporated biscuits can attenuate PPGR, and fermented BSG incorporation conferred a further PPGR controlling benefit.

Keywords: dietary fibre; dietary protein; brewery by-product; *Rhizopus oligosporus*; subjective appetite; postprandial lipid panel response

1. Introduction

Postprandial hyperglycaemia is an important contributor to type 2 diabetes mellitus (T2DM)-related complications, such as cardiovascular disease (CVD) [1]. Moreover, it has been reported that postprandial glucose response (PPGR) could serve as a more effective predictor of cardiovascular events compared to fasting glucose concentration [2]. Another factor closely associated with the development and progression of T2DM and CVD is metabolic syndrome (MetS) [3,4], a pathophysiological clustering of chronic disease risk factors including abdominal obesity, hypertension, hyperglycaemia and dyslipidaemia [5]. Individuals with MetS exhibit a higher risk of impaired PPGR, primarily due to insulin resistance, a key characteristic of MetS [6,7]. Consequently, establishing interventions that regulate PPGR in this population is crucial.

Biscuits are widely consumed due to their ease of consumption, long shelf life, high affordability, and palatability [8,9]. However, they are commonly formulated with refined grains and tend to cause a rise in PPGR [10]. Among the various factors affecting PPGR, fibre has been proven by numerous studies to modulate PPGR by slowing gastric emptying, promoting insulin release, and decreasing glucagon release [11–14]. Brewer's spent grains (BSG), a fibre-rich by-product of the beer-brewing process, contains 41–59% dietary fibre (dry weight basis) [15,16], thus, incorporating BSG into biscuits may improve the nutritional profiles in biscuits and promote a desirable PPGR after consumption [17,18]. Moreover, previous studies have also indicated that the extract of dietary protein and phenolic compounds from BSG exhibits potential in regulating PPGR [19–21]. However, since the primary dietary fibre in BSG is insoluble fibre, the dietary protein and phenolic compounds are physically trapped in the insoluble dietary fibre matrix [22]. This may reduce the bioaccessibility of these nutrients, thereby limiting the effectiveness of BSG in regulating PPGR after consumption.

Our recent study showed that the solid-state fermentation of BSG with *Rhizopus oligosporus* (RO) could break down insoluble dietary fibre into soluble dietary fibre [18]. Soluble dietary fibre can be further fermented by gut microbiota to promote the secretion of short-chain fatty acids [23], which in turn will regulate glucose metabolism by stimulating the secretion of glucagon-like peptide-1 and peptide tyrosine tyrosine, and upregulating the expression of glucose transporter-4 [24–26]. The consumption of fermented BSG-incorporated biscuits may therefore provide a further improvement on PPGR.

At present, limited clinical research has been conducted to verify the potential benefits of BSG and fermented BSG-containing biscuits consumption on PPGR regulation, particularly in individuals with MetS. Hence, the aim of the current study was to recruit individuals with MetS into a crossover randomised controlled trial (RCT) to explore the PPGR and postprandial insulin response (PPIR) after consuming BSG and fermented BSG-containing biscuits. The effect on the postprandial lipid panel, breath H_2 and CH_4 concentration and subjective appetite response were also examined as secondary outcomes in this research.

2. Materials and Methods

2.1. BSG Fermentation and Biscuits Baking

The fermentation of BSG and formulation of biscuits were carried out as per our previous work [18] and briefly elaborated below. BSG was provided by Brewerkz Brewing Co. (Singapore) and sterilised at 121 °C for 15 min to produce autoclaved BSG. Ragi tempe starter (PT. Aneka Fermentasi Industri, Bandung, Indonesia), a mixture of RO spores and rice flour, was added to autoclaved BSG cooled to room temperature at 4% of wet autoclaved BSG basis and subsequently incubated at 37 °C for 72 h to obtain fermented BSG. All autoclaved BSG and fermented BSG were freeze-dried (Lyovapor L-300, BUCHI, Singapore) to a water content of less than 5% and stored at −20 °C until use.

The control biscuits (Control), autoclaved BSG-containing biscuits (ABSG) and fermented BSG-containing biscuits (FBSG) were made with the same base recipe, with ABSG and FBSG made with a 30% wheat flour substitution of autoclaved or fermented BSG powder, respectively. The base recipe contained 180 g wheat flour, 10 g corn flour, 50 g fine sugar, 1 g salt, 1 g baking soda, 55 g sunflower oil and 70 g whole egg. Additionally, 15 g water was added to the ABSG and FBSG recipe to improve dough binding. The whole egg, sunflower oil and fine sugar were mixed at speed 2 for 1 min using a stand mixer flat beater (KitchenAid Classic, Battle Creek, MI, USA) before the remaining ingredients were mixed at speed 1 for 90 s. The dough was sheeted to a 3 mm thickness, sliced into 55 mm squares, and baked at 160 °C in a convection oven (Anna, Unox, Cadoneghe, Italy) for 20 min. Upon cooling to room temperature, all the biscuits were stored at 4 °C until use.

The nutritional profile of the test biscuits was measured and detail methods are described in previous work [18]. The results were converted to wet basis and tabulated in Table 1.

Table 1. Nutrient composition of test biscuits.

Nutrient (per 90 g Biscuits)	Control	ABSG	FBSG
Energy (kcal)	455 ± 5	440 ± 5	428 ± 5
Fat (g)	18.9 ± 0.0	20.0 ± 0.2	19.6 ± 0.2
Protein (g)	9.0 ± 0.1	10.4 ± 0.0	10.7 ± 0.0
ACHO (g)	61.2 ± 1.2	49.2 ± 0.9	47.1 ± 1.4
TDF (g)	1.9 ± 0.4	10.7 ± 0.4	10.4 ± 0.1
IDF (g)	1.2 ± 0.1	9.9 ± 0.3	9.3 ± 0.1
SDF (g)	0.7 ± 0.2	0.8 ± 0.2	1.1 ± 0.1
Phytic acid (g)	0.7 ± 0.0	0.9 ± 0.0	0.9 ± 0.0
Ash (g)	0.4 ± 0.2	1.0 ± 0.1	1.0 ± 0.2

Values are presented as mean ± SD. Control: control biscuit; ABSG: autoclaved brewers' spent grain-containing biscuit; FBSG: fermented brewers' spent grain-containing biscuit; ACHO: available carbohydrates; TDF: total dietary fibre; IDF: insoluble dietary fibre; SDF: soluble dietary fibre.

2.2. *Study Design and Subject Recruitment*

The study was registered at clinicaltrials.gov (NCT05421780) and approved by the National University of Singapore Institutional Review Board (NUS-IRB-2022-089) on 25 May 2022.

The recruitment of subjects with MetS aged 35–85 years old took place from June to December 2022 in Singapore by placing posters, sending emails, and advertising through word-of-mouth. The criteria for MetS was in accordance to the National Cholesterol Education Program Adult Treatment Panel III (NCEP-ATP III), where subjects that meet any three of the following five criteria were defined as with MetS: (1) waist circumference > 102 cm (male), >88 cm (female), and for the Asian population, >90 cm (male), >80 cm (female); (2) fasting glucose concentration ≥ 5.6 mmol/L or on known medication for blood glucose control; (3) triglyceride (TG) ≥ 1.7 mmol/L or on known medication for TG control; (4) high-density lipoprotein cholesterol (HDL-C) < 1.0 mmol/L (male), <1.3 mmol/L (female); (5) systolic or diastolic blood pressure > 130/85 mmHg or on known medication for blood pressure control. The exclusion criteria to reject subjects whose current lifestyle may impact the outcomes of interest was as follows: (1) significant change in weight (≥3 kg) over the past 3 months; (2) allergy to barley, wheat, corn, egg, or any other ingredients found inside the biscuits; (3) acute illness at study baseline; (4) exercising vigorously over the past 3 months; (5) following any restricted diet (e.g., vegetarian); (6) smoking; (7) have a daily intake of more than 2 alcoholic drinks per day; (8) prescribed and taking antihypertensive/cholesterol-lowering/T2DM medication for less than 3 years prior to the intervention participation; (9) taking fermented food regularly or any probiotic/prebiotic supplements; (10) consumption of antibiotics over past 3 months; (11) pregnant, lactating, or planning pregnancy in the next 6 months; (12) current daily intake of brown rice or wholemeal products comprising more than 6 servings; (13) current daily intake of vegetables comprising more than 2 servings.

Sixteen subjects were recruited and provided their written informed consent. Following the consent, subjects were randomly assigned to the intervention order using STATA (Version 13, StataCorp LC, College Station, TX, USA) by an unblinded staff. Fifteen subjects finished the entire intervention and one subject dropped out due to personal reasons. The CONSORT flow diagram for our study is as shown in Figure 1.

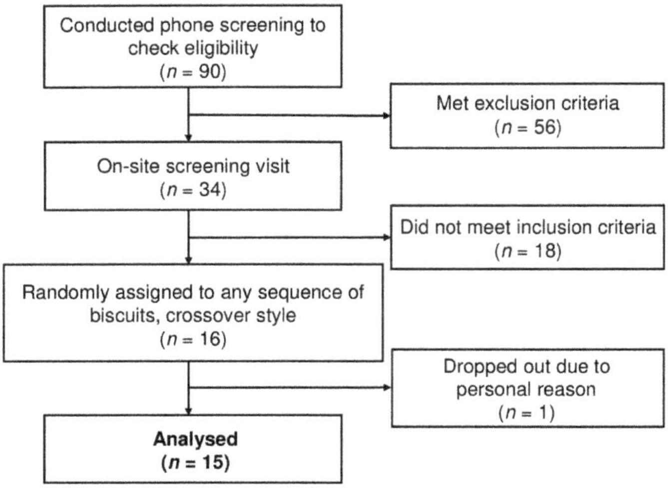

Figure 1. Flow chart of the clinical trial enrolment, randomisation, allocation, and analysis.

2.3. Study Design

The study procedure is depicted in Figure 2. This was a randomised, crossover, double-blinded study with three visits and a 7 day washout period was set between each visit. Prior to the test visit, subjects were instructed to fast for more than 10 h and to avoid taking high-fibre meals and alcohol the day before. Moreover, anti-diabetic, cholesterol-lowering and anti-hypertensive medications were also stopped. After the subjects arrived at the National University Health System Investigational Medicine Unit, anthropometric measurements and blood pressure were taken by trained staff. Subjects lay in a phlebotomy bed for the entirety of the visit, and blood collection was completed through an indwelling venous cannula inserted into the antecubital vein on the forearm, with a saline flush conducted every time a blood aliquot was taken. After fasting, blood and breath were taken (time = 0 min) and subjects were asked to consume a pack of biscuits (90 g) within 15 min. Additionally, a cup of plain water (250 mL) was provided to facilitate the consumption of the biscuits, although complete consumption of water was not obligatory. As the main aim of this study was to mimic a typical eating pattern incorporating whole BSG, the biscuits provided were standardised based on weight (90 g) instead of matching available carbohydrates (ACHO). The timer was started on their first bite, and blood samples were taken at time = 15, 30, 45, 60, 90, 120, 180 and 240 min. A heating pad was used to warm the subjects' forearm throughout the entire visit duration to avoid coagulation in the cannula.

2.4. Anthropometric and Blood Pressure Measurement

The height and weight were measured using an electronic scale stadiometer (BSM370, Biospace Co., Ltd., Seoul, Korea), with body mass index (BMI) calculated by weight/height2. The waist circumference (WC) was taken as the midpoint between the lower margin of the last rib and top of the iliac crest according to the World Health Organisation STEPwise Approach to Surveillance protocol [27]. Subjects lay on the phlebotomy bed for 5 min before taking resting blood pressure (Omron HEM-7121, Kyoto, Japan). All the measurements were taken in duplicates.

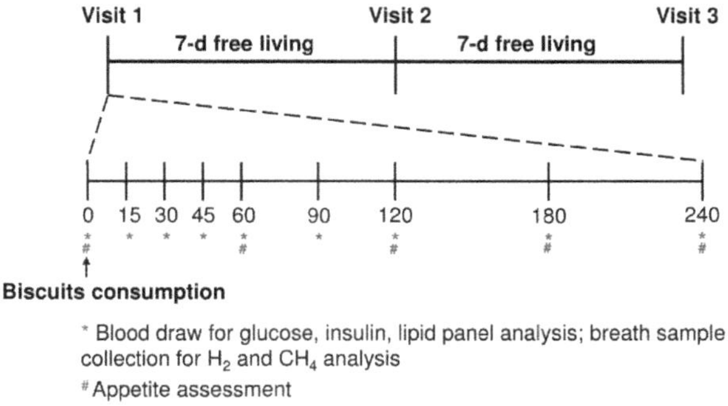

Figure 2. Study design timeline of a crossover design, 3-arm randomised controlled trial. Fifteen subjects were randomly assigned to the intervention order of consuming Control, FBSG and ABSG. Control: control biscuit; ABSG: autoclaved brewers' spent grain-containing biscuit; FBSG: fermented brewers' spent grain-containing biscuit.

2.5. Blood Sample Processing and Biochemical Analysis

All blood samples were drawn by certified phlebotomists. At each timepoint, 2 mL and 3 mL of blood was collected in a fluoride tube and serum separator tube respectively (Becton Dickinson, Franklin Lakes, NJ, USA). Fluoride tubes and serum separator tubes were kept at 4 °C and outsourced to Quest Laboratories Pte Ltd. (Singapore) for glucose, insulin and lipid panel analysis. The glucose and lipid panel was analysed using ADVIA 1800 and ADVIA Chemistry XPT (Siemens, Munich, Germany). Glucose concentration was determined using the enzymatic method with hexokinase and glucose-6-phosphate [28]. Insulin concentration was measured through electro-chemiluminescence immunoassay using a Cobas 6000 (e601) analyser (Roche, Basel, Switzerland). Total cholesterol (TC) and HDL-C concentrations were quantified using an enzymatic method based on the Trinder reaction and the Fossati three-step enzymatic reaction, while a Trinder endpoint was used to determine TG concentration [29]. After determining the TC, HDL-C, and TG, low-density lipoprotein cholesterol (LDL-C) was calculated using the Friedwald formula [30].

2.6. Breath Analysis

The breath samples were collected by trained staff. Subjects were required to continuously blow into a vented polyethylene bag with a medium-resistance leak at time = 0, 15, 30, 45, 60, 90, 120, 180 and 240 min. Subsequently, a 25 mL breath sample was drawn steadily into a syringe. Twenty-mL was injected into a Quintron breath analyser (BioMedix Singapore Pte Ltd., Singapore) to analyse the concentration of hydrogen (H_2) and methane (CH_4) present to assess gastrointestinal bacteria fermentation.

2.7. Appetite Sensation

Subjective appetite was evaluated via the Visual Analogue Scale (VAS) of appetite proposed by Flint [31]. Responses were collected every 60 min (time = 0, 60, 120, 180 and 240 min). The VAS contained questions regarding subjects' feelings of hunger, desire to eat, prospective consumption and fullness. For each query, answers were collected through a scale represented as a 100 mm long horizontal line, with no gradation. The subjects were asked to mark a cross at the point of the scale to assess their sensation at each timepoint. After collection, a ruler measurement from the left end of the line (0 mm) to the right end (100 mm) was performed to obtain the appetite score.

2.8. Power Calculation and Statistical Analysis

The primary outcome of interest in this research is the difference in postprandial glucose concentration after consuming BSG-containing biscuits, compared to the Control. Recent research reported that the plasma glucose concentration area under the curve$_{0\text{–}120min}$

($AUC_{0–120min}$) was lower with a BSG-containing high fat meal (2613 ± 193 a.u., mean ± SD), compared to the same meal without BSG (2936 ± 139 a.u., mean ± SD) in the animal model [32]. In addition, an in vitro digestion study reported that total glucose release $AUC_{0–360\ min}$ was lower with RO fermented okara-containing biscuits (166.8 ± 7.7 mg/min, mean ± SD) compared to the same biscuits without okara (187.9 ± 17.5 mg/min, mean ± SD) [33]. For the current study, presuming the proposed experiment yields similar results as previous studies, ≥6 subjects will provide ≥90% power at $\alpha = 0.05$ (two-tailed) to statistically confirm a similar difference. Due to the different study types (animal model vs. in vitro vs. human) and study design, a minimum recruitment of 15 subjects is required, with a maximum recruitment of 19 subjects to account for a 20% dropout.

Two researchers were involved in data collation to ensure accuracy. A blinded researcher proceeded with the data analysis and was only unblinded after the completion of statistical analysis. The change from the baseline of serum glucose, insulin and lipid lipoprotein concentration was plotted against time. Subsequently, the positive incremental area under the curve (iAUC) of glucose, insulin and TG were calculated with GraphPad Prism 9.3.0 software (GraphPad Software Inc., La Jolla, CA, USA) by using the trapezoidal rule formula. As TC, HDL-C and LDL-C may decrease before rising over the entire postprandial duration, negative iAUC for the postprandial lipid lipoprotein response was also calculated. The subjective appetite assessment score was also plotted against time. All data were checked for normality distribution using a Shapiro–Wilk test and were logarithmised if necessary. Two-way repeated measures ANOVA were used to examine the effect of time, intervention and their interaction (time × intervention). The Wilcoxon test considering Bonferroni correction for multiple comparisons was used to determine the simple main effect when the interaction effects were significant. One-way repeated-measures ANOVA and pairwise comparison with the Bonferroni test was conducted to assess the effect of intervention among groups in iAUC. The figures were drawn using GraphPad Prism 9.3.0 software (GraphPad Software Inc., La Jolla, CA, USA). IBM SPSS Statistics 20 (SPSS Inc., Chicago, IL, USA) was used for all statistical analysis, and a p value < 0.05 was considered statistically significant. The data of the clinical trial are presented as mean ± SEM.

3. Results

3.1. Subjects' Baseline Characteristics

Subjects' baseline characteristics are shown in Table 2. Fifteen middle-aged and older adults (mean age: 63 ± 10 years old, 10 males and 5 females) participated in this study and all of them were diagnosed as MetS. Among all subjects, 8, 6 and 1 subjects met 3, 4, 5 MetS components respectively.

3.2. Postprandial Glucose and Insulin Response

The changes in postprandial glucose and insulin concentration from 0 min and positive iAUC are shown in Figure 3. For PPGR (Figure 3A), a significant interaction effect was observed ($P_{interaction} = 0.013$) for glucose time-points concentration. Pairwise comparison showed that at time-points 90 and 120 min, FBSG resulted in a significantly lower glucose concentration compared to the Control (FBSG vs. Control: 90 min: 2.15 ± 0.44 mmol/L vs. 3.42 ± 0.62 mmol/L, $p = 0.010$; 120 min: 2.28 ± 0.47 mmol/L vs. 3.31 ± 0.72 mmol/L, $p = 0.041$). Moreover, at 180 min, both ABSG and FBSG showed a significantly lower glucose concentration compared to the Control (ABSG vs. Control: 1.57 ± 0.57 mmol/L vs. 2.83 ± 0.67 mmol/L, $p = 0.010$; FBSG vs. Control: 1.54 ± 0.46 mmol/L vs. 2.83 ± 0.67 mmol/L, $p = 0.002$), with no significant difference between ABSG and FBSG. As compared to the Control consumption, FBSG consumption resulted in a significantly lower positive iAUC for glucose (FBSG vs. Control: 413.9 ± 73.8 mmol/L × min vs. 602.0 ± 123.4 mmol/L × min, $p = 0.028$) while no difference between ABSG and the Control was observed ($p > 0.05$) (Figure 3B).

Table 2. Participant baseline characteristics.

Characteristics	Total	Male	Female
Number of subjects (*n*)	15	10	5
Age (years)	63 ± 10	61 ± 10	66 ± 12
WC (cm)	94.5 ± 12.0	96.6 ± 13.7	90.4 ± 6.5
HDL (mmol/L)	1.3 ± 0.1	1.1 ± 0.1	1.8 ± 0.2
Fasting glucose (mmo/L)	6.2 ± 0.4		
TG (mmol/L)	1.3 ± 0.2		
SBP (mmHg)	130.5 ± 18.5		
DBP (mmHg)	71.2 ± 11.2		
Insulin (mU/L)	10.0 ± 1.4		
HOMA-IR	2.61 ± 0.4		
T2DM medication (*n*)	5	3	2
Cholesterol-lowering medication (*n*)	9	6	3
Antihypertension medication (*n*)	6	3	3
The number of MetS criteria met			
3 (*n*)	8		
4 (*n*)	6		
5 (*n*)	1		

Values are presented as mean ± standard deviation or *n*. WC: waist circumference; HDL: high-density lipoprotein cholesterol; TG: triglyceride; SBP: systolic blood pressure; DBP: diastolic blood pressure; T2DM: type 2 diabetes mellitus; MetS: metabolic syndrome.

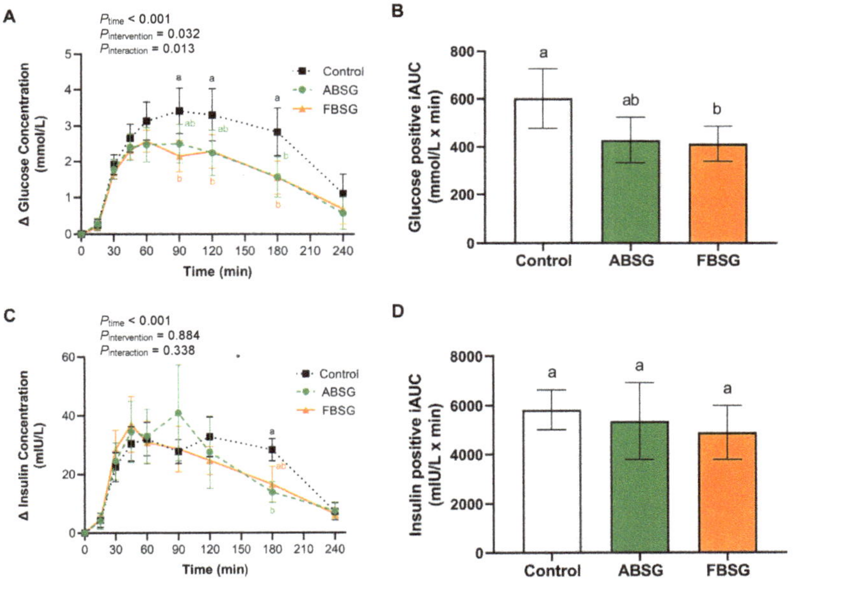

Figure 3. (**A**) Change from baseline of postprandial glucose concentration. (**B**) Positive iAUC of glucose. (**C**) Change from baseline of postprandial insulin concentration. (**D**) Positive iAUC of insulin. *p* values were determined using two-way repeated measures ANOVA (**A**,**C**) or one-way repeated measures ANOVA (**B**,**D**). Groups with different letters indicate a statistical difference. Control: control biscuit; ABSG: autoclaved brewers' spent grain-containing biscuit; FBSG: fermented brewers' spent grain-containing biscuit.

Regarding the PPIR (Figure 3C), no interaction effect was observed ($P_{\text{interaction}} = 0.34$). At 180 min, the concentration of insulin was significantly higher in the Control than ABSG (Control vs. ABSG: 28.32 ± 3.81 mIU/L vs. 13.91 ± 3.56 mIU/L, $p < 0.001$), and FBSG tended to have a lower insulin concentration compared to the Control (FBSG vs. Control:

16.55 ± 6.08 mIU/L vs. 28.32 ± 3.81 mIU/L, $p = 0.051$). However, no difference was noted in positive iAUC for insulin ($p > 0.05$) (Figure 3D).

3.3. Postprandial Lipid Panel Response

The responses in the postprandial concentration of TC, HDL-C, LDL-C, and TG concentrations from baseline are shown in Supplemental Figure S1 and the negative iAUC of TC, HDL-C and LDL-C, as well as positive iAUC of TG, are summarised in Supplemental Table S1. No difference was observed for any of the outcomes.

3.4. Breath Analysis

The changes in postprandial breath H_2 and CH_4 are shown in Supplemental Figure S2 and no difference was observed for postprandial breath H_2 and CH_4 after consuming the 3 types of biscuits ($p > 0.05$).

3.5. Subjective Appetite Assessment

The effect on the postprandial subjective sensation of hunger, desire to eat, prospective consumption and fullness are shown in Supplemental Figure S3 and there was no difference on those measurements between the Control, ABSG and FBSG ($p > 0.05$) at all time-points.

4. Discussion

While BSG and fermented BSG could serve as fibre-rich ingredients to be incorporated into food products to enhance their nutritional value, research into its effect on PPGR is lacking. In addition, there are limited studies exploring the potential of diet modifications in improving PPGR in the MetS population, who experience an impaired glucose metabolism. Our findings suggest that the consumption of biscuits with a 30% substitution of wheat flour with BSG improves PPGR in individuals with MetS and a further benefit on PPGR was offered by FBSG-containing biscuits consumption.

A significant difference was found in PPGR between the Control and BSG-containing biscuits consumption. Both ABSG and FBSG showed a more than 16% lower in glucose iAUC (29% and 31%, respectively), which is considered clinically relevant [34], although the decrease in ABSG is statistically insignificant. The decrease in the PPGR in the FBSG and ABSG could be attributed to the lesser ACHO content in the BSG-containing biscuits (20% in ABSG and 23% in FBSG). However, a previous study that substituted whole maize flour into biscuits which resulted in an ACHO reduction (18%), similar to our study here, only showed a 14% decrease in glucose AUC [35]. This implies that the presence of other compounds within BSG may also contribute to the decrease in PPGR observed. In a prior study by Ullah, et al. [36], the consumption of BSG extract-based breadsticks with matched ACHO showed an improvement in PPGR at 90 and 120 min compared to the consumption of plain breadsticks in individuals with impaired glucose tolerance. As a high-fibre food by-product, BSG-containing biscuits showed a significantly higher total dietary fibre content [18] and this dietary fibre is able to increase the viscosity of gastric contents and slow down gastric emptying, leading to a delayed and reduced glucose absorption [37]. Additionally, the antioxidants present in BSG, such as phenolic compounds and bioactive peptides, also contribute to the improvement in PPGR regulation. These antioxidants function as α-glucosidase inhibitor, which prevents the breakdown of complex carbohydrates into glucose [18]. On the other hand, these antioxidants could promote glucose consumption and glycogen synthesis, enhancing the utilisation of glucose [38].

Moreover, FBSG demonstrated a further improved PPGR regulating effect, exhibited through a significantly lower positive iAUC compared to Control, whereas ABSG did not. This outcome is attributed to a portion of the insoluble dietary fibre being broken down to soluble dietary fibre during the fermentation process by certain extracellular carbohydrases secreted by RO, such as cellulase, endocellulase, and hemicellulose [33]. Soluble dietary fibre demonstrates a greater ability to inhibit glucose absorption in the intestine compared to insoluble dietary fibre because of its high viscosity and gel-forming properties [39]. A previous RCT evidenced that incorporating β-glucan, a soluble fibre also present in BSG, into tortillas improved PPGR [40].

Nevertheless, a recent study found that there was no difference in PPGR among healthy subjects who consumed either 100 g of 40% autoclaved or RO fermented okara-substituted biscuits, which is also a dietary fibre-rich food by-product, compared to those who consumed 100 g of control biscuits [41]. The difference in findings may be caused by the different health conditions of the subjects involved in each study. The dietary fibre rich food by-products may be more effective in managing PPGR in individuals with MetS because they have impaired glucose tolerance and insulin resistance, which can lead to higher PPGR [42]. This finding is supported by the result of a systematic review and meta-analysis, which concluded that metabolically impaired individuals showed a more significant improvement in PPGR after consuming a low glycaemic index breakfast compared to healthy individuals [6].

In our results, ABSG and FBSG showed a lower insulin concentration at 180 min in comparison to Control, and the iAUC was lower although statistically insignificant. Dietary fibre is extensively credited as the main reason for the improvement of PPIR, as its consumption is associated with increased insulin sensitivity [43]. Additionally, phenolic compounds present in BSG could also play a supporting role. Reactive oxygen species generated from glucose metabolism could trigger the secretion of glucose-stimulated insulin [44] and phenolic compounds may combat these reactive oxygen species. Despite this, an overall insignificant change in insulin iAUC is likely due to the high dietary protein content in ABSG and FBSG compared to the Control. Dietary protein is recognised as insulinotropic since it stimulates insulin secretion by producing insulin-related amino acids during digestion, such as branched-chain amino acids [45]; therefore, this effect may offset the benefit provided by dietary fibre and phenolic compounds.

We also examined the postprandial lipid-panel response after biscuits consumption and no effect on postprandial TG response is observed, even though consumption of soluble fibre has been shown to enhance postprandial TG response [46]. That being said, another study conducted by Bourdon, et al. [47], also showed no difference in postprandial TG response after consumption of pasta made with a 40% barley flour substitution compared to pasta made with wheat flour. The reason for such conflicting results could be due to the amount of soluble dietary fibre present in the biscuits. Previous studies have shown that a minimum of 10 g of soluble dietary fibre is required to reduce postprandial TG response [48], while the soluble dietary fibre content in our biscuits is noted to be below 1.2 g [18]. In addition, the 4-h duration of the postprandial study could also explain the discrepancy in results, as lipid metabolism requires a longer postprandial duration of around 6-h [46].

We anticipated that the BSG-containing biscuits would increase the concentration of breath H_2 and CH_4, as BSG-containing biscuits contain more non-digestible carbohydrates, such as cellulose and hemicellulose, which can be fermented by gut microbiota [49]. However, in contrast, no difference was observed and it is possibly explained by the relatively short postprandial duration in this study. As the primary dietary fibre in the BSG is insoluble, it cannot be rapidly fermented by gut microbiota [50]. According to a previous study by Belobrajdic et al. [51], it takes at least 270 min for 121 g of wholegrain bread to reach the large intestine. In line with this study, results from O'Connor and Campbell [52] found that after consuming a 15 g insoluble dietary fibre-enriched snack bar, H_2 levels increased after 240 min, and a significant difference with a plain snack bar was only observed after 360 min.

A previous study demonstrated that arabinoxylans, the main oligosaccharide in BSG (21–28% on dry weight basis), exhibit a great water holding capacity, thereby increasing digested food viscosity and prolonging stomach distension [53]. With these characteristics, it was expected that the consumption of BSG-containing biscuits will suppress appetite; however, no difference was observed in this study. This may be attributed to the fact that the subjects enrolled into this study are relatively older adults and may exhibit a slower gastric emptying due to increased pyloric motility, greater gastric antral area, and decreased perception of gastric distension [54]. Moreover, older adults experience a decrease in appetite due to the elevated secretion of appetite-suppressing peptides in conjunction with a decreased secretion of appetite-stimulating peptides [55]. Hence, any changes in postprandial appetite for the older adults may be less obvious when reported through a subjective appetite assessment.

The novelty of the study lies in the incorporation of BSG or fermented BSG into food products and the subsequent assessment of its PPGR and PPIR controlling effects. These findings provide clinical evidence to facilitate the development of functional food with BSG that can assist in improving glucose metabolism while simultaneously addressing the environmental challenge of BSG disposal. Another key strength of this study lies in the recruitment of individuals with MetS, as these individuals specifically require dietary strategies for PPGR regulation. However, present limitations in our study should also be noted. Although 15 subjects provided more than 90% statistical power, the sample size remains relatively small. Therefore, future studies should consider enlarging the study population to enhance the generalizability of the findings across broader demographic groups. Additionally, as the main aim of this study was to mimic a typical eating pattern, the biscuits provided were standardised based on weight, resulting in varying ACHOs which may cause differences in PPGR. Therefore, it would be valuable to run future studies matching the ACHO content in the biscuits. Moreover, although a postprandial duration of 4 h is sufficient to observe differences in our primary outcomes such as PPGR and PPIR, a longer duration may be necessary to detect variations in other parameters such as postprandial breath and lipid panel response. Lastly, future studies which may explore the underlying mechanism responsible for the improvement of PPGR following the consumption of BSG-containing biscuits, as well as the effect over long-term consumption, are required.

5. Conclusions

In conclusion, BSG-containing biscuits have a higher nutritive value compared to commercial wheat-based biscuits, which can help to regulate postprandial glycaemic response in individuals with MetS. This study posits BSG as a promising source of value-added ingredient, and fermentation as a potential method for biovalorising food by-products to enhance their health benefits.

Supplementary Materials: The following supporting information can be downloaded at: https://www.mdpi.com/article/10.3390/nu16060909/s1. Figure S1: Change from baseline of (A) postprandial TC concentration. (B) Postprandial HDL-C concentration. (C) Postprandial LDL-C concentration. (D) Postprandial TG concentration; Figure S2: Change from baseline of (A) H_2 concentration. (B) CH_4 concentration; Figure S3: VAS score of subjective satiety assessment: (A) Hunger. (B) Desire to eat. (C) Prospective consumption. (D) Fullness; Table S1: Normalised AUC of TC, HDL-C and LDL-C, and positive iAUC and net iAUC of TG.

Author Contributions: Conceptualisation, Y.X. and J.E.K.; methodology, Y.X.; formal analysis, Y.X.; investigation, Y.X., Z.N.L., W.Z., X.J. and J.W.K.; writing—original draft preparation, Y.X.; writing—review and editing, J.E.K.; supervision, G.C.T.C. and J.E.K.; project administration, Y.X.; funding acquisition, J.E.K. All authors have read and agreed to the published version of the manuscript.

Funding: This work was supported by Ministry of Education, Singapore (A-0008420-00-00).

Institutional Review Board Statement: The study was registered at clinicaltrials.gov (NCT05421780), was conducted according to the guidelines of the Declaration of Helsinki and was approved by the National University of Singapore Institutional Review Board (NUS-IRB-2022-089) on 25 May 2022.

Informed Consent Statement: Informed consent was obtained from all subjects involved in the study.

Data Availability Statement: The data presented in this study will be made available on request from the corresponding author. The data are not publicly available due to restrictions of informed consent and institutional guidelines.

Acknowledgments: We thank Xianfang Wang's contribution in biscuit development. We appreciate the support of Amelia Chng, and Denise Tan throughout the clinical visits. We are grateful for the kind assistance from our phlebotomist team at the Metabolic Phenotyping Unit. We also acknowledge Brewerkz Brewing Co. (Singapore), who kindly provided the raw brewer's spent grain for this study.

Conflicts of Interest: The authors declare no conflicts of interest.

References

1. Gin, H.; Rigalleau, V. Post-prandial hyperglycemia. Post-prandial hyperglycemia and diabetes. *Diabetes Metab.* **2000**, *26*, 265–272.
2. Balkau, B.; Shipley, M.; Jarrett, R.J.; Pyörälä, K.; Pyörälä, M.; Forhan, A.; Eschwège, E. High blood glucose concentration is a risk factor for mortality in middle-aged nondiabetic men: 20-year follow-up in the Whitehall Study, the Paris Prospective Study, and the Helsinki Policemen Study. *Diabetes Care* **1998**, *21*, 360–367. [CrossRef]
3. Esser, N.; Legrand-Poels, S.; Piette, J.; Scheen, A.J.; Paquot, N. Inflammation as a link between obesity, metabolic syndrome and type 2 diabetes. *Diabetes Res. Clin. Pract.* **2014**, *105*, 141–150. [CrossRef]
4. Niedowicz, D.M.; Daleke, D.L. The role of oxidative stress in diabetic complications. *Cell Biochem. Biophys.* **2005**, *43*, 289–330. [CrossRef]
5. Expert Panel on Detection, Evaluation, and Treatment of High Blood Cholesterol in Adults. Executive summary of the third report of the National Cholesterol Education Program (NCEP) expert panel on detection, evaluation, and treatment of high blood cholesterol in adults (Adult Treatment Panel III). *JAMA* **2001**, *285*, 2486–2497. [CrossRef] [PubMed]
6. Toh, D.W.K.; Koh, E.S.; Kim, J.E. Lowering breakfast glycemic index and glycemic load attenuates postprandial glycemic response: A systematically searched meta-analysis of randomized controlled trials. *Nutrition* **2020**, *71*, 110634. [CrossRef] [PubMed]
7. Pladevall, M.; Singal, B.; Williams, L.K.; Brotons, C.; Guyer, H.; Sadurni, J.; Falces, C.; Serrano-Rios, M.; Gabriel, R.; Shaw, J.E. A single factor underlies the metabolic syndrome: A confirmatory factor analysis. *Diabetes Care* **2006**, *29*, 113–122. [CrossRef]
8. Olawoye, B.; Adeniyi, D.M.; Oyekunle, A.O.; Kadiri, O.; Fawale, S.O. Economic evaluation of cookie made from blend of brewers' spent grain (BSG), groundnut cake and sorghum flour. *Open Agric.* **2017**, *2*, 401–410. [CrossRef]
9. Petrović, J.S.; Pajin, B.S.; Kocić-Tanackov, S.D.; Pejin, J.D.; Fišteš, A.Z.; Bojanić, N.Đ.; Lončarević, I.S. Quality properties of cookies supplemented with fresh brewer's spent grain. *Food Feed Res.* **2017**, *44*, 57–63. [CrossRef]
10. Jenkins, A.L.; Jenkins, D.J.; Wolever, T.M.; Rogovik, A.L.; Jovanovski, E.; Božikov, V.; Rahelić, D.; Vuksan, V. Comparable postprandial glucose reductions with viscous fiber blend enriched biscuits in healthy subjects and patients with diabetes mellitus: Acute randomized controlled clinical trial. *Croat. Med. J.* **2008**, *49*, 772. [CrossRef] [PubMed]
11. Carlson, J.J.; Eisenmann, J.C.; Norman, G.J.; Ortiz, K.A.; Young, P.C. Dietary fiber and nutrient density are inversely associated with the metabolic syndrome in US adolescents. *J. Am. Diet. Assoc.* **2011**, *111*, 1688–1695. [CrossRef] [PubMed]
12. Galisteo, M.; Duarte, J.; Zarzuelo, A. Effects of dietary fibers on disturbances clustered in the metabolic syndrome. *J. Nutr. Biochem.* **2008**, *19*, 71–84. [CrossRef] [PubMed]
13. Dikeman, C.L.; Murphy, M.R.; Fahey, G.C. Dietary Fibers Affect Viscosity of Solutions and Simulated Human Gastric and Small Intestinal Digesta. *J. Nutr.* **2006**, *136*, 913–919. [CrossRef] [PubMed]
14. Garcia, A.L.; Otto, B.; Reich, S.C.; Weickert, M.O.; Steiniger, J.; Machowetz, A.; Rudovich, N.N.; Möhlig, M.; Katz, N.; Speth, M.; et al. Arabinoxylan consumption decreases postprandial serum glucose, serum insulin and plasma total ghrelin response in subjects with impaired glucose tolerance. *Eur. J. Clin. Nutr.* **2007**, *61*, 334–341. [CrossRef]
15. Fărcaş, A.; Tofană, M.; Socaci, S.; Mudura, E.; Scrob, S.N.A.; Salanţă, L.; Mureşan, V. Brewers' spent grain–A new potential ingredient for functional foods. *J. Agroaliment. Process Technol.* **2014**, *20*, 137–141.
16. Lynch, K.M.; Steffen, E.J.; Arendt, E.K. Brewers' spent grain: A review with an emphasis on food and health. *J. Inst. Brew.* **2016**, *122*, 553–568. [CrossRef]
17. Lamas, D.L.; Gende, L.B. Valorisation of brewers' spent grain for the development of novel beverage and food products. *Appl. Food Res.* **2023**, *3*, 100314. [CrossRef]
18. Wang, X.; Xu, Y.; Teo, S.Q.; Heng, C.W.; Lee, D.P.S.; Gan, A.X.; Kim, J.E. Impact of solid-state fermented Brewer's spent grains incorporation in biscuits on nutritional, physical and sensorial properties. *LWT* **2023**, *182*, 114840. [CrossRef]
19. Nuttall, F.Q.; Mooradian, A.D.; Gannon, M.C.; Billington, C.; Krezowski, P. Effect of Protein Ingestion on the Glucose and Insulin Response to a Standardized Oral Glucose Load. *Diabetes Care* **1984**, *7*, 465–470. [CrossRef]
20. Gutierrez-Barrutia, M.B.; Cozzano, S.; Arcia, P.; del Castillo, M.D. In Vitro Digestibility and Bioaccessibility of Nutrients and Non-Nutrients Composing Extruded Brewers' Spent Grain. *Nutrients* **2022**, *14*, 3480. [CrossRef]
21. Farago, C.V.; de Melo, G.B.; Escher, G.B.; Marcon, M.V.; Granato, D.; Danesi, E.D.G. Cereal bars made from brewers' spent grain, apple and Spirulina platensis: Antioxidant activity and antihyperglycaemic effects. *Res. Soc. Dev.* **2021**, *10*, 5.
22. Schettino, R.; Verni, M.; Acin-Albiac, M.; Vincentini, O.; Krona, A.; Knaapila, A.; Cagno, R.D.; Gobbetti, M.; Rizzello, C.G.; Coda, R. Bioprocessed Brewers' Spent Grain Improves Nutritional and Antioxidant Properties of Pasta. *Antioxidants* **2021**, *10*, 742. [CrossRef] [PubMed]
23. Prasad, K.N.; Bondy, S.C. Dietary fibers and their fermented short-chain fatty acids in prevention of human diseases. *Bioact. Carbohydr. Diet. Fibre.* **2019**, *17*, 100170. [CrossRef]

24. Utzschneider, K.M.; Kratz, M.; Damman, C.J.; Hullarg, M. Mechanisms Linking the Gut Microbiome and Glucose Metabolism. *J. Clin. Endocrinol. Metab.* **2016**, *101*, 1445–1454. [CrossRef] [PubMed]
25. Van Den Hoek, A.M.; Heijboer, A.C.; Corssmit, E.P.; Voshol, P.J.; Romijn, J.A.; Havekes, L.M.; Pijl, H. PYY3–36 reinforces insulin action on glucose disposal in mice fed a high-fat diet. *Diabetes* **2004**, *53*, 1949–1952. [CrossRef] [PubMed]
26. He, J.; Zhang, P.; Shen, L.; Niu, L.; Tan, Y.; Chen, L.; Zhao, Y.; Bai, L.; Hao, X.; Li, X.; et al. Short-Chain Fatty Acids and Their Association with Signalling Pathways in Inflammation, Glucose and Lipid Metabolism. *Int. J. Mol. Sci.* **2020**, *21*, 6356. [CrossRef]
27. World Health Organization. *WHO STEPS Surveillance Manual: The WHO STEPwise Approach to Chronic Disease Risk Factor Surveillance*; Report No.: 9241593830; World Health Organization: Geneva, Switzerland, 2005.
28. Bergmeyer, H. Determination with hexokinase and glucose-6-phosphate dehydrogenase. *Methods Enzym. Anal.* **1997**, *3*, 1196–1201.
29. Fossati, P.; Prencipe, L. Serum triglycerides determined colorimetrically with an enzyme that produces hydrogen peroxide. *Clin. Chem.* **1982**, *28*, 2077–2080. [CrossRef]
30. Friedewald, W.T.; Levy, R.I.; Fredrickson, D.S. Estimation of the concentration of low-density lipoprotein cholesterol in plasma, without use of the preparative ultracentrifuge. *Clin. Chem.* **1972**, *18*, 499–502. [CrossRef]
31. Flint, A. Reproducibility, power and validity of visual analogue scales in assessment of appetite sensations in single test meal studies. *Int. J. Obes.* **2000**, *24*, 38–48. [CrossRef]
32. Raza, G.S.; Maukonen, J.; Makinen, M.; Niemi, P.; Niiranen, L.; Hibberd, A.A.; Poutanen, K.; Buchert, J.; Herzig, K.-H. Hypocholesterolemic Effect of the Lignin-Rich Insoluble Residue of Brewer's Spent Grain in Mice Fed a High-Fat Diet. *J. Agric. Food Chem.* **2019**, *67*, 1104–1114. [CrossRef] [PubMed]
33. Lee, D.P.S.; Gan, A.X.; Kim, J.E. Incorporation of biovalorised okara in biscuits: Improvements of nutritional, antioxidant, physical, and sensory properties. *LWT* **2020**, *134*, 109902. [CrossRef]
34. de Carvalho, C.M.; de Paula, T.P.; Viana, L.V.; Machado, V.M.; de Almeida, J.C.; Azevedo, M.J. Plasma glucose and insulin responses after consumption of breakfasts with different sources of soluble fiber in type 2 diabetes patients: A randomized crossover clinical trial. *Am. J. Clin. Nutr.* **2017**, *106*, 1238–1245. [CrossRef]
35. Luhovyy, B.L.; Mollard, R.C.; Yurchenko, S.; Nunez, M.F.; Berengut, S.; Liu, T.T.; Smith, C.E.; Pelkman, C.L.; Anderson, G.H. The Effects of Whole Grain High-Amylose Maize Flour as a Source of Resistant Starch on Blood Glucose, Satiety, and Food Intake in Young Men. *J. Food Sci.* **2014**, *79*, H2550–H2556. [CrossRef]
36. Ullah, H.; Esposito, C.; Piccinocchi, R.; De Lellis, L.F.; Santarcangelo, C.; Minno, A.D.; Baldi, A.; Buccato, D.G.; Khan, A.; Piccinocchi, G.; et al. Postprandial Glycemic and Insulinemic Response by a Brewer' Spent Grain Extract-Based Food Supplement in Subjects with Slightly Impaired Glucose Tolerance: A Monocentric, Randomized, Cross-Over, Double-Blind, Placebo-Controlled Clinical Trial. *Nutrients* **2022**, *14*, 3916. [CrossRef] [PubMed]
37. Reynolds, A.N.; Akerman, A.P.; Mann, J. Dietary fibre and whole grains in diabetes management: Systematic review and meta-analyses. *PLoS Med.* **2020**, *17*, e1003053. [CrossRef] [PubMed]
38. Deng, N.; Zheng, B.; Li, T.; Liu, R.H. Assessment of the Phenolic Profiles, Hypoglycemic Activity, and Molecular Mechanism of Different Highland Barley (*Hordeum vulgare* L.) Varieties. *Int. J. Mol. Sci.* **2020**, *21*, 1175. [CrossRef]
39. Steemburgo, T.; Dall'Alba, V.; Almeida, J.; Zelmanovitz, T.; Gross, J.; De Azevedo, M. Intake of soluble fibers has a protective role for the presence of metabolic syndrome in patients with type 2 diabetes. *Eur. J. Clin. Nutr.* **2009**, *63*, 127–133. [CrossRef]
40. Ames, N.; Blewett, H.; Storsley, J.; Thandapilly, S.J.; Zahradka, P.; Taylor, C. A double-blind randomised controlled trial testing the effect of a barley product containing varying amounts and types of fibre on the postprandial glucose response of healthy volunteers. *Br. J. Nutr.* **2015**, *113*, 1373–1383. [CrossRef]
41. Lee, D.P.S.; Gan, A.X.; Sutanto, C.N.; Toh, K.Q.X.; Khoo, C.M.; Kim, J.E. Postprandial glycemic and circulating SCFA concentrations following okara- and biovalorized okara-containing biscuit consumption in middle-aged and older adults: A crossover randomized controlled trial. *Food Funct.* **2022**, *13*, 9687–9699. [CrossRef]
42. Alberti, K.G.; Zimmet, P.; Shaw, J.; IDF Epidemiology Task Force Consensus Group. The metabolic syndrome—A new worldwide definition. *Lancet* **2005**, *366*, 1059–1062. [CrossRef] [PubMed]
43. Liese, A.D.; Schulz, M.; Fang, F.; Wolever, T.M.; D'Agostino, R.B., Jr.; Sparks, K.C.; Mayer-Davis, E.J. Dietary glycemic index and glycemic load, carbohydrate and fiber intake, and measures of insulin sensitivity, secretion, and adiposity in the Insulin Resistance Atherosclerosis Study. *Diabetes Care* **2005**, *28*, 2832–2838. [CrossRef] [PubMed]
44. Serrano, J.C.; Martín-Gari, M.; Cassanye, A.; Granado-Serrano, A.B.; Portero-Otín, M. Characterization of the post-prandial insulinemic response and low glycaemic index of a soy beverage. *PLoS ONE* **2017**, *12*, e0182762. [CrossRef] [PubMed]
45. Rietman, A.; Schwarz, J.; Tome, D.; Koke, F.J.; Mensink, M. High dietary protein intake, reducing or eliciting insulin resistance? *Eur. J. Clin. Nutr.* **2014**, *68*, 973–979. [CrossRef] [PubMed]
46. Lee, D.P.S.; Low, J.H.M.; Chen, J.R.; Zimmermann, D.; Actis-Goretta, L.; Kim, J.E. The influence of different foods and food ingredients on acute postprandial triglyceride response: A systematic literature review and meta-analysis of randomized controlled trials. *Adv. Nutr.* **2020**, *11*, 1529–1543. [CrossRef] [PubMed]
47. Bourdon, I.; Yokoyama, W.; Davis, P.; Hudson, C.; Backus, R.; Richter, D.; Knuckles, B.; Schneeman, B.O. Postprandial lipid, glucose, insulin, and cholecystokinin responses in men fed barley pasta enriched with β-glucan. *Am. J. Clin. Nutr.* **1999**, *69*, 55–63. [CrossRef] [PubMed]
48. Khossousi, A.; Binns, C.; Dhaliwal, S.; Pal, S. The acute effects of psyllium on postprandial lipaemia and thermogenesis in overweight and obese men. *Br. J. Nutr.* **2008**, *99*, 1068–1075. [CrossRef] [PubMed]

49. Le Marchand, L.; Wilkens, L.R.; Harwood, P.; Cooney, R.V. Use of breath hydrogen and methane as markers of colonic fermentation in epidemiologic studies: Circadian patterns of excretion. *Environ. Health Perspect.* **1992**, *98*, 199–202. [CrossRef]
50. Glitsø, L.V.; Gruppen, H.; Schols, H.A.; Højsgaard, S.; Sandström, B.; Bach Knudsen, K.E. Degradation of rye arabinoxylans in the large intestine of pigs. *J. Sci. Food Agric.* **1999**, *79*, 961–969. [CrossRef]
51. Belobrajdic, D.P.; Regina, A.; Klingner, B.; Zajac, I.; Chapron, S.; Berbezy, P.; Bird, A.R. High-Amylose Wheat Lowers the Postprandial Glycemic Response to Bread in Healthy Adults: A Randomized Controlled Crossover Trial. *J. Nutr.* **2019**, *149*, 1335–1345. [CrossRef]
52. O'Connor, L.E.; Campbell, W.W. A novel fiber composite ingredient incorporated into a beverage and bar blunts postprandial serum glucose and insulin responses: A randomized controlled trial. *Nutr. Res.* **2016**, *36*, 253–261. [CrossRef] [PubMed]
53. Marciani, L.; Pritchard, S.E.; Hellier-Woods, C.; Costigan, C.; Hoad, C.; Gowland, P.A.; Spiller, R.C. Delayed gastric emptying and reduced postprandial small bowel water content of equicaloric whole meal bread versus rice meals in healthy subjects: Novel MRI insights. *Eur. J. Clin. Nutr.* **2013**, *67*, 754–758. [CrossRef] [PubMed]
54. Giezenaar, C.; Chapman, I.; Luscombe-Marsh, N.; Feinle-Bisset, C.; Horowitz, M.; Soenen, S. Ageing Is Associated with Decreases in Appetite and Energy Intake—A Meta-Analysis in Healthy Adults. *Nutrients* **2016**, *8*, 28. [CrossRef]
55. Akimoto, S.; Miyasaka, K. Age-associated changes of appetite-regulating peptides. *Geriatr. Gerontol. Int.* **2010**, *10*, S107–S119. [CrossRef] [PubMed]

nutrients

MDPI

Article

Factors Impacting the Reduction in Neophobia Prevalence in Phenylketonuria Patients

Meda-Ada Bugi [1,2,3], Iulius Jugănaru [2,4,5], Raluca Isac [6,7], Iulia-Elena Simina [8], Andrei-Ioan Munteanu [2,4,5], Niculina Mang [2,4,5], Georgiana-Flavia Brad [2,4], Delia-Maria Nicoară [1,4], Daniela Cîrnatu [9,*] and Otilia Mărginean [2,4,5]

1 Ph.D. School Department, 'Victor Babeş' University of Medicine and Pharmacy of Timisoara, 300041 Timisoara, Romania; bugi.ada@umft.ro (M.-A.B.); nicoara.delia@umft.ro (D.-M.N.)
2 Department of Pediatrics I, Children's Emergency Hospital 'Louis Turcanu', 300011 Timisoara, Romania; juganaru.iulius@umft.ro (I.J.); munteanu.andrei@umft.ro (A.-I.M.); mang.nina@umft.ro (N.M.); brad.giorgiana@umft.ro (G.-F.B.); marginean.otilia@umft.ro (O.M.)
3 Department of Pharmacy, University of Medicine and Pharmacy 'Vasile Goldis', 310025 Arad, Romania
4 Department XI Pediatrics, Discipline I Pediatrics, 'Victor Babeş' University of Medicine and Pharmacy of Timisoara, 300041 Timisoara, Romania
5 Department XI Pediatrics, Discipline I Pediatrics, Disturbances of Growth and Development in Children–BELIVE, 300011 Timisoara, Romania
6 Department of Pediatrics II, Children's Emergency Hospital 'Louis Turcanu', 300011 Timisoara, Romania; isac.raluca@umft.ro
7 Department of Pediatrics, IIIrd Pediatric Clinic, 'Victor Babeș' University of Medicine and Pharmacy of Timișoara, 300041 Timisoara, Romania
8 Department of Genetics, Center of Genomic Medicine, 'Victor Babeş' University of Medicine and Pharmacy of Timișoara, 300041 Timisoara, Romania; simina.iulia@umft.ro
9 Department of Medicine, University of Medicine and Pharmacy 'Vasile Goldis', 310025 Arad, Romania
* Correspondence: daniela.cirnatu@insp.gov.ro

Abstract: Food neophobia (FN), the fear of sampling new foods, can have a significant impact on children's eating habits. Children with phenylketonuria (PKU), a hereditary condition that inhibits the body's capacity to metabolize phenylalanine, should take this attitude with caution. Patients with PKU must follow a rigorous phenylalanine (Phe)-restricted diet to avoid brain malfunction that can include intellectual disability, seizures, and behavioral difficulties. The novelty of our work stems from the fact that we explored the origins of this incorrect intake pattern, which exacerbates PKU patients' already fragile health. We conducted a cross-sectional study on 34 previously diagnosed phenylketonuria patients and a control group ranging in age from 7 months to 40 years, with a sex ratio of M/F 2:1. The Food Neophobia Scale (FNS) was used to determine neophobia. We used JASP (version 0.18.1) statistical analysis to examine the relationship between neophobia and PKU condition, age and nutritional status at the time of study, diet compliance, parental educational level, period from birth to PKU diagnosis, and environmental (rural/urban) provenience of PKU patients. According to the data, 61.76% of patients with PKU were neophobic, as were 70.57% of the control group. Food neophobia was associated with PKU patients' present age, the period from birth to PKU diagnosis, and parental educational level.

Keywords: neophobia; phenylketonuria; nutrition

1. Introduction

Phenylketonuria (PKU; OMIM 261600) is a rare autosomal recessive genetic disorder triggered by mutations in the phenylalanine hydroxylase (*PAH*) gene, leading to reduced catalytic activity and affecting the breakdown of phenylalanine (Phe) [1]. The excess of Phe causes severe and irreversible intellectual disability, along with autistic behaviors, motor impairments, rashes, and seizures [2]. At present, there is no curative treatment for PKU. The primary approach to managing the condition predominantly involves strict dietary control. Dietary Phe intake should be minimized to meet the essential requirements for normal growth, supplemented with specially created metabolic formulas. The initiation of this regimen is recommended in infancy, ideally within the five days post-birth [3]. Adherence to the prescribed diet is important as failure to do so may render the severe neurological impairment irreversible [4]. The highest reported prevalence of PKU was found in Turkey; 38 per 100,000 neonates were diagnosed with PKU, based on data from 46 research studies conducted between 1964 and 2017. Conversely, the lowest prevalence was obtained in Thailand; 0.3 per 100,000 neonates suffer from PKU [5]. Based on the previously mentioned sources, it has been stated that the worldwide incidence rate of the disease stands at 6 cases per 100 infants. The forecast prevalence rate in Romania is 1 case per 100,000 individuals [6]. Dietary restrictions observed in PKU patients' diets bring to the surface the most common attitudes related to novel food acceptance: neophobia, picky/fussy, and neophilia. Food neophobia (FN), defined as the aversion or reluctance to consume unfamiliar foods, is a psychological phenomenon that must be distinguished from pickiness, which is the refusal to consume familiar foods that are detested [7–9]. To date, our comprehension of the relationship between food neophobia in individuals with PKU and various clinical and treatment-related factors, such as age, blood Phe control, anthropometric measurements, lifestyle choices, breastfeeding during infancy, and the timing of introducing solid foods, remains limited [10]. Within the confines of these dietary restrictions, it is essential to promote normal consumption behavior and a wide variety of flavors in order to maximize food options [11]. The early administration of L-amino acid supplements with a bitter flavor to neonates with PKU may influence the development of food preferences [12]. In addition, there is evidence that such early flavor experiences have a lasting impact on flavor preferences [12–14]. Therefore, the precise impact of food neophobia in PKU has yet to be thoroughly investigated, especially in relation to the introduction of L-amino acid supplements and their potential long-term effects on food preferences [12–15]. Food neophobia significantly impacts the dietary choices and overall daily routines of children and caregivers [16]. The phenomenon in question holds evolutionary relevance, yet it can give rise to avoidance habits and present challenges for youngsters who adhere to dietary restrictions [17,18]. It falls under avoidant/restrictive food intake disorder and affects psychological well-being due to taste familiarity and parental attitudes [18,19]. Multiple factors contribute to food neophobia, necessitating countermeasures against its influence on children's preferences and nutritional imbalance [19]: dietary diversity, health, development, body mass index (BMI), age, diagnostic moment, and parent's level of education [20]. The objective of this study was to examine the relationship between neophobia and socioeconomic status, familial background, or dietary habits among patients with PKU.

2. Materials and Methods

2.1. Study Population

We conducted a cross-sectional study between 2018 and 2023 using self-applied online questionnaires on individuals previously diagnosed with phenylketonuria, including those under the care of the "Louis Turcanu" Emergency Hospital for Children in Timisoara or affiliated with the PKU Life Romania Association or their parents. The inclusion criteria were a prior diagnosis of PKU. Of the 70 patients who met the inclusion criteria, 34 agreed to participate and were analyzed in this study as the PKU group (PKUG). The sample comprised an equal number of non-PKU controls, carefully selected to match in terms of both age and sex. The control group (CG) consisted of patients admitted to the Pediatric Ward of the Timisoara "Louis Turcanu" Emergency Children's Hospital. Regarding age, the patients were further divided into five categories: <2 years (8.82%), 2–7.9 years (44.11%), 8–13.9 years (17.64%), 14–17.9 years (17.64%), and ≥18 years (11.76%).

The research adhered to the guidelines outlined in the Declaration of Helsinki and Resolution 466/2012 of the National Health Council. The research protocol received approval from the Ethics Committee of the "Victor Babes" University of Medicine and Pharmacy

(No.60/12.11.2018) and from the hospital's Ethics Committee (No.3392/24.02.2023). Additionally, each participant provided a signed informed consent.

2.2. Data Collection

An interview questionnaire partly developed by the researcher after reviewing the related literature and translated into the Romanian language was used to carry out the current study and offered to the participants. The respondents filled in the online survey using the Google Forms platform. The questionnaire consisted of three parts: the participant's characteristics, familial characteristics, and eating behavior. The first part comprised the subject's identifying information, including age, sex, anthropometric measurements at birth and during data collection, PKU diagnosis details (year and method of diagnosis—neonatal screening through dry blood spot or other method), and Phe level in µmol/L at the diagnosis. To assess the nutritional status of pediatric patients, percentiles were utilized, and specific terminology was employed, interpreted by referencing World Health Organization girls' percentiles expanded tables [21] and boys' percentiles expanded tables [22]. The second part included familial characteristics, such as the socio-economic status of the family, the educational level of the parents or guardians, and specifications regarding the person taking care of the child and preparing/supervising the diet. The third part comprised details regarding the eating behavior, in which the child's eating habits were assessed using the Food Neophobia Scale (FNS) [6]. The specific measurement of food neophobia in adults commonly employs the FNS, a validated questionnaire consisting of ten items and developed by Pliner and Hobden. Comprising 10 items, the scale includes 5 related to neophobic behavior and 5 related to neophilic behavior. A higher FNS value indicates a greater inclination toward neophobia. This scale proves valuable for measuring the willingness to try new food items, investigating the acceptance of exotic cuisines, and exploring expectations regarding novel food items [23]. We chose to apply this questionnaire to both adults and children, translating the 10 statements to Romanian and requesting parents to help their children or to answer on their behalf when/if necessary. The questionnaire was applied to a CG, respecting the sex ratio and distribution by age groups. The typical Likert scales consist of seven choices. However, this configuration poses certain challenges. The inclusion of a "neutral" option may lead survey respondents to easily bypass the question, potentially choosing this middle ground without much deliberation. In contrast, a six-point scale encourages participants to approach the question with greater consideration, compelling them to make a choice that leans either positively or negatively [24]. Given that our perceptions often lean away from neutrality, we considered the seven-point scale to better capture the nuances of the subjective experiences of our respondents. They were asked to answer each statement on a Likert scale of 1–7, where 1 stood for "strongly disagree", while 7 stood for "strongly agree". Regarding statement translation, we have replaced the term "ethnic" with "specific" as it includes references to restaurants or cuisines of different cultures and countries, proven to influence neophobia [23–26].

The individual total scores were calculated by summing the values assigned to each scale item, ranging from 1 to 7, resulting in a total score range of 10 to 70 points. The statements used in this study and their corresponding English translations are noted in Supplementary Table S1. Therefore, an increase in the total individual score signifies a food neophobia level. Similar to previous research studies [25–28], the participants were categorized into three groups based on their FNS total scores, using cutoff points defined as the FNS total scores ± standard deviation: neophilic, neutral, and neophobic groups. Despite the lack of a standardized approach for setting FN cutoffs, we categorized patients as neutral (FN score = 30–35), neophobic (FN score > 35), or neophilic (FN score < 30). Additionally, the results of the FNS were compared between the two groups.

2.3. Data Analysis

The distributions of collected data analyzed with the Shapiro–Wilk test were shown to be not normally distributed for birth height (cm), current age (month), BMI, current height (cm), current level of Phe (µmol/L), current weight (kg), year of diagnostic phenylketonuria, diagnostic year (month), and level of Phe at birth (µmol/L). Consequently, a nonparametric Mann–Whitney U test was used for intergroup comparison, while Pearson's r test was used for correlations. For the Mann–Whitney test, the effect size was given by the rank biserial correlation. All statistical analyses were performed using the JASP (Version 0.18.1 [29]).

2.4. Hypothesis

In order to assess the factors impacting the reduction in neophobic prevalence in phenylketonuria patients, the following hypotheses were formulated:

H1. *There is a distribution variation for FN score among PKU and control groups.*

H2. *There is a relationship between the patient's current BMI and the development of neophobia.*

H3. *There is a relationship between diet compliance and neophobia.*

H4. *There is a relationship between a patient's Phe level at the diagnostic moment and the development of neophobia.*

H5. *There is a correlation between parental educational level and the patient's development of neophobia.*

H6. *There is a correlation between the patient's age and the onset of neophobia.*

H7. *There is a relationship between the length of the period from birth to diagnosis and neophobia.*

H8. *There is a variation of distribution for FN scores among urban and rural patients.*

3. Results

Sixty-eight participants (34 PKU patients and 34 controls) were included in the study. The sex ratio M/F = 2:1, and the age groups between the PKU group and the CG were comparable (Table 1). Patients with PKU were diagnosed in the following ways: 29.41% within the first 15 days of life, 44.11% between days 16 and 30, 14.70% between day 31 and three months, 2.94% by the end of the first year, and 8.82% beyond the first year. Of the patients, 88.23% were diagnosed using screening. The individuals within the PKUG had a substantial prevalence of malnutrition, with 35.30% displaying underweight, 23.53% being overweight, and 23.53% displaying severe obesity.

Table 1. Statistical description of the lot and results analyses.

Parameters	PKUG (n = 34)	CG (n = 34)	p-Value
Sex % (n)			
Males	67.6 (23)	67.6 (23)	1
Females	32.4 (11)	32.4 (11)	
Age at diagnosis in days (median, IQR)	21(15, 41)		
Age in years (median, IQR)	7.05 (4.9, 15.2)	5.75 (3.8, 14.87)	0.615

Abbreviations: PKUG, phenylketonuria group; CG, control group; IQR, inter-quartile range.

Table 2 provides a summary of the *PAH* gene mutations observed in patients for whom genetic results were available. Notably, all patients exhibited pathogenic mutations, with

29.4% identified as heterozygotes. Among these individuals, three manifested classic or moderate phenotypes of PKU, while only one displayed a mild phenotype of the disorder.

Table 2. *PAH* gene mutations and phenotypic associations.

N0.	*PAH* Variant	Zygosity Status	ACMG [30] Classification	PKU Phenotype
001	NM_000277.3:c.1066-11G>A	Heterozygous	Pathogenic	Classic or moderate
001	NM_000277.3:c.1222C>T NP_000268.1:p.(Arg408Trp)	Heterozygous	Pathogenic	Classic
002	NM_000277.3:c.472C>T NP_000268.1:p.Arg158Trp	Heterozygous	Pathogenic	Classic
002	NM_000277.3:c.1222C>T NP_000268.1: p.Arg408Trp	Heterozygous	Pathogenic	Classic
003	NM_000277.3:c.1222C>T NP_000268.1: p.Arg408Trp	Homozygous	Pathogenic	Classic
004	NM_000277.3:c.1222C>T NP_000268.1:p.Arg408Trp	Homozygous	Pathogenic	Classic
005	NM_000277.3:c.754C>T NP_000268.1: p.Arg252Trp	Heterozygous	Pathogenic	Classic
005	NM_000277.3:c.782G>A NP_000268.1:p.Arg261 GLN	Heterozygous	Pathogenic	Classic or moderate
006	NM_000277.3:c.1222C>T NP_000268.1:p.Arg408Trp	Heterozygous	Pathogenic	Classic
006	NM_000277.3:c.782G>A NP_000268.1:p.Arg261Gln	Heterozygous	Pathogenic	Classic or moderate
007	NM_000277.3:c.1222C>T NP_000268.1:p.(Arg408Trp)	Homozygous	Pathogenic	Classic
008	NM_000277.3:c.1222C>T NP_000268.1:p.Arg408Trp	Homozygous	Pathogenic	Classic
009	NM_000277.3:c.1315+1G>A	Heterozygous	Pathogenic	Classic
009	NM_000277.3:c.533A>G NP_000268.1:p.Glu178Gly	Heterozygous	Pathogenic	Mild
010	NM_000277.3:c.1222C>T NP_000268.1:p.(Arg408Trp)	Homozygous	Pathogenic	Classic
011	NM_000277.3:c.1222C>T NP_000268.1:p.(Arg408Trp)	Homozygous	Pathogenic	Classic
012	NM_000277.3:c.1222C>T NP_000268.1:p.(Arg408Trp)	Homozygous	Pathogenic	Classic
013	NM_000277.3:c.1222C>T NP_000268.1:p.(Arg408Trp)	Homozygous	Pathogenic	Classic
014	NM_000277.3:c.1222C>T NP_000268.1:p.(Arg408Trp)	Homozygous	Pathogenic	Classic
015	NM_000277.3:c.1222C>T NP_000268.1:p.(Arg408Trp)	Homozygous	Pathogenic	Classic
016	NM_000277.3:c.1222C>T NP_000268.1:p.(Arg408Trp)	Homozygous	Pathogenic	Classic
017	NM_000277.3:c.1222C>T NP_000268.1:p.(Arg408Trp)	Homozygous	Pathogenic	Classic

Abbreviations: N0., patient ID; ACMG, American College of Medical Genetics and Genomics guidelines; PKU, phenylketonuria.

After the questionnaire's data were analyzed, we observed that an important percentage of both PKUG and CG respondents had FNS scores exceeding 35. For instance, in the over-18-year-old group, the percentage was 100% for all the subjects. Table 3 provides a summary of the data for each age group, with further division based on sex. There were no statistical differences between incidences of neophobia by sex.

Table 3. Tertiles of FNS score results for both study groups.

Age Group (Years)	Group and Sex	FNS < 30 (%)	31–35 FNS (%)	>35 FNS (%)
Under 2	PKUG—male	33.33	33.33	33.33
	PKUG—female	0.00	0.00	0.00
	Total PKUG	33.33	33.33	33.33
	CG male	33.33	0.00	66.67
	CG female	0.00	0.00	0.00
	Total CG	33.33	0.00	66.67
2–7.9	PKUG—male	18.18	9.09	72.73
	PKUG—female	25.00	25.00	50.00
	Total PKUG	14.29	14.29	71.43
	CG male	0.00	8.33	91.67
	CG female	0.00	0.00	100.00
	Total CG	0.00	6.67	93.33
8–13.9	PKUG—male	0.00	0.00	100.00
	PKUG—female	0.00	50.00	50.00
	Total PKUG	0.00	16.67	83.33
	CG male	0.00	0.00	100.00
	CG female	0.00	0.00	100.00
	Total CG	0.00	0.00	100.00
14–17.9	PKUG—male	0.00	25.00	75.00
	PKUG—female	50.00	0.00	50.00
	Total PKUG	16.67	16.67	66.67
	CG male	25.00	0.00	75.00
	CG female	0.00	0.00	100.00
	Total CG	16.67	0.00	83.33
Over 18	PKUG—male	0.00	0.00	100.00
	PKUG -female	0.00	0.00	100.00
	Total PKUG	0.00	0.00	100.00
	CG male	0.00	0.00	100.00
	CG female	0.00	0.00	100.00
	Total CG	0.00	0.00	100.00

Abbreviations: PKUG, phenylketonuria group; CG, control group; FNS, Food Neophobia Scale.

The answers to each question revealed significant disparities for several questions—Q4, Q6, or Q9—even though the FNS scores were rather close. The outcome of each question is shown in Figure 1 for both groups.

The results of the statistical tests employed for hypothesis testing are presented in Figure 2. The data show statistically significant correlations between the score on the FNS (Food Neophobia Scale), the current age, and the period between birth and diagnosis. Consequently, as age advanced, the FNS evaluation score tended to be higher, indicating an increased risk of neophobia with age. A correlation between the two was observed regarding the duration between birth and diagnosis, such that an earlier diagnosis was associated with a lower risk of obtaining a high score on the FNS evaluation scale.

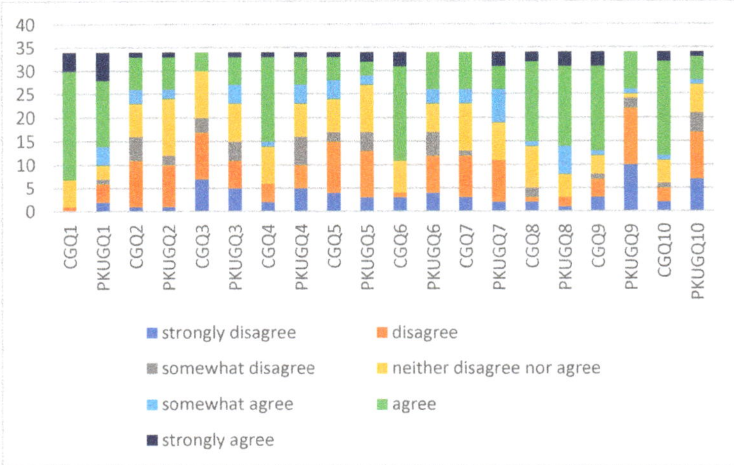

Figure 1. Responses to the food neophobia questionnaire using the Likert scale.

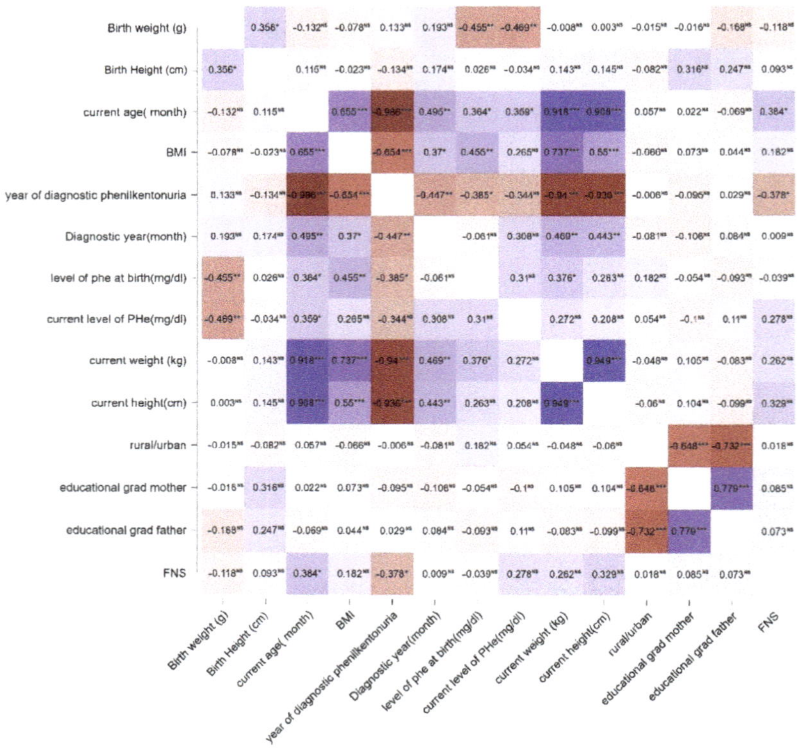

Figure 2. Statistical correlations among analyzed variables. Significance levels: *** $p < 0.001$, ** $p < 0.01$, * $p < 0.05$. Nonsignificant correlations are marked as 'NS'. The color gradient in the figure reflects the strength of significance, with deeper shades representing more substantial associations, and lighter tones suggesting comparatively lower statistical significance. Positive correlation coefficients denote positive associations, while negative correlation coefficients indicate negative associations.

We applied additional tests for the observed correlations to corroborate the findings and provide an enhanced statistical data analysis. The parents' levels of education and the distribution of the PKU group's scores were the subjects of the further tests. Table 4 presents a comparison of data between PKUG and CG, encompassing factors such as development, FNS, the primary caregiver's educational level, and the level of Phe at birth and currently. The results revealed statistically significant differences in the primary caregiver's educational level and, to some extent, in the FNS scores between groups.

Table 4. PKUG in comparison with CG statistical analyses.

Parameters	PKUG (n = 34)	CG (n = 34)	p-Value
Development median (IQR)	80 (43.7, 99.9)	75.5 (22.5, 97)	0.473
Primary caretaker % (n)			
Mother	91.2 (31)	41.2 (14)	0.003
Other	8.8 (3)	58.8 (20)	
Mother education % (n)			
Primary school	23.5 (8)	23.5 (8)	
Vocational school	14.7 (5)	14.7 (5)	0.634
High school	14.7 (5)	26.5 (9)	
University	47.1 (16)	35.2 (12)	
Father education % (n)			
Primary school	17.6 (6)	20.6 (7)	
Vocational school	17.6 (6)	38.2 (13)	0.279
High school	23.5 (8)	14.7 (5)	
University	41.2 (14)	26.5 (9)	
FNS median (IQR)	38.5 (30, 46)	45 (40, 49)	0.013
Level of Phe at birth median (IQR)	1089.6 (484.3, 1785.8)		
Level of Phe at present median (IQR)	260.3 (183.4, 381.3)		

Abbreviations: PKUG, phenylketonuria group; CG, control group; FNS, Food Neophobia Scale. Phe levels are measured in μmol/L. Statistically significant differences, with $p < 0.05$, are represented in bold.

Given the findings that highlighted a significant variation in the parents' educational levels based on whether they came from urban or rural environments, a follow-up inquiry was conducted. A comparison of the outcomes of the correlation between the FNS score and each parameter investigated by PKUG based on the environment of origin (rural/urban) is presented in Table 5. The results showed no significant differences for any of the investigated items.

Synthesizing the results of the statistical analyses from the eight proposed hypotheses, only three were confirmed: those related to the correlations between FNS and age and the correlations between FNS and the interval between the moment of birth and the diagnosis of PKU, as seen in Table 6.

Table 5. FNS score based on the environment of origin (rural/urban).

Independent Samples t-Test Rural/Urban	W	p-Value	Hodges–Lehmann Estimate	Rank-Biserial Correlation
FNS	141.000	0.931	-3.769×10^{-5}	−0.021
Birth weight (g)	146.500	0.945	2.580×10^{-5}	0.017
Birth height (cm)	157.500	0.651	1.000	0.094
Current age (months)	134.500	0.756	−4.331	−0.066
BMI	155.000	0.717	0.330	0.076
Time to diagnosis (years)	145.000	0.986	4.815×10^{-6}	0.007

Table 5. *Cont.*

Independent Samples *t*-Test Rural/Urban	W	*p*-Value	Hodges–Lehmann Estimate	Rank-Biserial Correlation
Age at diagnosis (months)	157.500	0.653	2.685	0.094
Level of Phe at birth (μmol/L)	107.500	0.313	−5.000	−0.210
Current level of Phe (μmol/L)	126.500	0.772	−0.500	−0.063
Current weight (kg)	152.000	0.796	1.577	0.056
Current height (cm)	154.000	0.743	2.451	0.069

Abbreviations: FNS, Food Neophobia Scale; BMI, body mass index; Phe, phenylalanine.

Table 6. The results of tested hypotheses.

No	Hypothesis	Result
H1	There is a variation of distribution for FN score among PKU group and control group.	Rejected
H2	There is a relationship between the patients' current BMI and the development of neophobia.	Rejected
H3	There is a relationship between diet compliance and neophobia.	Rejected
H4	There is a relationship between a patient's Phe level at the diagnostic moment and the development of neophobia.	Rejected
H5	There is a correlation between parental educational level and the patient's development of neophobia.	Accepted
H6	There is a correlation between the patient's age and the onset of neophobia.	Accepted
H7	There is a relationship between length of period from birth to diagnostic and neophobia.	Accepted
H8	There is a variation of distribution for FN score among urban and rural patients.	Rejected

4. Discussion

While food neophobia unquestionably shapes the initial encounter with novel foods, it is crucial to recognize that the decision to consistently include such foods in one's diet is influenced by a complex interaction of additional factors [31]. The concept of "picky/fussy" behavior is distinct from food neophobia [8]; however, measures regarding these attitudes are still in development [32]. "Picky/fussy" individuals are particularly children who consume an insufficient variety of foods by rejecting both familiar and unfamiliar foods [33]. Concurrently, the term "neophilic" refers to a characteristic in which an individual or group demonstrates a strong propensity and willingness to experiment with new foods. Neophilic individuals or cultures are inclined to embrace culinary innovations; investigate a variety of ingredients, flavors, and dishes; and be willing to experiment with unconventional food pairings [34]. This disposition contributes to the variety of culinary practices, the adaptability of the diet, and the incorporation of novel nutritional sources. Of the three attitudes toward new foods, neophobia has the most significant impact on the dietary choices of PKU patients. In this study, a high prevalence of food neophobia was observed in both the PKU patient group and the control group. This conclusion was not surprising given that the majority of the participants in our study were children. Previous research has consistently shown a higher prevalence of neophobic behavior among parents of children in both the PKU group and the control group, compared with adults [3,35]. According to our results, food neophobia does not seem to be associated with BMI, the phenylalanine level at the time of diagnosis, parental education level, or the patient's background. Instead, food neophobia appears to be more influenced by the patient's age and the year of PKU diagnosis. Our study found no association between patient's sex and FNS, consistent with previous research [36]. A study from 2016 carried out in Germany that focused on adolescents found no significant difference in food phobia levels between boys and girls [37]. However, the researchers highlighted that food neophobia levels in relation to sex were influenced by age. Despite achieving optimal metabolic control, the current Phe level, characterized by maintaining phenylalanine levels within the established reference range, did not play a significant role in fostering a higher aversion or reluctance toward experimenting with novel foods. Similar results were revealed in a previous study by Tonon et al. [10]. With the introduction of a modified diet in 1951, newborn screening for PKU was initiated, allowing affected individuals to live normal lives [5]. Despite the difficulty of adherence, phenylalanine restriction in the diet has been the primary treatment for phenylketonuria for over 60 years [7]. Diets low in phenylalanine and novel drug mechanisms [14] keep blood phenylalanine concentrations within the target range. However, the diet's lifetime adherence is unknown, and the organism's protein status may be compromised [9]. Compliance with the restrictive diet is challenging, and abandonment can lead to a decline in academic performance and psychosocial problems [9]. The management of PKU requires adherence to a phenylalanine-restricted diet and routine clinic visits, presenting patients with ongoing challenges and negatively impacting their quality of life [12]. Increasing natural protein intake while maintaining metabolic control may improve outcomes, but the optimal protein intake is unknown [13]. The strict dietary regimen substantially impacts the quality of life, causing a substantial proportion of patients to discontinue treatment. Dietary cessation can worsen symptoms and raise phenylalanine levels [27]. This dietary management approach necessitates significantly reducing high-protein foods, like meat, fish, eggs, cheese, nuts, seeds, and pulses. To compensate for the restricted protein intake, individuals with PKU are supplemented with bitter-tasting Phe-free L-amino acids. Consequently, the diet naturally becomes high in carbohydrates, particularly from plant and cereal sources, to fulfill energy requirements [9]. From an evolutionary standpoint, food neophobia might have conferred a selective advantage by safeguarding against harmful foods in patients with PKU or limits to dietary diversity and causes of malnutrition. Considering the significance of the phenylalanine-free diet in the progression of the disease, adherence to it is crucial for the lifelong treatment of PKU patients. Critical to the progression of the disease is the early detection of the medical condition at birth, continuous monitoring of patients, and providing support to both patients and their families. The neonatal screening system in Romania is based on collecting a single drop of blood 48–72 h after birth. Despite management guidelines for phenylketonuria recommending the initiation of dietary treatment between 7 and 10 days of life, our study indicates that a significant number of patients commence treatment at a later stage. In 2022, the number of children with PKU monitored was 342. Patients with phenylketonuria in Romania who are enrolled in the National Program for Women and Children have access to the following foods: protein substitutes, infant formulas without phenylalanine, and hypo protein dietary products with reduced phenylalanine content (such as flour, cereals, pasta, biscuits, semolina, egg replacer, and cooking mixes) are available for those with phenylketonuria. Foods intended for special medical use in the case of phenylketonuria are administered under strict medical supervision and are tailored to individual cases, tolerance, weight, age, and body size. The neonatal screening situation in Romania can be compared with that of other European countries and the rest of the world in the EAEC regarding coverage, the number of conditions screened for, and screening methods. In France, for instance, an extended test is administered for approximately 50 conditions, including congenital diseases of metabolism, endocrine disorders, and hematological disease [4]. In the United Kingdom, the national neonatal screening program searches for a variety of conditions, including congenital hypothyroidism, phenylketonuria, and cystic fibrosis.

Analyzing the interval between birth and diagnosis, we observed that one of the subjects received a diagnosis at the age of 10 years. For statistical interpretation, this value with an abnormal distribution was excluded. The mean interval until the diagnosis was 21 days, whereas guidelines recommend an interval of up to 10 days. Our analysis aimed to explore whether the age of patients with PKU can indicate a tendency toward food neophobia. Establishing a diet low in phenylalanine from birth, diversifying appropriately to prevent micronutrient deficiencies, and maintaining a regimen meeting the individual's micronutrient requirements throughout life are all necessary. From this perspective, an early diagnosis and the implementation of an appropriate diet to reduce the risk of metabolic imbalance are essential. However, another nutritional condition, pickiness, exists in patients younger than 5 years old [38]. High scores on the FNS suggest a reduced anticipation of liking unfamiliar foods, limited familiarity with foreign cuisines [1], and a decreased willingness to try new and unfamiliar foods. Pliner described how differences in measured food neophobia did not significantly correlate with actual hedonic responses when participants tasted unfamiliar foods. However, other studies have indicated that individuals with higher levels of food neophobia tended to express less liking for unfamiliar foods compared with those with lower neophobia [5,6]. It has been observed that early exposure to a particular food significantly enhances the willingness to eat that food again, regardless of the individual's level of food neophobia [4]. This suggests that food neophobia primarily influences responses to unfamiliar foods rather than familiar ones [1,6]. It is evident that food neophobia plays a role in shaping the initial tasting experience of unfamiliar foods. However, the decision to continue consuming such foods is influenced by a combination of other factors as well [6].

Nutritional support is required for the duration of a person's life; however, support is also required to perceive treatment as a positive factor, particularly during adolescence [39]. Lifelong maintenance of blood phenylalanine concentrations within the therapeutic range is a challenging objective. It is not uncommon for adults with PKU to discontinue their dietary management [40]. During adolescence and adulthood, when many individuals with PKU find the restricted diet disagreeable, difficult to follow, and a hindrance to social relationships [41], therapeutic phenylalanine levels may be difficult to achieve. Previous research has demonstrated that a restrictive diet does not hinder physical development [42]. All children who switch from specific infant feeding to foods consumed by family members might have difficulties embracing new foods, not just PKU patients, as a potential natural developmental stage [17].

Despite recommendations to maintain blood phenylalanine concentrations in the therapeutic range throughout life, it is not uncommon for adults with PKU to discontinue dietary management of their disorder. An early diagnosis was associated with a reduced need for special education or other special services, and continuous treatment was associated with decreased psychological co-morbidities. In our study, preschooler-age children's average food neophobia score was 35.3, which was higher than the results obtained in another study, 23.73 ± 4.45 (25). In the same study, parental modeling (β: −0.470; 95%CI: −0.732, −0.207) and the frequency of children eating with their families at home (β: −0.407; 95%CI: −0.707, −0.108) were negatively associated with children's food neophobia scores [43]. A previous study showed there was a statistically significant correlation between the neophobia that children develop and the neophobic characteristics of their parents' dietary behavior [44]. Pliner found that variations in quantified food neophobia did not correlate significantly with actual hedonic responses when participants sampled unfamiliar foods [1]. In contrast, other studies [45,46] have suggested that individuals with elevated levels of food neophobia tend to have a lower preference for unfamiliar cuisines than those with lower levels of neophobia. This observation suggests that psychological and sensory factors interact intricately to determine food preferences [19]. In addition, research has shown that early exposure to specific foods increases a person's willingness to ingest them in the future, regardless of their initial level of food aversion [47]. This indicates that food neophobia affects responses to unfamiliar cuisines more than familiar ones [1,6]. For all the participants, the FNS score was calculated by adding the average score for each statement belonging to the FNS. The average FNS score was 30.35 on the PKUG and 44.09 in CG in this study and was compared with the FNS scores from other countries: 38.5 for

Indonesia, 37.4 for the Philippines, 39.3 for Malaysia, 34.1 for Vietnam, 26.1 for the UK, and 34.7 for Australia [48]. It must be mentioned that our subjects are primarily children, which could explain the high numbers on the FNS. The incidence of obesity in our study group was comparable to that in the general population, as demonstrated by a previous meta-analysis that found no evidence to support the notion that a Phe-restricted diet is a risk factor for overweight or obesity [4]. One longitudinal study found that food neophobia at 1 year was positively associated with the later introduction of dairy products, the use of ready-prepared baby foods, and the use of ready-prepared adult foods [49]. In our study, diet adherence did not affect the FNS, which is similar to age, BMI, and sex. Identical results were discovered by other researchers as well [50–52].

To ascertain whether the genetic phenotype had an impact on the FNS score, we analyzed the available genetic mutations in our patients. The majority of patients exhibited classic phenotypes of PKU, marked by the presence of either heterozygous or homozygous mutations within the *PAH* gene. Consistent with findings in the literature [53,54], the p.Arg408Trp mutation emerged as the most prevalent among these cases, highlighting its role in the pathogenesis of PKU. Our study also revealed a subset of individuals presenting with moderate or mild phenotypes of the disorder, but their number was too small to be able to make assumptions regarding the influence of the genetic phenotype on the FNS score.

When considering each question in the form, regarding Q1, the control group had a more significant percentage of people who "strongly agree" and "agree" to try new foods (79.41%) than the PKU group (58.82%). This variation might occur because individuals in the control group have fewer dietary limitations than those in the PKU group. Being on a specialized diet, the latter group must exercise greater caution in their food choices, contributing to this distinction. Analyzing respondents who would "strongly disagree" and "disagree" with trying new foods, the percentage is 2.94% in the control group and 17.65% in the PKU group. This reinforces the reluctance of PKU patients to try new foods. Previous investigations in the PKU group found that neophobia was a significant factor in food refusal [35]. Regarding the PKU group, however, many respondents attempt new foods less frequently or even avoid them. This is well known in the literature even from 2007, when researchers [9] pointed to food neophobia as an evolutionary safeguard to prevent humans from consuming potentially hazardous food. Therefore, it is not unexpected that within our study, individuals with PKU exhibited neophobic tendencies [55]. Variable responses to continually trying new foods may indicate divergent approaches to accepting and incorporating food diversity into a limited diet among PKU patients. It is essential to recognize that strict diets and the need to monitor phenylalanine levels can affect the willingness to try novel foods.

For the second question, the PKU group showed a higher proportion (35.29%) of individuals who expressed a neutral stance (neither disagree nor agree) toward trying new foods compared with the control group (20.59%). This variance may be attributed to the dietary constraints imposed on individuals with phenylketonuria (PKU), necessitating adherence to a stringent diet. As a result, their curiosity toward unfamiliar foods must be regulated to prevent health complications. Many respondents lack confidence in novel foods, indicating a lack of faith in their quality or safety. The propensity of PKU patients to rely on foods with known and monitored phenylalanine content may be reflected by a lack of confidence in ingesting novel foods. This reaction can be interpreted as a preventative measure to avoid accidentally consuming foods that could alter the nutritional balance specific to PKU.

The divergence in responses for Q3 between the control group and the PKU group regarding trying food without knowledge of its ingredients highlights distinct perspectives shaped by different priorities. Within the control group, a significant 50% expressed strong disagreement or disagreement with the notion, suggesting a prevalent willingness to explore food even without comprehensive ingredient knowledge. This openness reflects a curiosity and willingness to embrace new culinary experiences. Conversely, in the PKU group, where strict dietary control is crucial due to health concerns, approximately 32.35%

disagreed with trying food without prior knowledge of its ingredients. This inclination aligns with the necessity for careful dietary management among individuals with PKU. These contrasting percentages underscore the impact of health considerations and diverse attitudes on food choices. While a considerable portion of the general population remains open to culinary exploration despite limited ingredient awareness, the PKU group's response emphasizes the importance of health-related caution in food consumption. The majority of participants appear to avoid foods whose ingredients are unknown. This may be indicative of a circumspect approach to unfamiliar ingredients. Observing that unfamiliar foods should be avoided has significant implications for PKU patients. This behavior may be determined by the need to know the exact amount of phenylalanine in the food ingested to avoid excessive consumption.

In the study's Question 4 responses, it becomes evident that a more significant percentage of individuals in the control group (55.88%) agreed with trying food from other countries compared with the PKU group (20.59%). This disparity can be attributed to the absence of strict dietary restrictions within the control group, allowing them the freedom to explore and sample a more comprehensive array of foods from various countries. The majority of participants did not appear to be interested in foreign cuisine. However, a significant number of participants were either less or more willing to try foods from diverse cultures.

In Question 5, a higher percentage of individuals in the control group (44.12%) expressed disagreement and strong disagreement toward trying foreign food due to its perceived extreme dissimilarity, compared with the PKU group (38.24%), which also demonstrated a noteworthy percentage. This variance might be clarified by the tendency of individuals in the control group to emphasize the appearance aspect of a new food. Patients with phenylketonuria may have a range of preferences and comfort levels regarding culinary exploration, as indicated by this diversity of attitudes. The interest in cuisines from other countries suggests that PKU patients may be positioned favorably toward opportunities for dietary diversity. Due to their familiarity with dietary restrictions, their appreciation of cultural diversity may reflect a desire to investigate alternative and safe sources of nutrients. A substantial proportion of participants were unwilling to sample ethnic foods, possibly because they viewed them as "strange" or "unusual". Patients with phenylketonuria are conditioned to avoid novel and unusual foods based on their phenylalanine content, which may contribute to their negative perceptions of ethnic food. This can result in an unwillingness to experiment with foreign flavors and ingredients.

When discussing the exploration of new foods (Q6), particularly at parties, the contrast between the PKU group and the control group was striking. In the control group, a considerable 67.65% expressed agreement or strong agreement with the idea of trying new foods. This suggests a prevalent openness and willingness to indulge in culinary experiences. Conversely, within the PKU group, only 23.53% shared this sentiment, indicating a substantial divergence in attitude. This discrepancy underscores the necessity for individuals with PKU to be exceptionally cautious and vigilant when navigating unfamiliar foods. The stringent dietary restrictions imposed by PKU make it imperative for these individuals to exercise careful consideration and scrutiny regarding the ingredients before trying new dishes, particularly in social settings like parties. These percentages underscore the distinct challenges faced by the PKU group, emphasizing the need for heightened caution and conscientiousness due to the health-related dietary constraints imposed by their condition. This stark response highlights the unique considerations shaping the relationship between individuals' health conditions and their approach to trying new foods, especially within social contexts. It appeared that participants were more willing to sample new foods at dinner parties, indicating a greater social willingness to experience food. The ability to overcome food neophobia is significantly influenced by social factors, as evidenced by the willingness to sample new foods at dinner parties. The social context can facilitate the development of a framework of approval and encouragement for culinary experimentation.

In the seventh question, noteworthy percentages in the control group were 29.41% expressing neutrality, while 23.53% provided a neutral response in the PKU group. This contrast might be elucidated by the comparative absence of stringent phenylalanine intake regulations in the control group, affording them greater assurance when encountering unfamiliar studies. Significant numbers of participants and their children feared eating unfamiliar foods, indicating a certain degree of apprehension in the face of novel culinary experiences. PKU patients' constant awareness of the potential impact on their phenylalanine levels may contribute to their irrational aversion to foreign cuisines. This demonstrates a delicate balance between the desire to try novel foods and the need to maintain strict dietary control.

In assessing children's food selectivity, both the control and PKU groups showed a 50% similarity, indicating comparable discernment levels. However, this alignment is not solely due to PKU-related dietary restrictions. In the control group, selectivity might be due to taste preferences or habits, not just health concerns. Differences arose in the middle ground: 26.47% in the control group vs. 14.71% in PKU. Interestingly, while only 2.94% in the control group somewhat agreed, 17.65% in the PKU group leaned toward agreement. This suggests a wider openness among PKU individuals to potential food choices compared with the control group, emphasizing varied influences on their preferences. Participants' food preferences appeared to span a broad spectrum, from moderate selectivity to greater openness. The stringent need to adhere to dietary restrictions can justify demanding food choices. Patients with PKU are aware of the significance of nutritional balance and limiting phenylalanine intake so they can make more prudent food selections.

The responses for Q9 to the statement "My kid eats almost everything" revealed distinct attitudes toward children's eating habits between the control group and the PKU group. In the control group, a significant 61.76% strongly agreed or agreed with this statement, indicating a prevailing belief that children within this group had versatile eating habits and were open to a wide variety of foods. This suggests a general perception of less selectivity in dietary preferences among children in this cohort. Conversely, within the PKU group, a notably lower 23.53% strongly agreed or agreed with the statement. This stark contrast likely stems from the stringent dietary restrictions imposed by PKU, leading to a more selective approach to food choices among individuals managing this condition. The higher percentage of disagreement in the PKU group (64.71%) further underscores the impact of these dietary limitations on their perception of a child's eating habits. These considerable percentage differences between the two groups highlight the divergent attitudes toward children's food acceptance. The control group's higher agreement suggests a broader dietary openness among children. In comparison, the lower agreement in the PKU group underscores the influence of health-related dietary constraints on a child's food choices and the heightened selectivity necessitated by the condition. Most respondents indicated they were willing to consume almost any variety of food, indicating that their palates were adaptable. The willingness to attempt new foods can be interpreted as a desire to consume all safe and permissible PKU diet foods. Within the limitations imposed by their medical condition, this approach reflects a favorable attitude toward nutritional diversity.

The marked contrast in responses between the control and PKU groups regarding their child's preference for food at specific restaurants (Q10) showed significant divergence in perceptions and preferences. Within the control group, a substantial 64.71% expressed agreement (agree or strongly agree) with the statement, indicating a prevalent belief that children in this cohort were inclined to enjoy food at ethnic restaurants. This suggested an openness to cultural culinary experiences and a positive attitude toward diverse cuisines. On the contrary, in the PKU group, only 17.65% agreed with the statement. This substantial difference of 47.06 percentage points indicated a markedly lower inclination toward their child's preference for specific restaurant food. The lower agreement percentage within the PKU group aligns with the dietary limitations associated with managing PKU, possibly resulting in a more cautious approach to diverse cuisines and ethnic restaurant foods. Additionally, the contrast in disagreement percentages is noteworthy. While 14.71% in the control group disagreed with the statement, a higher percentage, 50%, within the PKU group expressed disagreement. This significant discrepancy of 35.29 percentage points highlights a more pronounced opposition to their child preferring food at ethnic restaurants among the PKU respondents. These striking disparities underscore the profound impact of health-related dietary constraints on the perceived preferences of children in different culinary settings. The higher inclination toward ethnic restaurant food in the control group compared with the PKU group emphasizes the need to consider health conditions and their associated dietary limitations when evaluating preferences for diverse cuisines. The attendees' willingness to try new specific restaurants demonstrates their appreciation for international culinary experiences. In the context of the restrictive PKU diet, dining at ethnic restaurants could be considered a pursuit for alternative dining experiences. In this context, specific restaurants can provide patients with the opportunity to experience a variety of flavors and ingredients in a controlled environment. Regarding the impact of food neophobia in patients with phenylketonuria, we can see how the constant focus on phenylalanine control can contribute to the tendency to avoid unfamiliar foods. This may provide a level of nutritional security, but it may limit opportunities to experience variety and culinary innovation. It is essential to investigate these relationships further and determine how to balance specific dietary requirements with the desire to diversify diets and try new foods safely.

Building upon our FNS results, which indicated that the primary caretaker was the mother in 91% of cases and that 47.1% of mothers had a university education level, our study highlights the significance of family dynamics and the educational level of caregivers in managing neophobia among patients with PKU. To address these factors, we propose targeted interventions. First, personalized guidance and resources should be provided to empower mothers, particularly those with higher education levels, to actively engage in dietary management strategies. Leveraging their knowledge and skills, mothers can play a pivotal role in optimizing dietary adherence and overcoming neophobic behaviors in their child with PKU. Furthermore, we advocate for the encouragement of active participation from both parents, with a focus on the primary caretaker (typically the mother), in meal planning, preparation, and food exploration activities. This collaborative approach within the family unit can foster a supportive environment and facilitate the successful implementation of dietary strategies to mitigate neophobia and promote overall well-being in patients with PKU.

As a notable observation, the FNS scores were higher in the control group compared with PKU patients. This discrepancy could be attributed to the study's timing during the COVID-19 pandemic, leading to increased home-based eating habits among children. Additionally, concerns about accidentally consuming allergens in foods may have contributed to heightened fears among parents or children in the control group. Unfortunately, comprehensive information about the control group is limited to the details provided in the questionnaires.

The results have to be viewed in light of certain study limitations. First, the control group included children with different pathologies. Adult patients were enrolled from the nutrition clinic, indicating individuals in need of dietary habit changes. Another limitation of the study may be the diversity of ages accepted in the research since food neophobia at young ages can be confused with typical food preferences or whims. Additionally, our study is constrained by the limited availability of protein substitutes in Romania and the lack of data regarding the incorporation of glycomacropeptides into the diet within the Romanian National Phenylketonuria Program, as these were introduced in 2022. Specifically, for each age group, there are at most two options with similar characteristics, with a maximum difference of 5 g protein equivalent per 100 g product. Our study is also limited by the small number of participants and the short time frame during which they were observed. It is plausible for FN values to fluctuate with age. Therefore, multicenter prospective studies are warranted to enhance the generalizability of our findings and contribute to a more comprehensive understanding of the relationship between age and food neophobia in this population.

5. Conclusions

Our study revealed a prevalent occurrence of neophobia both among patients diagnosed with PKU and the control group. Among PKU patients, neophobia exhibited correlations with various factors, including the patient's age, the time lapse between birth and disease diagnosis, and the educational attainment of the parents. These findings underscore the multifaceted nature of neophobia in individuals with PKU, shedding light on potential factors influencing its prevalence and manifestation. Also, by correlating the educational level of the parent or carer with neophobia incidence, we believe that offering clear and strong information to parents at the time of diagnosis and providing nutritional guidance can decrease the occurrence of this eating behavior disorder. The study also concluded that diagnosing as soon as feasible after birth and implementing a suitable diet are factors that are correlated with reducing the occurrence of neophobia.

Supplementary Materials: The following supporting information can be downloaded at https://www.mdpi.com/article/10.3390/nu16060768/s1: Table S1: Adapted version of the Food Neophobia Scale [7].

Author Contributions: Conceptualization, M.-A.B., D.C. and O.M.; methodology, M.-A.B., D.C. and I.-E.S.; software, D.C. and D.-M.N.; validation, M.-A.B. and D.C.; formal analysis, M.-A.B. and R.I. investigation, M.-A.B., D.C. and I.J.; resources, M.-A.B., G.-F.B. and N.M.; data curation M.-A.B., D.C., A.-I.M. and D.-M.N.; writing—original draft preparation, M.-A.B., D.-M.N. and D.C.; writing—review and editing, M.-A.B., D.C. and O.M.; visualization, M.-A.B., D.-M.N., D.C., I.-E.S. and O.M.; supervision, M.-A.B., D.C. and O.M. All authors have read and agreed to the published version of the manuscript.

Funding: This research received no external funding.

Institutional Review Board Statement: The study was conducted in accordance with the Declaration of Helsinki. The research protocol received approval from the Ethics Committee of the "Victor Babes" University of Medicine and Pharmacy (No.60/12 November 2018) and from the hospital's Ethics Committee (No.3392/24 February 2023).

Informed Consent Statement: Informed consent was obtained from all subjects involved in the study.

Data Availability Statement: The data presented in this study are available on request from the corresponding author. The data are not publicly available due to ethical restrictions.

Conflicts of Interest: The authors declare no conflicts of interest.

References

1. Zurflüh, M.R.; Zschocke, J.; Lindner, M.; Feillet, F.; Chery, C.; Burlina, A.; Stevens, R.C.; Thöny, B.; Blau, N. Molecular genetics of tetrahydrobiopterin-responsive phenylalanine hydroxylase deficiency. *Hum. Mutat.* **2008**, *29*, 167–175. [CrossRef] [PubMed]
2. Al Hafid, N.; Christodoulou, J. Phenylketonuria: A review of current and future treatments. *Transl. Pediatr.* **2015**, *4*, 304–317. [CrossRef]
3. Evans, S.; Daly, A.; Chahal, S.; Ashmore, C.; MacDonald, J.; MacDonald, A. The influence of parental food preference and neophobia on children with phenylketonuria (PKU). *Mol. Genet. Metab.* **2018**, *14*, 10–14. [CrossRef] [PubMed]
4. Rodrigues, C.; Pinto, A.; Faria, A.; Teixeira, D.; van Wegberg, A.M.J.; Ahring, K.; Feillet, F.; Calhau, C.; MacDonald, A.; Moreira-Rosário, A.; et al. Is the Phenylalanine-Restricted Diet a Risk Factor for Overweight or Obesity in Patients with Phenylketonuria (PKU)? A Systematic Review and Meta-Analysis. *Nutrients* **2021**, *13*, 3443. [CrossRef] [PubMed]
5. Shoraka, H.R.; Haghdoost, A.A.; Baneshi, M.R.; Bagherinezhad, Z.; Zolala, F. Global prevalence of classic phenylketonuria based on Neonatal Screening Program Data: Systematic Review and Meta-Analysis. *Clin. Exp. Pediatr.* **2020**, *63*, 34–43. [CrossRef]
6. Raport Privind Procedurile Utilizate, Indicatorii de Performanță, Standarde, și Modele de Raportare Periodică in Cadrul Institutului Național de Endocrinologie C.I. Parhon, Ministerului Sanătății (Programul Național de Tratament Dietetic pentru Boli Rare) și Institutului Național de Sănătate Publică (Programul Național de Management Al Registrelor Naționale): Modulul 9. Available online: https://acad.ro/proiecteFSE/OPERE_SCRISE-SIPOCA13/3.1/3.1_Modulul_9.pdf (accessed on 17 September 2023).
7. Pliner, P.; Hobden, K. Development of a scale to measure the trait of food neophobia in humans. *Appetite* **1992**, *19*, 105–120. [CrossRef]
8. Galloway, A.T.; Lee, Y.; Birch, L.L. Predictors and consequences of food neophobia and pickiness in young girls. *J. Am. Diet. Assoc.* **2003**, *103*, 692–698. [CrossRef]
9. Knaapila, A.; Tuorila, H.; Silventoinen, K.; Keskitalo, K.; Kallela, M.; Wessman, M.; Peltonen, L.; Cherkas, L.F.; Spector, T.D.; Perola, M. Food neophobia shows heritable variation in humans. *Physiol. Behav.* **2007**, *91*, 573–578. [CrossRef]

10. Tonon, T.; Martinez, C.; Poloni, S.; Nalin, T.; MacDonald, A.; Schwartz, I.V.D. Food neophobia in patients with phenylketonuria. *J. Endocrinol. Metab.* **2019**, *9*, 108–112. [CrossRef]
11. MacDonald, A.; Rylance, G.; Asplin, D.; Booth, I.W. Abnormal feeding behaviors in phenylketonuria. *J. Hum. Nutr. Diet.* **1997**, *10*, 163–170. [CrossRef]
12. Menella, J.A.; Kennedy, J.M.; Beauchamp, G.K. Vegetable acceptance by infants: Effects of formula flavours. *Early Hum. Dev.* **2006**, *82*, 463–468. [CrossRef]
13. Beauchamp, G.K.; Mennella, J.A. Flavor perception in human infants: Development and functional significance. *Digestion* **2011**, *83* (Suppl. S1), 1–6. [CrossRef]
14. Mennella, J.A.; Beauchamp, G.K. Flavour experiences during formula feeding are related to preferences during childhood. *Early Hum. Dev.* **2002**, *68*, 71–82. [CrossRef]
15. Beauchamp, G.K.; Mennella, J.A. Early flavor learning and its impact on later feeding behavior. *J. Pediatr. Gastroenterol. Nutr.* **2009**, *48* (Suppl. S1), 25–30. [CrossRef]
16. Ramalho, C.; Sampaio, M.; Rocha, N.; Poínhos, R. Food Neophobia among Primary School Children and Their Caregivers. *Acta Port. Nutr.* **2016**, *7*, 10–13.
17. Białek-Dratwa, A.; Szczepańska, E.; Szymańska, D.; Grajek, M.; Krupa-Kotara, K.; Kowalski, O. Neophobia—A Natural Developmental Stage or Feeding Difficulties for Children? *Nutrients* **2022**, *14*, 1521. [CrossRef] [PubMed]
18. Łoboś, P.; Januszewicz, A. Food neophobia in children. *Pediatr. Endocrinol. Diabetes Metab.* **2019**, *25*, 150–154. [CrossRef] [PubMed]
19. Gedrich, K. Determinants of nutritional behavior: A multitude of levers for successful intervention? *Appetite* **2003**, *41*, 231–238. [CrossRef] [PubMed]
20. Green, A. The First Treatment for PKU: The Pioneers-Birmingham 1951. *Int. J. Neonatal Screen.* **2021**, *7*, 19. [CrossRef] [PubMed]
21. World Health Organization. *Child Growth Standards*; World Health Organization: Geneva, Switzerland, 2006; ISBN 924154693X.
22. Canadian Paediatric Society. A health professional's guide for using the new WHO growth charts. *Paediatr. Child Health* **2010**, *15*, 84–98. [CrossRef]
23. Barrena, R.; Sánchez, M. Neophobia, personal consumer values and novel food acceptance. *Food Qual. Prefer.* **2013**, *27*, 72–84. [CrossRef]
24. Taherdoost, H. What Is the Best Response Scale for Survey and Questionnaire Design; Review of Different Lengths of Rating Scale/Attitude Scale/Likert Scale. *Int. J. Acad. Res. Manag.* **2019**, *8*, ffhal-02557308f.
25. D'Antuono, L.F.; Bignami, C. Perception of typical Ukrainian foods among an Italian population. *Food Qual. Prefer.* **2012**, *25*, 1–8. [CrossRef]
26. Choe, J.Y.; Cho, M.S. Food neophobia and willingness to try non-traditional foods for Koreans. *Food Qual. Prefer.* **2011**, *22*, 671–677. [CrossRef]
27. Olabi, A.; Najm, N.E.O.; Baghdadi, O.K.; Morton, J.M. Food neophobia levels of Lebanese and American college students. *Food Qual. Prefer.* **2009**, *20*, 353–362. [CrossRef]
28. Ribeiro de Andrade Previato, H.D.; Behrens, J.H. Translation and validation of the Food Neophobia Scale (FNS) to the Brazilian Portuguese. *Nutr. Hosp.* **2015**, *32*, 925–930. [CrossRef]
29. JASP Team. *JASP, version 0.18.1*; JASP Team: Amsterdam, The Netherlands, 2023.
30. Richards, S.; Aziz, N.; Bale, S.; Bick, D.; Das, S.; Gastier-Foster, J.; Grody, W.W.; Hegde, M.; Lyon, E.; Spector, E.; et al. Standards and guidelines for the interpretation of sequence variants: A joint consensus recommendation of the American College of Medical Genetics and Genomics and the Association for Molecular Pathology. *Genet. Med.* **2015**, *17*, 405–424. [CrossRef]
31. Pliner, P.; Salvy, S.-J. Food neophobia in humans. In *The Psychology of Food Choice*; Shepherd, R., Raats, M., Eds.; CABI Publishing: Oxfordshire, UK, 2006; pp. 75–92.
32. Kauer, J.; Rozin, P.; Pelchat, M.L. Adulty picky eating: Is there more to it. *Soc. Study Ingestion Behav.* **2002**.
33. Smith, A.M.; Roux, S.; Naidoo, N.R.; Venter, D.J. Food choices of tactile defensive children. *Nutrition* **2005**, *21*, 14–19. [CrossRef]
34. MacDonald, A.; Rylance, G.W.; Asplin, D.A.; Hall, K.; Harris, G.; Booth, I.W. Feeding problems in young PKU children. *Acta Paediatr.* **1994**, *407*, 73–74. [CrossRef]
35. Evans, S.; Daly, A.; Chahal, S.; MacDonald, J.; MacDonald, A. Food acceptance and neophobia in children with phenylketonuria: A prospective controlled study. *J. Hum. Nutr. Diet.* **2016**, *29*, 427–433. [CrossRef] [PubMed]
36. Cole, N.C.; An, R.; Lee, S.Y.; Donovan, S.M. Correlates of picky eating and food neophobia in young children: A systematic review and meta-analysis. *Nutr. Rev.* **2017**, *75*, 516–532. [CrossRef] [PubMed]
37. Rossbach, S.; Foterek, K.; Schmidt, I.; Hilbig, A.; Alexy, U. Food neophobia in German adolescents: Determinants and association with dietary habits. *Appetite* **2016**, *101*, 184–191. [CrossRef] [PubMed]
38. Addessi, E.; Galloway, T.A.; Visalberghi, E.; Birch, L.L. Specific social influences on the acceptance of novel foods in 2–5-year-old children. *Appetite* **2005**, *45*, 264–271. [CrossRef] [PubMed]
39. Levy, H.L.; Waisbren, S.E. PKU in adolescents: Rationale and psychosocial factors in diet continuation. *Acta Paediatr. Suppl.* **1994**, *407*, 92–97. [CrossRef]
40. Kenneson, A.; Singh, R.H. Natural history of children and adults with phenylketonuria in the NBS-PKU Connect registry. *Mol Genet. Metab.* **2021**, *134*, 243–249. [CrossRef] [PubMed]
41. Rose, A.M.; Grosse, S.D.; Garcia, S.P.; Bach, J.; Kleyn, M.; Simon, N.J.E.; Prosser, L.A. The financial and time burden associated with phenylketonuria treatment in the United States. *Mol. Genet. Metab.* **2019**, *21*, 100523. [CrossRef] [PubMed]

42. Matic, J.; Zeltner, N.A.; Häberle, J. Normal Growth in PKU Patients Under Low-Protein Diet in a Single-Center Cross-Sectional Study. *JIMD Rep.* **2019**, *43*, 1–6. [CrossRef]
43. Yong, C.; Kuang, X.; Liu, Y.; Xiang, C.; Xi, Y.; Huo, J.; Liang, J.; Zou, H.; Lin, Q. Parental food neophobia, feeding practices, and preschooler's food neophobia: A cross-sectional study in China. *Appetite* **2023**, *185*, 106547. [CrossRef]
44. Pliner, P. Development of measures of food neophobia in children. *Appetite* **1994**, *23*, 147–163. [CrossRef]
45. Tuorila, H.; Meiselman, H.L.; Bell, R.; Cardello, A.V.; Johnson, W. Role of sensory and cognitive information in the enhancement of certainty and liking for novel and familiar foods. *Appetite* **1994**, *23*, 231–246. [CrossRef]
46. Pliner, P.; Lähteenmäki, L.; Tuorila, H. Correlates of human food neophobia. *Appetite* **1998**, *30*, 93. [CrossRef]
47. Tuorila, H.; Lähteenmäki, L.; Pohjalainen, L.; Lotti, L. Food neofobia among the Finns and related responses to familiar and unfamiliar foods. *Food Qual. Prefer.* **2001**, *12*, 29–37. [CrossRef]
48. Kim, S.-J.; Park, H.-J.; Lee, K.-H. Comparison of Food Neophobia Scale and Food Involvement Scale Between Koreans and East-South Asians. *J. Korean Soc. Food Cult.* **2011**, *26*, 429–436. [CrossRef]
49. Yuan, W.L.; Rigal, N.; Monnery-Patris, S.; Chabanet, C.; Forhan, A.; Charles, M.A.; de Lauzon-Guillain, B. Early determinants of food liking among 5y-old children: A longitudinal study from the EDEN mother–child cohort. *Int. J. Behav. Nutr. Phys. Act.* **2016**, *13*, 20. [CrossRef] [PubMed]
50. Kozioł-Kozakowska, A.; Piórecka, B.; Schlegel-Zawadzka, M. Prevalence of food neophobia in pre-school children from southern Poland and its association with eating habits, dietary intake and anthropometric parameters: A cross-sectional study. *Public Health Nutr.* **2018**, *21*, 1106–1114. [CrossRef] [PubMed]
51. Maratos, F.A.; Staples, P. Attentional biases towards familiar and unfamiliar foods in children. The role of food neophobia. *Appetite* **2015**, *91*, 220–225. [CrossRef] [PubMed]
52. Faith, M.S.; Heo, M.; Keller, K.L.; Pietrobelli, A. Child food neophobia is heritable, associated with less compliant eating, and moderates familial resemblance for BMI. *Obesity* **2013**, *21*, 1650–1655. [CrossRef] [PubMed]
53. van Wegberg, A.M.J.; MacDonald, A.; Ahring, K.; Bélanger-Quintana, A.; Blau, N.; Bosch, A.M.; Burlina, A.; Campistol, J.; Feillet, F.; Giżewska, M.; et al. The complete European guidelines on phenylketonuria: Diagnosis and treatment. *Orphanet J. Rare Dis.* **2017**, *12*, 162. [CrossRef] [PubMed]
54. Brooks, D.L.; Whittaker, M.N.; Said, H.; Dwivedi, G.; Qu, P.; Musunuru, K.; Ahrens-Nicklas, R.C.; Alameh, M.G.; Wang, X. A base editing strategy using mRNA-LNPs for in vivo correction of the most frequent phenylketonuria variant. *HGG Adv.* **2024**, *5*, 100253. [CrossRef]
55. Falciglia, G.A.; Couch, S.C.; Gribble, L.S.; Pabst, S.M.; Frank, R. Food neophobia in childhood affects dietary variety. *J. Am. Diet. Assoc.* **2000**, *100*, 1474–1481. [CrossRef] [PubMed]

nutrients

MDPI

Article

Metabolic Dysfunction-Associated Steatotic Liver Disease in a Dish: Human Precision-Cut Liver Slices as a Platform for Drug Screening and Interventions

Mei Li [1], Frederik T. Larsen [2], Marius C. van den Heuvel [3], Konstanze Gier [1], Alan R. Gorter [1], Dorenda Oosterhuis [1], Johan Bijzet [4,5], Vincent E. de Meijer [6], Kim Ravnskjaer [2], Anika Nagelkerke [7,*] and Peter Olinga [1,*]

1 Department of Pharmaceutical Technology and Biopharmacy, Groningen Research Institute of Pharmacy, University of Groningen, 9713 AV Groningen, The Netherlands; mei.li@rug.nl (M.L.); k.gier@rug.nl (K.G.); a.r.gorter@rug.nl (A.R.G.); d.oosterhuis@rug.nl (D.O.)
2 Department of Biochemistry and Molecular Biology, University of Southern Denmark, 5230 Odense M, Denmark; tibert@bmb.sdu.dk (F.T.L.); ravnskjaer@bmb.sdu.dk (K.R.)
3 Department of Pathology and Medical Biology, Pathology Section, University Medical Center Groningen, 9700 RB Groningen, The Netherlands; m.c.van.den.heuvel@umcg.nl
4 Amyloidosis Center of Expertise, University Medical Center Groningen, 9713 GZ Groningen, The Netherlands; j.bijzet@umcg.nl
5 Department of Laboratory Medicine, Division of Medical Immunology, University Medical Center Groningen, 9713 GZ Groningen, The Netherlands
6 Department of Surgery, Division of Hepato-Pancreato-Biliary Surgery and Liver Transplantation, University Medical Center Groningen, 9713 AV Groningen, The Netherlands; v.e.de.meijer@umcg.nl
7 Department of Pharmaceutical Analysis, Groningen Research Institute of Pharmacy, University of Groningen, 9713 AV Groningen, The Netherlands
* Correspondence: a.p.nagelkerke@rug.nl (A.N.); p.olinga@rug.nl (P.O.)

Abstract: Metabolic dysfunction-associated steatotic liver disease (MASLD) is a growing healthcare problem with limited therapeutic options. Progress in this field depends on the availability of reliable preclinical models. Human precision-cut liver slices (PCLSs) have been employed to replicate the initiation of MASLD, but a comprehensive investigation into MASLD progression is still missing. This study aimed to extend the current incubation time of human PCLSs to examine different stages in MASLD. Healthy human PCLSs were cultured for up to 96 h in a medium enriched with high sugar, high insulin, and high fatty acids to induce MASLD. PCLSs displayed hepatic steatosis, characterized by accumulated intracellular fat. The development of hepatic steatosis appeared to involve a time-dependent impact on lipid metabolism, with an initial increase in fatty acid uptake and storage, and a subsequent down-regulation of lipid oxidation and secretion. PCLSs also demonstrated liver inflammation, including increased pro-inflammatory gene expression and cytokine production. Additionally, liver fibrosis was also observed through the elevated production of pro-collagen 1a1 and tissue inhibitor of metalloproteinase-1 (TIMP1). RNA sequencing showed that the tumor necrosis factor alpha (TNFα) signaling pathway and transforming growth factor beta (TGFβ) signaling pathway were consistently activated, potentially contributing to the development of inflammation and fibrosis. In conclusion, the prolonged incubation of human PCLSs can establish a robust ex vivo model for MASLD, facilitating the identification and evaluation of potential therapeutic interventions.

Keywords: precision-cut liver slices (PCLSs); long-term incubation; metabolic dysfunction-associated steatotic liver disease (MASLD); non-alcoholic fatty liver disease (NAFLD); hepatic steatosis; liver fibrosis

1. Introduction

Metabolic dysfunction-associated steatotic liver disease (MASLD), previously termed non-alcoholic fatty liver disease (NAFLD), is a chronic and progressively developing condition impacting approximately 38% of the global population [1]. It is closely linked to metabolic syndromes, such as obesity and type 2 diabetes mellitus. MASLD results from fat accumulation within the liver and encompasses a spectrum of disease stages, ranging from an isolated lipid buildup or steatosis (metabolic dysfunction-associated steatotic liver, MASL), to its more active inflammatory manifestation, metabolic dysfunction-associated steatohepatitis (MASH), formerly known as nonalcoholic steatohepatitis (NASH). The persistence of MASH can ultimately lead to the development of fibrosis, cirrhosis, and even liver cancer. Although several drugs are in advanced stages of development, there is currently no approved cure for MASLD [2]. Long-term lifestyle modification with a focus on healthy diet, weight loss, and regular exercise remain the cornerstone of therapy in children and adults [3], yet willpower and persistence are required. So far, a major barrier to the development of therapies for MASLD is the lack of preclinical models of disease that are appropriately validated to represent the biology and outcomes of human liver diseases [4]. Though many advances in the development of preclinical models for MASLD have been made that have provided valuable insights on disease pathogenesis, only a few models recapitulate the key elements needed to be representative for human liver disease [4].

Therefore, the ideal MASLD models should mimic human disease as closely as possible if the objective is to enhance the likelihood that a drug that improves MASLD in the preclinical models also improves the disease in humans [4]. A relatively simple model, monolayer cultures of primary human hepatocytes or liver-based cell lines, fails to capture the interactions with other cell types seen in the liver [5]. Co-cultures of hepatocytes and non-parenchymal cells like Kupffer cells or fibroblasts and sandwich cultures of primary hepatocytes between two layers of collagen type 1 have been developed as well, but these planar culture systems have failed in representing the complex architecture of hepatic tissue in vivo. Therefore, here, we aimed to create an ex vivo model utilizing PCLSs which preserve cell–cell contacts and the extracellular matrix (ECM). This model serves as a valuable tool for investigating liver diseases, given its capacity to closely mimic the in vivo context [6].

To date, several studies have utilized PCLSs for the research of MASLD. Healthy murine PCLSs exposed to steatotic conditions, such as high concentration of fructose, glucose, insulin and/or fatty acids for a maximum of 48 h incubation could develop the phenotype of MASLD, represented by the increased positive area of Oil Red O staining and triglyceride (TG) content [7,8]. Still, these organotypic models did not instigate an inflammatory response [8] or only showed a modest increase in *Il1β* expression [9]. Similar treatments of rat [10] and human [11,12] PCLSs also led to lipid deposition, TG secretion, and/or lipotoxicity. However, these PCLS models either show no evidence of MASH or a late stage of MASLD. In this study, we prolonged the incubation time of healthy human PCLSs beyond the previously reported 48 h, under conditions that induce steatotic liver disease. We investigated the progression of MASLD including steatosis, steatohepatitis, and liver fibrosis. The aim of this study was to ascertain whether an extended incubation period could lead to the advancement of MASLD beyond the initial stage of simple hepatic steatosis.

2. Materials and Methods

2.1. Collection of Human Liver Tissue and Preparation of PCLSs

The use of human material was approved by the Medical Ethical Committee of the University Medical Centre Groningen (UMCG) according to Dutch legislation and the Code of Conduct for dealing responsibly with human tissue in the context of health research (www.federa.org), refraining the need of written consent for "further use" of coded-anonymous human tissue. We collected tissues and prepared PCLSs as previously published [13]. Briefly, human liver tissues were sourced from individuals undergoing partial hepatectomy or from organ donors. These tissues were conserved in a University of Wisconsin (UW) preservation solution (Cat. BUWC, Bridge to Life Ltd., London, UK) at 4 °C. Biopsies with a 6 mm diameter were sectioned and subsequently sliced with a

Krumdieck tissue slicer (Alabama Research and Development, Munford, TN, USA), which was filled with an ice-cold Krebs–Henselheit buffer, following the protocol described by de Graaf et al. [14]. PCLSs were estimated to be approximately 250 μm in thickness and were stored at 4 °C in UW preservation solution. The number of PCLSs and analyses conducted were constrained by the size of tissue available from donations. For each analytical procedure, tissues from a minimum of five distinct donor livers were utilized. Livers that failed to exhibit adenosine triphosphate (ATP) production after 1 h of culture were considered non-viable and consequently excluded from the study.

2.2. PCLS Culture and Induction of Steatosis

PCLSs were cultured in Williams medium E with GlutaMAX™ (Cat. 32551020, Invitrogen, Bleiswijk, The Netherlands), supplemented with gentamycin (50 μg/mL; Cat. 15750-037, Invitrogen) and glucose (25 mM; Cat. 1.08342.1000, Merck, Darmstadt, Germany), hereinafter referred to as "WEGG". To induce steatosis, additional glucose (36 mM), fructose (5 mM, Cat. F3510-100G, Sigma-Aldrich, St. Louis, MO, USA), human insulin (1 nM, Cat. I9278, Sigma-Aldrich), palmitic acid (240 μM, Cat. P0500-10G, Sigma-Aldrich), and oleic acid (480 μM, Cat. 75096-1L, Sigma-Aldrich) were added, a combination denoted as "GFIPO". Palmitic acid and oleic acid were solubilized in 0.04% bovine serum albumin (BSA; Cat. A2153-100G, Sigma-Aldrich). The fatty acids were initially dissolved in 0.1 M sodium hydroxide (Cat. 1.06469.5000, Merck) at 70 °C, followed by combination with pre-warmed BSA in water at 55 °C. It was observed that the final concentrations of BSA and sodium hydroxide did not influence the pH of the medium nor affect the PCLSs' viability. The medium preparation and final concentrations were based on previous experiments [10]. PCLSs were incubated at 37 °C in an atmosphere containing 20% O_2 and 5% CO_2 for periods ranging from 24 to 96 h. The medium was renewed at 24 h intervals, and samples were collected at the same frequency.

2.3. ATP Measurement

Metabolic activity and viability of PCLSs were assessed by measuring ATP content using the ATP Bioluminescence Assay Kit CLS II (Cat. 11699695001, Roche Diagnostics, Mannheim, Germany), as previously described [14]. ATP was measured per slice and corrected for total protein content.

2.4. Triglyceride Quantitation

The triglyceride (TG) levels in PCLSs and the medium were quantified utilizing the Trig/GB kit (Cat. 11877771, Roche Diagnostics), according to the protocol provided by the manufacturer. In short, PCLSs were snap-frozen and preserved at −80 °C pending analysis. They were subsequently homogenized in a Tris-buffered saline solution, and lipids were extracted using the Bligh and Dyer method [15]. Media samples, pooled from 12 PCLSs derived from the same donor and identical conditions, were stored at −20 °C until further processing. The TG values were determined by measuring the absorption at 540 nm after 1 h and adjusting for the total protein content [10].

2.5. Total Protein Measurement

The total protein in the PCLSs was assessed via the Lowry assay (Cat. 5000113, Cat. 5000114; BioRad DC Protein Assay, Hercules, CA, USA) and a BSA calibration curve [16]. The quantified protein concentrations were used to normalize the ATP and TG measurements as previously published [10,13].

2.6. RNA Sequencing and Quantitative Real-Time PCR

Transcriptomic alterations were assessed through total RNA sequencing and quantitative Real-Time PCR (RT-qPCR). For each experimental condition, three to nine PCLSs from a single donor were combined, snap-frozen in liquid nitrogen with Qiazol buffer (Cat. No. 79306, Qiagen, Venlo, The Netherlands) added, and, subsequently, stored at

−80 °C pending further analysis. For homogenization and lysis, PCLSs were processed using Qiazol buffer, followed by vigorous mixing with one-fifth the volume of chloroform (Cat. 1.02445.4000, Sigma-Aldrich). Phase separation was achieved using extraction tubes, and RNA was isolated via the RNeasy Lipid Tissue Mini Kit (Ref. 74106, Qiagen) [10,13]. The integrity and concentration of the extracted RNA were determined using the RNA Screen Tape assay on the TapeStation 4200 System (Agilent, Santa Clara, CA, USA).

For next-generation sequencing (NGS), the construction of libraries was performed using the NEBNext Ultra RNA Library Prep Kit for Illumina (New England Biolabs, San Diego, CA, USA) according to the manufacturer's protocol. RNA was paired-end sequenced using the NovaSeq™ 6000 platform (Illumina, San Diego, CA, USA). Reads were aligned with STAR (v. 2.7.3a) [17] to the human genome assembly (GRCh38, Ensembl release 101). FeatureCounts (v.2.0) [18] was employed for exon read quantification. The quality of raw sequencing was assessed using FastQC (v.0.11.8) [19] and MultiQC (v.1.10.1) [20]. NGS was performed on seven liver donors per condition.

DESeq2 (v. 1.40.2) [21] was used to identify differentially expressed genes (DEGs) by fitting a negative binominal model to the raw counts. A Wald test with α-error accumulation correction using Benjamini–Hochberg implemented in the DESeq2 package was employed to determine genes as differentially expressed. Adjusted p-values < 0.05 were considered statistically significant. Regularized log transformation was used to normalize raw gene counts for further analysis. The regulation of biological processes was assessed through a gene set enrichment analysis (GSEA) using GSEA software (v. 4.3.2) [22,23].

The expression levels of critical genes associated with inflammation and fibrosis were quantified using RT-qPCR as previously published [10,13]. The process of the reverse transcription of RNA into complementary DNA (cDNA) was facilitated using the Reverse Transcription System (Promega, Leiden, The Netherlands). Subsequent RT-qPCR analysis was conducted on a ViiA 7 Real-Time PCR System (Applied Biosystems, Waltham, MA, USA), utilizing SYBR Green primers (Table S1) and SYBR Green reagent (Cat. No. 04913914001, Roche Diagnostics). The threshold cycle (Ct) values were normalized to the Ct values of reference gene *YWHAZ* (resulting in ΔCt values) and further compared to those of the control group to calculate the relative expression levels (ΔΔCt). The results were presented as the mean fold change in expression ($2^{-\Delta\Delta Ct}$), computed using the method delineated by Livak et al. [24]. Statistical analyses were subsequently performed using the ΔΔCt values.

2.7. Measurement of Excreted Proteins

Mediums from 12 PCLSs per condition (one 12-well plate) were pooled together and stored at −20 °C. Cytokines and chemokines were measured with Luminex® Multiplex Assay (LXSAHM-13, R&D Systems, Minneapolis, MN, USA) according to the manufacturer's instructions.

Pro-collagen Ia1 concentrations were quantitatively assessed in a 1:20 dilution of a pooled culture medium utilizing the Human Pro-Collagen I alpha 1 ELISA Kit (ab210966, Abcam, Cambridge, UK), following the manufacturer's protocol. The data are expressed as relative values compared to the control (WEGG, 24 h) [10,13].

ApoB100 concentration was measured in a 1:1 dilution using a Human Apolipoprotein B ELISA development kit (3715-1H-6, Mabtech, Nacka Strand, Sweden) according to instructions from the manufacturer.

2.8. Histology

PCLSs were fixed in 4% buffered formalin overnight and stored at 4 °C in 70% ethanol. Fixed PCLSs were embedded in paraffin and sectioned 4 µm thick. Tissue morphology and fibrosis were assessed through hematoxylin and eosin (H&E) staining (Hematoxyline: Cat. 4085.9002, Klinipath, Duiven, The Netherlands; Eosin: Cat. HT110232, Sigma-Aldrich) and Picrosirius Red (PSR) staining (ab150681, Abcam) according to the standard histological procedure. Stained tissue sections were scanned using a Nanozoomer Digital Pathology

Scanner (NDP Scan U10074-01, Hamamatsu Photonics K.K., Shizuoka, Japan) [10,13]. ImageJ was used to quantify the stained areas. The histopathologic percentages of necrosis and fat vesicles were determined by an experienced pathologist (MH).

2.9. Statistics

Statistical significance was identified by comparing each experimental condition to its respective control. Replicates included at least five distinct livers, with a minimum of three precision-cut liver slices (PCLSs) per condition sourced from each liver. The data are presented as mean ± standard error of the mean (SEM). Statistical analyses were conducted using GraphPad Prism version 9.2.0 (GraphPad Software Inc., Boston, MA, USA). Group comparisons were made using one-way ANOVA followed by Dunnett's multiple comparisons test, in addition to paired two-tailed *t*-tests. A *p*-value less than 0.05 was considered indicative of statistical significance [10,13].

3. Results

3.1. Characterization of PCLSs

3.1.1. Slices Retain Viability after Incubation

Slices were incubated in the GFIPO medium to induce the MASLD phenotype and in the WEGG medium as the control. Notably, slices were incubated in 20% O_2 as they exhibited stable viability up to 96 h compared to 80% O_2, and three out of four livers showed down-regulation in ATP production at 80% O_2 compared to 20% O_2 (Figure S1A). The analysis of the PCLS morphology using H&E staining revealed that the slices showed no obvious signs of the loss of cell viability (Figure 1A). However, over time, increased necrotic areas were observed in PCLSs after incubation in both WEGG and GFIPO, as compared to the 24 h incubation in WEGG (Figure 1B). This indicates that incubation could cause moderate cell death towards the end, regardless of the medium conditions used. Measurements of ATP showed that GFIPO resulted in lower ATP levels compared to WEGG, particularly after 72 h of incubation. As the source of energy at the cellular level, ATP in the GFIPO group might have been consumed to a larger extent than in the WEGG group for energy, in the processes including lipid transport and fat accumulation, leading to the decreased ATP levels in the PCLSs. However, ATP levels remained stable throughout the incubation period with GFIPO, with an average ATP content of 4 pmol/µg protein (Figure 1C), which could be considered as viable according to our previous studies [6,8,10].

Following these observations on PCLS viability, we further investigated gene expression using NGS to gain a deeper insight into the differences between WEGG and GFIPO.

3.1.2. Principal Component Analysis (PCA)

PCA was employed to ascertain the impact of experimental variables, namely, the medium and incubation time, on the data variance. The resulting two-dimensional plot, which was derived from the first two principal components (PC1 and PC2), is depicted in Figure 1D. Samples from the 0 h time point, obtained immediately after slicing, were separated along PC1, indicating an effect of incubation on liver slices compared to untreated slices. Despite the intrinsic human variability noted among the 0 h samples, this heterogeneity did not appear to substantially affect the variation of samples after incubation. All samples from 24 h incubation were separated together at the bottom part of the plot, regardless of whether they were in the GFIPO or WEGG medium. For the 48 h time point, the samples remained distinct from other time points, and, notably, they were also separated by different medium used. Upon reaching 72 h and 96 h, a clear separation between GFIPO and WEGG was observed. Overall, PCLSs can be differentiated based on both the medium and the duration of incubation time, suggesting that these two parameters could be the principal determinants influencing the effects observed in the PCLSs culture over time.

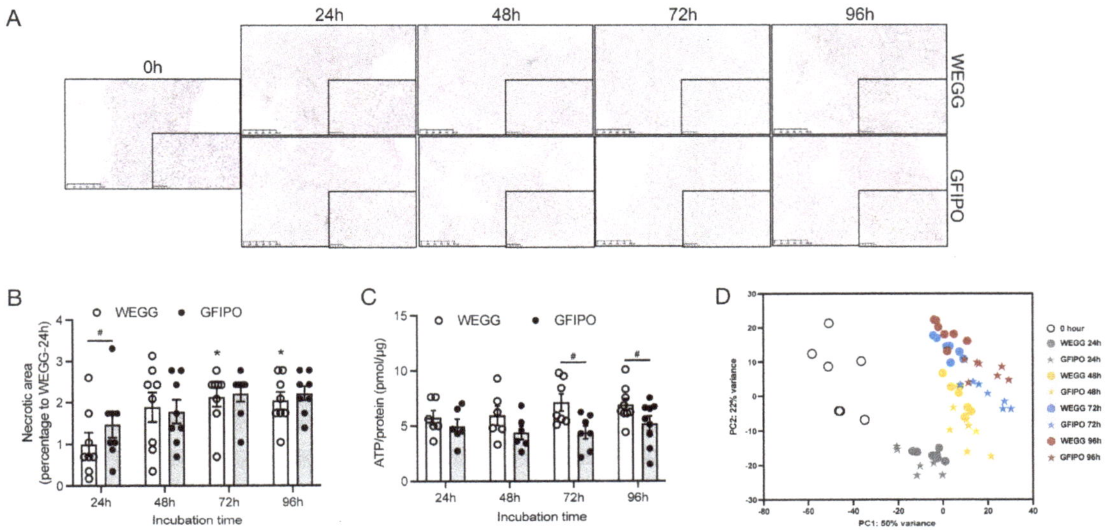

Figure 1. Initial characterization of human PCLSs, cultured for up to 96 h in WEGG or GFIPO. (**A**) Representative H&E staining images of PCLSs after long-term incubation in WEGG and GFIPO. Scale bar = 500 μm; inset: 100 μm. (**B**) Presence of necrotic areas compared to WEGG 24 h in H&E staining. (**C**) ATP/protein content in PCLSs after up to 96 h incubation in WEGG or GFIPO. (**D**) PCA showing the first two principal components. Each symbol corresponds to an individual patient sample composed of 3 slices, with a total of 7 patients (n = 7) assessed in each group. Data are presented as mean ± SEM. (#) denotes statistical differences between GFIPO and WEGG at each time point, while (*) denotes statistical differences in GFIPO or WEGG compared to their corresponding 24 h; $^{*(\#)}$ $p < 0.05$.

In assessing how PCLSs change with long-term incubation, we performed overrepresentation analysis on dysregulated genes in WEGG compared to the 0 h groups. It showed an increase in pro-inflammatory and pro-fibrotic processes (Table S2) and a decrease in metabolic processes upon incubation (Table S3). Therefore, our primary focus was to compare the influence of GFIPO on PCLSs to that of WEGG (control group) at each time point, aiming to assess if GFIPO would aggravate tissue dysfunction, inflammation, and fibrotic response in the context of MASLD progression.

3.1.3. Total Number of Differentially Expressed Genes

Next, we analyzed differentially expressed genes (DEGs) under various experimental conditions as shown in volcano plots (Figure 2A). At 24 h, we observed a relatively modest number of DEGs, comprising 59 induced genes and 33 repressed genes. However, a notable elevation in the number of DEGs was evident at 48 h, with 149 genes up-regulated and 159 genes down-regulated. This observation suggests that the influence of GFIPO may commence at around 48 h. Remarkably, at both 72 h and 96 h, we identified a substantial number of DEGs induced by GFIPO (175 induced and 507 repressed at 72 h, 164 induced and 435 repressed at 96 h). These findings align with our PCA analysis, reinforcing the notion that the impact of GFIPO builds up over time. Notably, our analysis revealed a larger number of down-regulated genes compared to up-regulated genes. This suggests that the influence of GFIPO may primarily entail the down-regulation of specific biological pathways.

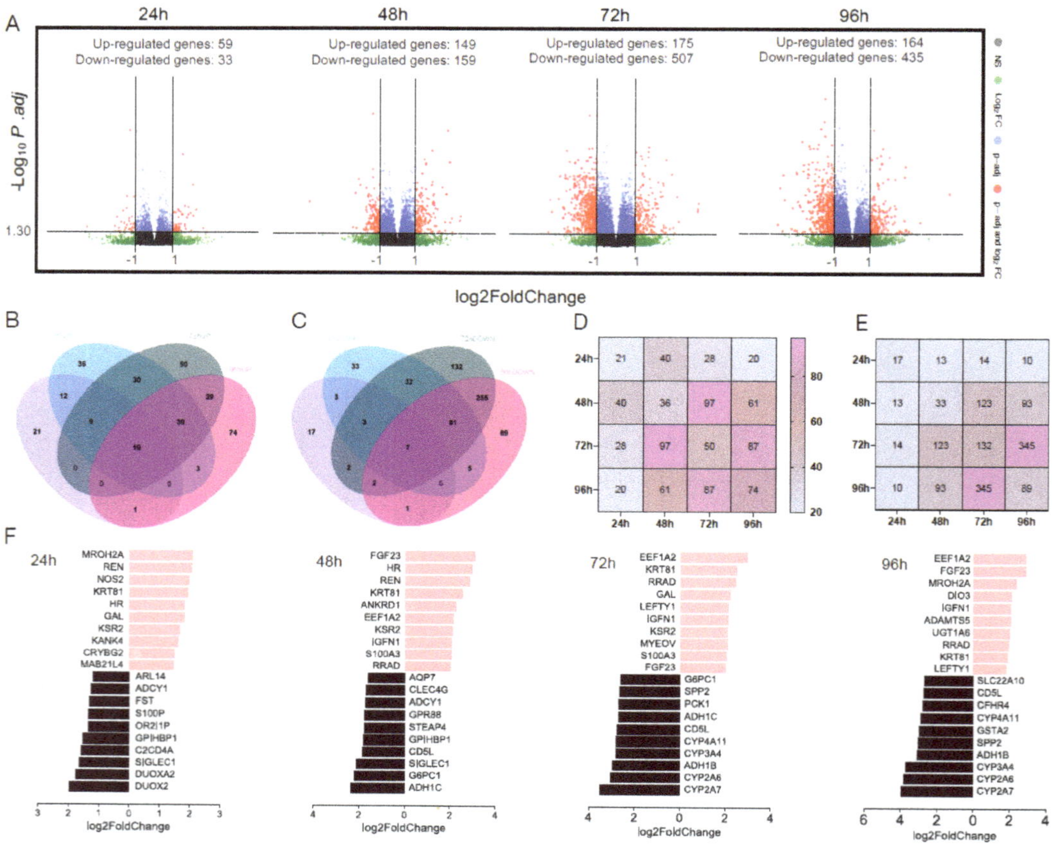

Figure 2. NGS of PCLSs, cultured for up to 96 h in WEGG or GFIPO. (**A**) Volcano plots representing the differentially expressed genes (red), with a threshold of log2FoldChange > 1 on the x-axis and p.adj < 0.05 (−Log10p.adj > 1.30) on the y-axis, comparing GFIPO and WEGG at each time point. WEGG serves as the control group. (**B**,**C**) Venn diagrams illustrating the distribution of up-regulated genes (**B**) and down-regulated genes (**C**) across all time points. (**D**,**E**) Heatmap showing the number of overlapped up-regulated (**D**) and down-regulated (**E**) DEGs between two timepoints; overlap of the same timepoint stands for the unique DEGs amount as shown in the Venn diagrams. (**F**) The top 10 up-regulated genes (highlighted in pink, right) and down-regulated genes (highlighted in black, left) at each time point.

3.1.4. Overlapping Differentially Expressed Genes

To examine transcriptional alterations within each comparison, we conducted a cross-referencing analysis of the DEGs to identify shared genes across different time points. Figure 2B,C illustrate a relatively limited overlap in DEGs shared in all time points. Up-regulated gene sets have relatively more in common (19 out of 323 up-regulated genes), while down-regulated genes only share 7 out of 662 genes. When assessing each time point, we observed a progressive increase in the number of uniquely up-regulated DEGs over time, while this trend reversed for down-regulated genes starting at 72 h. Considering that the total amount of both up-regulated genes and down-regulated genes peaked at 72 h compared to other time points, the rising number of uniquely up-regulated DEGs after 96 h indicates a divergence in transcriptional profiles at this time point compared to 72 h. Additionally, our analysis suggests that the primary change in up-regulated genes occurs between 48 and 72 h, while predominantly down-regulated genes are found at 72 h–96 h (Figure 2D,E).

3.1.5. Top 10 Significantly Differentially Expressed Genes

Next, we aimed to pinpoint the most significantly altered genes at each time point. The column charts in Figure 2F illustrate the top 10 up-regulated and down-regulated genes in GFIPO compared to WEGG, with the selection criteria based on thresholds (baseMean > 50, p.adj < 0.05, and log2FoldChange > 1). These genes were ranked according to their absolute log2FoldChange values. Table S4 provides additional information, including ensemble gene IDs and functions of encoded proteins, and a comprehensive list of all DEGs. The top 10 up-regulated genes exhibited substantial overlap across all time points and were implicated in diverse biological processes. These processes encompassed inflammatory response (*NOS2*), carbohydrate binding activity (*CRYBG2*), signaling transduction (*KSR2*), and endothelial cell activation (*ANKRD1*), which were predominantly observed in the early stages. Genes associated with cell cycle regulation (*S100A3*), cell survival (*FGF23*), cell adhesion (*IGFN1*), and cell migration (*ADAMTS5*) were prominent in later stages. On the other hand, down-regulated genes caused by GFIPO were primarily linked to metabolic processes (*DUOXA2, DUOX2, AQP7, STEAP4, G6PC1, ADH1C, PCK1, ADH1B, CFHR4*) and inflammation response (*C2CD4A, ADCY1, SIGLEC1, CLEC4G, CD5L*). Notably, a cluster of cytochrome P450-encoding genes (*CYP4A11, CYP3A4, CYP2A6, CYP2A7*) manifested alterations after 72 h of incubation. Similar trends were also observed in other studies where *CYP2E1, CYP2C19*, and *CYP2C8* showed down-regulation that followed the disease stage progression [25].

3.2. *Characterization of PCLSs*

3.2.1. Genes Related to Fatty Acid Metabolism

Next, we specifically examined the DEGs and pathways related to fatty acid metabolism, comparing the effects of GFIPO and WEGG at different time points. Figure 3 illustrates the up-regulated (in red) and down-regulated (in green) DEGs induced by GFIPO in comparison to WEGG over time (A–D). We focused on four processes known to contribute to TG accumulation: fatty acid uptake, de novo lipogenesis (DNL), TG synthesis, and very low-density lipoprotein (VLDL) secretion [26]. At 24 h (Figure 3A), GFIPO up-regulated the expression of the *GOT2* gene, which might lead to increased fatty acid uptake. It also up-regulated *ACSL5, DGAT1, DGAT2*, and *PNPLA3*, implying the promoted TG synthesis, leading to the accumulation of lipid droplets and lipid secretion. At 48 h (Figure 3B), some genes associated with TG accumulation and secretion, such as *APOB, CIDEC*, and *PLIN5*, started to be down-regulated. A lower production of ApoB100 was observed in the GFIPO medium compared to the corresponding WEGG medium, although the down-regulation started from 24 h (Figure S1B). However, the expression of *ACLY*, involved in DNL, was up-regulated, while *CPT1A*, which plays a role in mitochondrial β-oxidation, was down-regulated. At 72 h (Figure 3C), numerous genes related to fatty acid uptake, intracellular fatty acid transportation, TG synthesis, and TG secretion were down-regulated. Additionally, genes involved in fatty acid oxidation, such as *CPT1A* (mitochondrial β-oxidation) and *ACOX1* (peroxisomal β-oxidation), along with *CYP4A11* (α-oxidation), were also down-regulated. At 96 h, a similar pattern to 72 h was observed, although there were no dysregulated genes in mitochondrial β-oxidation.

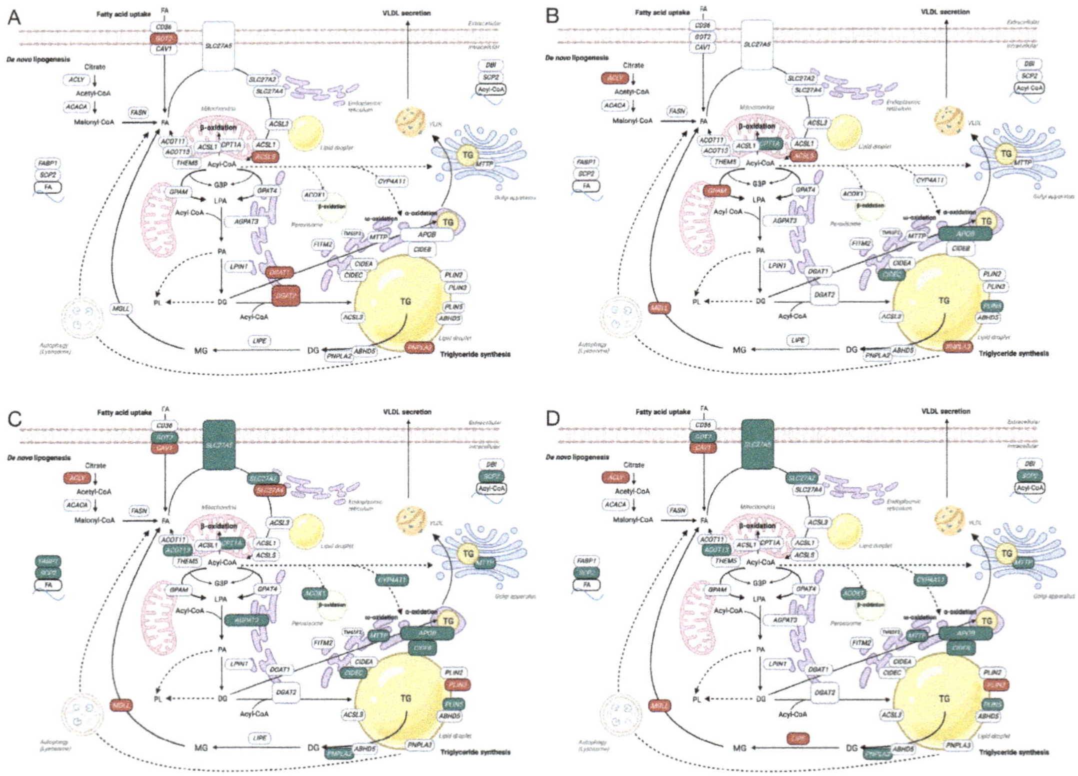

Figure 3. Gene regulation in fatty acid metabolism pathways by GFIPO compared to WEGG. (**A–D**) Up-regulated (in red) or down-regulated (in green) DEGs associated with fatty acid metabolism induced by GFIPO compared to WEGG at 24 h (**A**), 48 h (**B**), 72 h (**C**), and 96 h (**D**) of incubation.

3.2.2. Fatty Acid Metabolism Pathway

Next, we performed GSEA on hallmark pathways (Figure 4A–D). The results indicated that the fatty acid metabolism pathway was up-regulated at 24 h but subsequently down-regulated at 48 h and beyond. This aligns with the results shown in Figure 3A–D. The heatmaps in Figure 4A–D display the gene expression changes in the fatty acid metabolism pathway. The majority of genes involved in the breakdown of fatty acids, particularly through mitochondrial β-oxidation, showed up-regulation at 24 h. However, starting from 48 h and continuing to 96 h, there was a shift towards down-regulation, with more genes participating in this process. Additionally, several genes related to the metabolism of fatty acids were altered at 48 h, primarily showing down-regulation. Biological processes derived from Gene Ontology (GO) (Figure 5A) also indicated the up-regulation of fatty acid biosynthetic and metabolic process at 24 h and down-regulation at later time points. It has been shown that after the development of a more severe disease, lower levels of metabolic intermediates are needed to sustain growth compared with steatosis [25]. This may explain the down-regulation in metabolism. Meanwhile, fatty acid oxidation and fatty acid catabolic process were reduced as well from 48 h onwards.

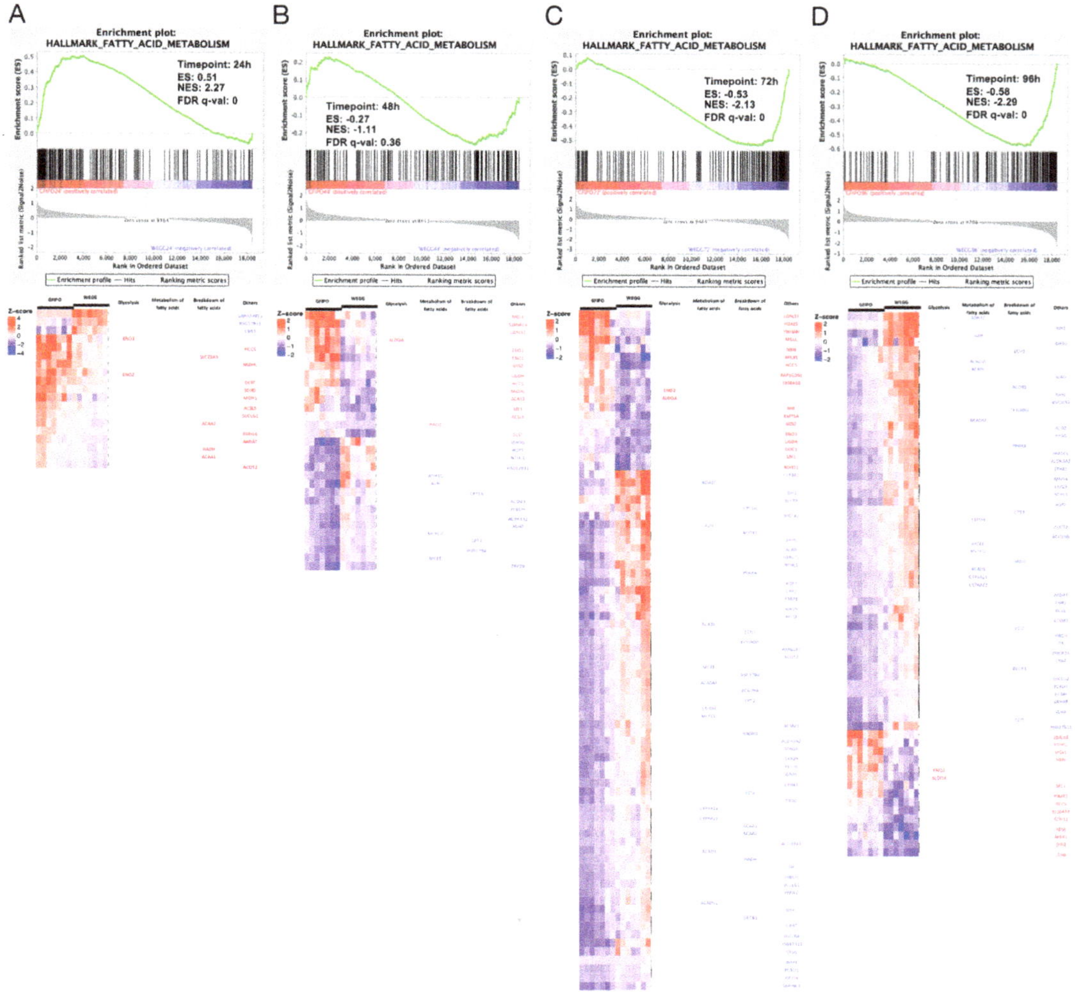

Figure 4. Gene regulation in fatty acid metabolism pathways by GFIPO compared to WEGG as identified by Gene Set Enrichment Analysis (GSEA). (**A–D**) GSEA plots of the hallmark fatty acid metabolism pathway at each time point: (**A**) 24 h, (**B**) 48 h, (**C**) 72 h, and (**D**) 96 h; below are significantly dysregulated genes in this pathway.

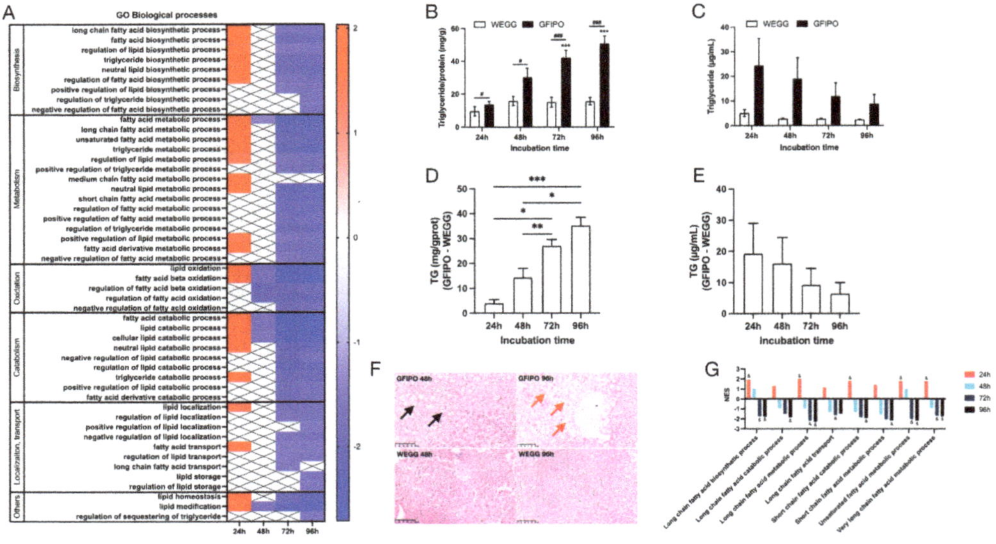

Figure 5. Fatty acid metabolism pathways by GFIPO compared to WEGG as identified by Gene Set Enrichment Analysis (GSEA) and phenotype of triglyceride regulation. (**A**) Altered biological processes associated with lipid, fatty acid, and TG metabolisms (nominal *p*. value < 0.01) at each time point, ranked by the normalized enrichment score (NES) value on the scale bar (NES > 0, up-regulated; NES < 0, down-regulated). (**B**,**C**) Increase in TG content relative to control at 0 h in PCLSs (**B**) or medium (**C**) after up to 96 h incubation in WEGG or GFIPO. D-E Differences in TG increase between GFIPO and WEGG in PCLSs (**D**) and medium (**E**). (**F**) Representative images of H&E staining on PCLSs (scale bar = 100 μM, black arrows indicate microvesicular steatosis, red arrows indicate macrovesicular steatosis). (**G**) Altered GO biological processes by GFIPO compared to WEGG in fatty acid with different chain lengths (& indicates significantly changed compared to the corresponding WEGG). Data are presented as mean ± SEM. (#) denotes statistical differences between GFIPO and WEGG at each time point, while (*) denotes statistical differences in GFIPO or WEGG compared to their corresponding 24 h; $^{*(\#)}$ $p < 0.05$, ** $p < 0.01$, $^{***(\#\#\#)}$ $p < 0.001$.

3.2.3. Phenotype of Fat Accumulation on PCLSs

To assess fat accumulation in slices, we measured TG content both in slices and in the medium. Figure 5B shows that TG in slices is significantly increased in GFIPO compared to WEGG, which indicates that GFIPO induces fat accumulation in slices more extensively after long-term incubation. The TG levels in PCLSs in WEGG remain constant. We found TG in the medium to be higher in GFIPO than in WEGG, although the levels decreased over time (Figure 5C). Figure 5D,E show the differences between GFIPO and WEGG at each time point. Overall, in slices, TG levels increased over time, but in the medium, they decreased. This difference may be due to the down-regulation of fatty acid transport (Figure 5A), which perhaps leads to suppressed VLDL export. H&E staining (Figure 5F) showed the presence of microvesicular steatosis in GFIPO after 48 h, and macrovesicular steatosis after 96 h, while WEGG did not induce steatosis. GFIPO also caused the down-regulation in genes involved in the metabolism of a very long-chain, long-chain, and unsaturated fatty acids after 72 h, and in short-chain fatty acid after 96 h at the gene level via GO analysis (Figure 5G).

3.3. *Inflammatory Response in PCLSs by GFIPO Compared to WEGG*

Next, we investigated the inflammatory response of PCLSs when exposed to GFIPO compared to WEGG. Hallmark pathway results showed that genes associated with inflammatory responses only started to be up-regulated after 48 h of incubation, in which

TNFα signaling via nuclear factor kappa-light-chain-enhancer of activated B cells (NFκB) played a major role with 24, 25, and 21 genes involved at 48 h, 72 h, and 96 h, respectively (Figure 6A–D). Moreover, the increase in the pro-inflammatory cytokines involved (*TNFα*, *IL-6*, and *IL-1β*) was confirmed at the gene expression level through RT-qPCR (Figure 6E–G). The gene expressions of pro-inflammatory cytokines (*TNF*, *IL6*, *IL1β*) were up-regulated in slices upon treatment with GFIPO compared to that with WEGG at each time point, although the differences found were not statistically significant. In addition, the expression of *TNF* increased over time in slices incubated with GFIPO. WEGG led to similar trends. However, the expression of *IL-6* and *IL-1β* showed the opposite effect, as well as the production of TNFα measured in the medium (Figures 6H and S1C). The results of cytokine and chemokine secretion further demonstrated the pro-inflammatory effects of GFIPO on PCLSs (Figure 6H). GFIPO significantly increased the production of TNFα, IL6, IL8, interleukin-1 receptor antagonist protein (IL-1RA), and C-X-C motif chemokine ligand 10 (CXCL10) in the medium after 48 h of incubation. As mentioned above, TNFα showed a significant increase in GFIPO compared to WEGG at the later stages, but the overall trend was a decrease after 48 h. This may indicate that the contribution of long-term incubation to the further development of liver inflammation is limited, as no additional inflammatory cell infiltration can be induced in this ex vivo model. Additionally, we also observed an up-regulation in cell adhesion pathways and the intercellular adhesion molecule 1 (*ICAM1*)'s gene expression through NGS after incubation in GFIPO (Figure S2A,B).

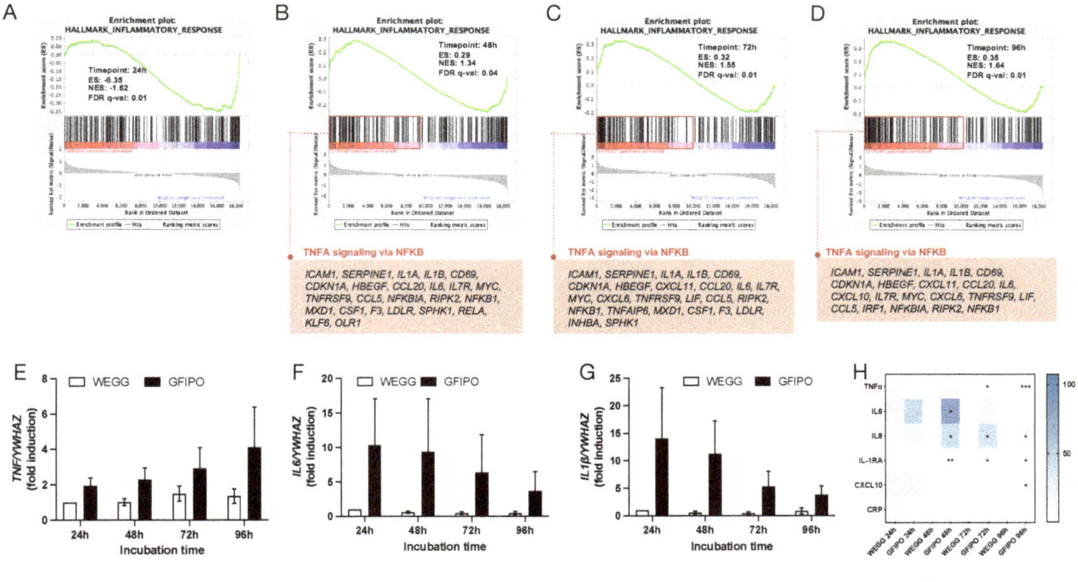

Figure 6. Gene regulation in inflammatory response of GFIPO compared to WEGG in PCLSs. (**A–D**) GSEA plots of hallmark inflammatory response pathway at each time point: (**A**) 24 h, (**B**) 48 h, (**C**) 72 h, and (**D**) 96 h; below are up-regulated genes involved in TNFα signaling via NFκB pathway. (**E–G**) mRNA expression of inflammatory biomarkers (*TNF*, *IL6*, *IL1β*) in PCLSs after up to 96 h of incubation. (**H**) Secretion of inflammatory cytokines by PCLSs after up to 96 h of incubation (100% to WEGG 24 h). Data are presented as mean ± SEM, (*) denotes statistical differences between GFIPO and WEGG at each time point; * $p < 0.05$, ** $p < 0.01$, *** $p < 0.001$.

3.4. Development of Liver Fibrosis in PCLSs by GFIPO Compared to WEGG

To assess the development of liver fibrosis on PCLSs, we conducted GSEA and found that the TGFβ signaling pathway was up-regulated by GFIPO compared to WEGG at all time points (Figure 7A–D). the reactome pathway results demonstrated that GFIPO activated many processes participating in TGFβ signaling, especially at 72 h of incubation (Figure 7E). Regarding the mRNA expression of *COL1A1* and *ACTA2* (Figure 7F,G), GFIPO led to an increasing trend, only having significant differences at 48 h in *ACTA2* expression. Previously, collagen genes, such as collagen type 1 alpha 1 chain (*COL1A1*), were shown to be consistently up-regulated with MASLD progression from steatosis to fibrosis [25]. Here, we also observed both an increasing trend and an increasing difference between GFIPO and WEGG. However, the production of pro-collagen 1a1 and TIMP1 protein was significantly up-regulated by GFIPO at the later stages (Figure 7H,I). The GSEA results also showed elevated focal adhesion and matrix metalloproteinases in WikiPathways, as well as ECM receptor interaction in Kyoto Encyclopedia of Genes and Genomes (KEGG) at the later stages (Figure S1D). Meanwhile, PSR staining revealed the up-regulation of positive areas in GFIPO compared to WEGG, indicating increased ECM accumulation by GFIPO (Figure 7J).

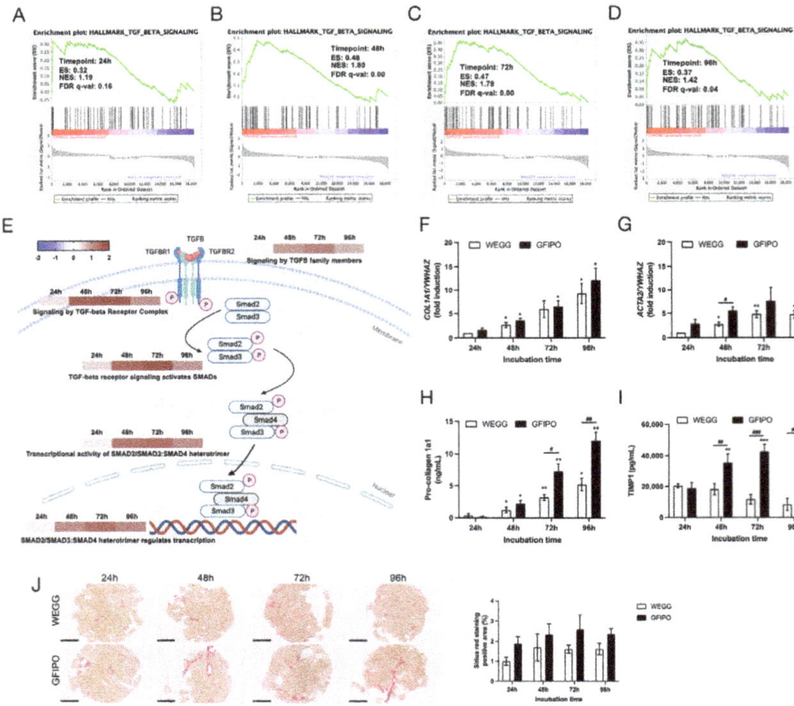

Figure 7. Development of liver fibrosis in PCLSs by GFIPO compared to WEGG. (**A–D**) GSEA plots of hallmark TGFβ signaling pathway at each time point: (**A**) 24 h, (**B**) 48 h, (**C**) 72 h, and (**D**) 96 h. (**E**) Dysregulation of reactome pathways ("Signaling by TGFβ family members", "Signaling by TGFβ Receptor Complex", "TGFβ receptor signaling activates SMADs", "Transcriptional activity of SMAD2/SMAD3:SMAD4 heterotrimer", and "SMAD2/SMAD3:SMAD4 heterotrimer regulates transcription") in TGFβ signaling, ranked by NES value on the scale bar (NES > 0, up-regulated; NES < 0, down-regulated). (**F,G**) mRNA expression of inflammatory biomarkers (*COL1A1*, *ACTA2*) in PCLSs after up to 96 h of incubation. (**H,I**) Secretion of pro-collagen 1a1 and TIMP1 from PCLSs after up to 96 h of incubation. (**J**) Representative images of Picro Sirius Red staining on PCLSs (scale bar = 1 mM) and quantitative analysis of positive areas (control to WEGG 24 h) using ImageJ. Data are presented as mean ± SEM. (#) denotes statistical differences between GFIPO and WEGG at each time point, while (*) denotes statistical differences in GFIPO or WEGG compared to their corresponding 24 h; $^{*(\#)}$ $p < 0.05$, $^{**(\#\#)}$ $p < 0.01$, $^{***(\#\#\#)}$ $p < 0.001$.

4. Discussion

MASLD is increasingly recognized as a major health challenge, necessitating advanced research models to elucidate its pathology [27]. Studies have identified a multifaceted etiology behind hepatic steatosis, inflammation, and fibrosis in MASLD [28]. While animal studies are informative [29], human-specific pathophysiological responses require more accurate models. Therefore, human PCLSs emerge as a superior ex vivo alternative, offering a preserved tissue microenvironment and controlled experimental variables [6]. This study utilized healthy human PCLSs to simulate MASLD onset, showing that GFIPO—rich in sugars, insulin, and fatty acids—induces macrovesicular steatosis, inflammation, and ECM accumulation compared to controls.

First, we evaluated the viability of human PCLSs for extended incubation as a model for MASLD. Our findings revealed that human PCLSs remained viable after 96 h in both GFIPO and the control medium, maintaining intact morphology and ATP levels. H&E staining indicated no significant toxicity from GFIPO. However, NGS highlighted significant gene-level differences between GFIPO and control conditions, affecting metabolic, inflammatory, and fibrogenic pathways. These findings support the use of human PCLSs as a valuable ex vivo model for prolonged studies under GFIPO conditions.

In this study, GFIPO significantly promoted TG accumulation in PCLSs, a key indicator of MASLD initiation. Notably, extended incubation with GFIPO also resulted in the formation of macrovesicles, as confirmed by H&E staining, a finding not previously reported in other ex vivo models.

Next, we investigated how GFIPO influences MASLD progression in human PCLSs with an initial focus on genes involved in fatty acid uptake and metabolism, which are key early MASLD processes [30]. GFIPO was found to increase the mRNA level of fatty acid transporters like *GOT2* (FA translocase/CD36) [31] at 24 h, potentially enhancing fatty acid binding to hepatocytes. The up-regulation of *CAV1* (Caveolin-1) may also play a role in fatty acid uptake, as it is essential for fatty acid translocase localization and function [32]. GO analysis suggested that fatty acid transport was initially up-regulated by GFIPO at 24 h, and then it started to be reduced after 72 h, probably due to cellular stress or damage. Consistent with the literature suggesting shifts from fatty acid uptake to synthesis as MASLD progresses [31], we noted increased DNL from 48 h onwards, implied by the elevated expression of *ACLY*, whose encoded protein is a key enzyme converting citrate to acetyl-CoA [33].

The assembly of TG molecules is a crucial process for storing and exporting fatty acids [26]. GFIPO initially increased the mRNA expression of GPAM and DGAT enzymes, key in TG synthesis, suggesting an early stimulation of lipid biosynthesis, which is also supported by the elevated GO pathways related to fatty acid biosynthesis. From 72 h, these biosynthetic pathways began to decrease. It aligns with a clinical trial where DGAT1 levels fell as liver disease progressed from MASH to cirrhosis [25]. Besides, in the same study, triacylglycerol (TAG) levels were uniformly elevated in patients with MASH and steatosis compared to healthy individuals but were lower in MASH than in steatosis alone. Specifically, MASH patients had reduced polyunsaturated fatty acids (PUFA)-TAGs but increased short and saturated fatty acyl chain-containing TAGs, with these levels decreasing further in cirrhosis [25]. Here, we also observed the down-regulation in pathways involved in very long-chain, long-chain, and unsaturated fatty acids after 72 h, and in short chain fatty acid after 96 h, which reflects the transition from steatosis to more severe liver conditions. Moreover, research in obese mice suggests that decreased TG synthesis would exacerbate liver damage in MASH, indicating that TG might help protect against liver injury [34]. Our NGS analysis implied a decrease in lipogenesis, which might link to liver damage and the progression of MASLD in our PCLSs model.

Furthermore, GFIPO decreased genes essential for VLDL assembly and secretion, alongside reduced lipid storage pathways, at 72 h and 96 h. This was mirrored by a decrease in ApoB100, a VLDL stabilizer [35], indicating suppressed lipid secretion. Malaguarnera et al. found that the diminished VLDL assembly and secretion during MASLD progression could result from lipotoxicity and cytokines activity affecting transcription factors for lipogenesis [36]. This may correlate with the lower lipogenesis and increased cytokine release observed in our study due to GFIPO. Interestingly, we also found a suppression in fatty acid oxidation implied by the decreased gene expression of key enzymes in mitochondrial β-oxidation (*CPT1A*), peroxisomal β-oxidation (*ACOX1*), and ω-oxidation or α-oxidation (*CYP4A11)*, as well as several reduced fatty acid oxidation pathways. Impaired β-oxidation leads to an abnormal TG accumulation and the development of MASLD [37]. Therefore, the suppression of lipid oxidation could potentially contribute to MASLD progression in PCLSs.

Overall, our results suggest that GFIPO exerts a temporal effect on lipid metabolism. In the initial stages, GFIPO increases genes involved in fatty acid uptake and storage. With prolonged exposure, there is a down-regulation of genes related to fatty acid oxidation and secretion, while genes involved in DNL are up-regulated, which may lead to increased lipid accumulation.

MASLD can progress from simple hepatic lipid accumulation to MASH, characterized by inflammation and fibrosis. Unlike current PCLSs models, which show only steatosis, lipid deposition, and lipotoxicity [10–12,38], human PCLSs under GFIPO after 96 h exhibited markers of liver inflammation, including up-regulated inflammatory genes (*TNFα*, *IL6*, and *IL1β*) and cytokines (TNFα, IL6, IL8), detectable up to 96 h. This was evidenced by RT-qPCR and cytokine assays, with NGS confirming the increase in inflammatory response over time. Noteworthy, IL1β, IL6, and IL8 showed earlier up-regulation than TNFα, suggesting that TNFα production might be activated by these pro-inflammatory cytokines, potentially through TNFα signaling via NFκB, as evidenced by the associated up-regulated genes from NGS. Additionally, MASH is also characterized by inflammatory cells' infiltration [39]. Although immune cell recruitment is not directly observed in PCLSs, increased cell adhesion pathways and intercellular adhesion molecule 1 (*ICAM1)*'s gene expression (Figure S2A,B) suggest a facilitated adhesion of immune cells to the endothelium, leading to inflammatory infiltration in MASH [40,41]. Thus, GFIPO-treated human PCLSs provide a model for studying the pro-inflammatory aspects of MASLD progression.

In addition to liver inflammation, liver fibrosis is also a crucial phase in MASH development, largely driven by TGFβ signaling [42]. In our study, NGS showed that GFIPO up-regulated this pathway, specifically at 48 h and 72 h. In chronic liver injury or inflammation, TGFβ binds to its receptors, triggering a signaling cascade, for example through SMAD families, and promoting fibrosis [43]. Our study showed that TGFβ-activated SMAD2/SMAD3:SMAD4 heterotrimer signaling pathways were significantly enhanced under GFIPO on the gene level, consistent with elevated pro-fibrotic biomarkers like Pro-Collagen 1a1. PSR staining and TIMP1 production exhibited a similar trend. The TGFβ-stimulated TIMP1 production could inhibit matrix degradation enzymes, leading to ECM accumulation [43–45]. It has also been reported that up-regulated TGFβ activates hepatic stellate cells (HSCs), driving their transformation into collagen-producing myofibroblasts [46]. GFIPO also appeared to activate HSCs, as evidenced by increased *ACTA2* expression, suggesting an early sign of fibrogenesis. Additionally, focal adhesion and matrix metalloproteinases in WikiPathways, as well as ECM receptor interaction in KEGG, were all up-regulated in later stages (Figure S1D), echoing gene patterns seen in cirrhosis [25]. Furthermore, a decline in metabolic genes, such as *MAT1A* and *GNMT*, could also be linked to severe MASLD (fibrosis stages 3–4) [47]. The inhibition of these genes leads to steatohepatitis, advanced liver fibrosis, and hepatocellular carcinoma in mouse models as well [47]. All of this supports the development of liver fibrosis induced by GFIPO at the later stages.

The progression of MASLD is understood to be driven by the intricate interplay of inflammatory stress and lipid accumulation, aided by pro-inflammatory cytokines like interleukins and TGFβ [44]. This study suggests that the GFIPO-induced activation of TNFα and TGFβ pathways, coupled with the previously mentioned disruption in lipid metabolism, is implicated in the advancement of MASLD, characterized by liver inflammation and the development of fibrosis.

Hepatocellular carcinoma (HCC) often marks the terminal phase of MASLD with potential oncogenic signals emerging during the MASH stage [25]. Metabolic dysregulation, oxidative damage, chronic inflammation, and a fibrotic environment in MASLD promote hepatocellular destruction and compensatory proliferation, increasing the risk of genetic aberrations [48,49]. Our study indicated heightened cell growth, evidenced by increased pathways associated with cell cycle (Figure S2C,D). Additionally, a persistent inflammatory status associated with the increased release of tumor necrosis factors could predispose to carcinogenesis specifically related to steatosis and steatohepatitis [50]. Moreover, fibrosis probably caused a poor oxygen exchange, leading to the increase in hypoxia (Figure S2D). Hypoxia might stimulate tumor angiogenesis via the increased gene expression of vascular endothelial growth factor (VEGF) and hypoxia-inducible factor (HIF)-1 (Figure S1E,F), which are identified as potential early indicators of HCC [51]. However, these results could also be triggered by inflammation or liver injury. Therefore, future studies could focus on genetic/epigenetic alterations to further elucidate the progression to HCC.

5. Conclusions

Our study demonstrates that human PCLSs treated with GFIPO for up to 96 h are a viable model for replicating the patient experience of MASLD, effectively simulating its progression, including MASH and fibrosis—stages that have been challenging to study in short-term PCLSs models. The use of NGS in our methodology sheds light on the complex processes of MASLD, allowing for a detailed exploration of its progression. To achieve an even more comprehensive pro-inflammatory environment, additional methods such as introducing pro-inflammatory stimuli or co-culturing with inflammatory cells could be employed. Diseased liver tissues from MASH cirrhosis patients or animal models could also be utilized for elongated incubation.

Supplementary Materials: The following are available online at https://www.mdpi.com/article/10.3390/nu16050626/s1, Figure S1: Altered ATP, ApoB100, pre-selected biological pathways, and gene expression in PCLSs; Figure S2: Altered biological processes or pathways and ICAM1 gene expression in PCLSs; Table S1: List of SYBR Green primers used for RT-Qpcr; Table S2: Overrepresented analysis results on up-regulated genes in WEGG compared to 0-hour groups; Table S3: Overrepresented analysis results on down-regulated genes in WEGG compared to 0-hour groups; Table S4: Additional information including ensemble gene IDs and functions of encoded proteins of all top 10 DEGs.

Author Contributions: Conceptualization, M.L., A.N. and P.O.; methodology, M.L.; software, M.L. and F.T.L.; formal analysis, M.L., F.T.L., M.C.v.d.H., K.G. and J.B.; investigation, M.L., F.T.L., A.R.G., D.O. and K.G.; resources, M.C.v.d.H., J.B., V.E.d.M. and K.R.; data curation, M.L. and P.O.; writing—original draft preparation, M.L.; writing—review and editing, A.N. and P.O.; visualization, M.L.; supervision, A.N. and P.O.; project administration, M.L., A.N. and P.O.; funding acquisition, P.O. and M.L. All authors have read and agreed to the published version of the manuscript.

Funding: This project was funded by the China Scholarship Council (grant number 201906210058) (M.L.).

Institutional Review Board Statement: The study was approved by the Medical Ethical Committee of the University Medical Centre Groningen (UMCG), according to Dutch legislation and the Code of Conduct for dealing responsibly with human tissue in the context of health research (www.federa.org), waiving the need for a written consent for "further use" of coded-anonymous human tissue.

Informed Consent Statement: Patient consent for "further use" of coded-anonymous human tissue was waived according to Dutch legislation and the Code of Conduct for dealing responsibly with human tissue in the context of health research (www.federa.org).

Data Availability Statement: The raw data supporting the conclusions of this article will be made available by the authors on request.

Acknowledgments: The authors thank W. Alfredo Ríos-Ocampo for his help with ApoB100 evaluation.

Conflicts of Interest: The authors declare no conflicts of interest.

References

1. Teng, M.L.; Ng, C.H.; Huang, D.Q.; Chan, K.E.; Tan, D.J.; Lim, W.H.; Yang, J.D.; Tan, E.; Muthiah, M.D. Global incidence and prevalence of nonalcoholic fatty liver disease. *Clin. Mol. Hepatol.* **2023**, *29*, S32–S42. [CrossRef]
2. Powell, E.E.; Wong, V.W.-S.; Rinella, M. Non-alcoholic fatty liver disease. *Lancet* **2021**, *397*, 2212–2224. [CrossRef] [PubMed]
3. Neuschwander-Tetri, B.A. Non-alcoholic fatty liver disease. *BMC Med.* **2017**, *15*, 45. [CrossRef] [PubMed]
4. Santhekadur, P.K.; Kumar, D.P.; Sanyal, A.J. Preclinical models of non-alcoholic fatty liver disease. *J. Hepatol.* **2018**, *68*, 230–237. [CrossRef]
5. Willebrords, J.; Pereira, I.V.A.; Maes, M.; Yanguas, S.C.; Colle, I.; Bossche, B.V.D.; Silva, T.C.D.; Oliveira, C.P.M.S.d.; Andraus, W.; Alves, V.A.; et al. Strategies, models and biomarkers in experimental non-alcoholic fatty liver disease research. *Prog. Lipid Res.* **2015**, *59*, 106–125. [CrossRef]
6. Graaf, I.A.M.d.; Groothuis, G.M.M.; Olinga, P. Precision-cut tissue slices as a tool to predict metabolism of novel drugs. *Expert. Opin. Drug Met.* **2007**, *3*, 879–898. [CrossRef]
7. Morán-Salvador, E.; López-Parra, M.; García-Alonso, V.; Titos, E.; Martínez-Clemente, M.; González-Périz, A.; López-Vicario, C.; Barak, Y.; Arroyo, V.; Clària, J. Role for PPARγ in obesity-induced hepatic steatosis as determined by hepatocyte- and macrophage-specific conditional knockouts. *FASEB J.* **2011**, *25*, 2538–2550. [CrossRef] [PubMed]
8. Prins, G.H.; Rios-Morales, M.; Gerding, A.; Reijngoud, D.J.; Olinga, P.; Bakker, B.M. The Effects of Butyrate on Induced Metabolic-Associated Fatty Liver Disease in Precision-Cut Liver Slices. *Nutrients* **2021**, *13*, 4203. [CrossRef]
9. Rius, B.; Duran-Guell, M.; Flores-Costa, R.; Lopez-Vicario, C.; Lopategi, A.; Alcaraz-Quiles, J.; Casulleras, M.; Lozano, J.J.; Titos, E.; Claria, J. The specialized proresolving lipid mediator maresin 1 protects hepatocytes from lipotoxic and hypoxia-induced endoplasmic reticulum stress. *FASEB J.* **2017**, *31*, 5384–5398. [CrossRef] [PubMed]
10. Prins, G.H.; Luangmonkong, T.; Oosterhuis, D.; Mutsaers, H.A.M.; Dekker, F.J.; Olinga, P. A Pathophysiological Model of Non-Alcoholic Fatty Liver Disease Using Precision-Cut Liver Slices. *Nutrients* **2019**, *11*, 507. [CrossRef]
11. Palma, E.; Doornebal, E.J.; Chokshi, S. Precision-cut liver slices: A versatile tool to advance liver research. *Hepatol. Int.* **2019**, *13*, 51–57. [CrossRef] [PubMed]
12. Shepherd, E.L.; Karim, S.; Newsome, P.N.; Lalor, P.F. Inhibition of vascular adhesion protein-1 modifies hepatic steatosis in vitro and in vivo. *World J. Hepatol.* **2020**, *12*, 931–948. [CrossRef]
13. Gore, E. *Integrating an Ex Vivo Model into Fibrosis Research*; University of Groningen: Groningen, The Netherlands, 2019.
14. Graaf, I.A.M.d.; Olinga, P.; de Jager, M.H.; Merema, M.T.; de Kanter, R.; van de Kerkhof, E.G.; Groothuis, G.M. Preparation and incubation of precision-cut liver and intestinal slices for application in drug metabolism and toxicity studies. *Nat. Protoc.* **2010**, *5*, 1540–1551. [CrossRef] [PubMed]
15. Bligh, E.G.; Dyer, W.J. A rapid method of total lipid extraction and purification. *Can. J. Biochem. Physiol.* **1959**, *37*, 911–917. [CrossRef]
16. Lowry, O.H.; Rosebrough, N.J.; Farr, A.L.; Randall, R.J. Protein measurement with the Folin phenol reagent. *J. Biol. Chem.* **1951**, *193*, 265–275. [CrossRef]
17. Dobin, A.; Davis, C.A.; Schlesinger, F.; Drenkow, J.; Zaleski, C.; Jha, S.; Batut, P.; Chaisson, M.; Gingeras, T.R. STAR: Ultrafast universal RNA-seq aligner. *Bioinformatics* **2013**, *29*, 15–21. [CrossRef] [PubMed]
18. Liao, Y.; Smyth, G.K.; Shi, W. featureCounts: An efficient general purpose program for assigning sequence reads to genomic features. *Bioinformatics* **2014**, *30*, 923–930. [CrossRef]
19. Andrews, S. FastQC. Available online: https://www.bioinformatics.babraham.ac.uk/projects/fastqc/ (accessed on 22 January 2024).
20. Ewels, P.; Magnusson, M.; Lundin, S.; Kaller, M. MultiQC: Summarize analysis results for multiple tools and samples in a single report. *Bioinformatics* **2016**, *32*, 3047–3048. [CrossRef]
21. Love, M.I.; Huber, W.; Anders, S. Moderated estimation of fold change and dispersion for RNA-seq data with DESeq2. *Genome Biol.* **2014**, *15*, 550. [CrossRef]
22. Subramanian, A.; Tamayo, P.; Mootha, V.K.; Mukherjee, S.; Ebert, B.L.; Gillette, M.A.; Paulovich, A.; Pomeroy, S.L.; Golub, T.R.; Lander, E.S.; et al. Gene set enrichment analysis: A knowledge-based approach for interpreting genome-wide expression profiles. *Proc. Natl. Acad. Sci. USA* **2005**, *102*, 15545–15550. [CrossRef]
23. Mootha, V.K.; Lindgren, C.M.; Eriksson, K.-F.; Subramanian, A.; Sihag, S.; Lehar, J.; Puigserver, P.; Carlsson, E.; Ridderstråle, M.; Laurila, E.; et al. PGC-1α-responsive genes involved in oxidative phosphorylation are coordinately downregulated in human diabetes. *Nat. Genet.* **2003**, *34*, 267–273. [CrossRef]
24. Livak, K.J.; Schmittgen, T.D. Analysis of Relative Gene Expression Data Using Real-Time Quantitative PCR and the $2^{-\Delta\Delta Ct}$ Method. *Methods* **2001**, *25*, 402–408. [CrossRef]
25. Gorden, D.L.; Myers, D.S.; Ivanova, P.T.; Fahy, E.; Maurya, M.R.; Gupta, S.; Min, J.; Spann, N.J.; McDonald, J.G.; Kelly, S.L.; et al. Biomarkers of NAFLD progression: A lipidomics approach to an epidemic. *J. Lipid Res.* **2015**, *56*, 722–736. [CrossRef] [PubMed]
26. Alves-Bezerra, M.; Cohen, D.E. Triglyceride metabolism in the liver. *Compr. Physiol.* **2017**, *8*, 1–8. [PubMed]
27. Oligschlaeger, Y.; Shiri-Sverdlov, R. NAFLD preclinical models: More than a handful, less of a concern? *Biomedicines* **2020**, *8*, 28. [CrossRef]
28. Friedman, S.L.; Neuschwander-Tetri, B.A.; Rinella, M.; Sanyal, A.J. Mechanisms of NAFLD development and therapeutic strategies. *Nat. Med.* **2018**, *24*, 908–922. [CrossRef] [PubMed]

29. Castro, R.E.; Diehl, A.M. Towards a definite mouse model of NAFLD. *J. Hepatol.* **2018**, *69*, 272–274. [CrossRef]
30. Postic, C.; Girard, J. The role of the lipogenic pathway in the development of hepatic steatosis. *Diabetes Metab.* **2008**, *34*, 643–648. [CrossRef] [PubMed]
31. Ge, J.; Walewski, J.; Anglade, D.; Berk, P. Regulation of Hepatocellular Fatty Acid Uptake in Mouse Models of Fatty Liver Disease with and without Functional Leptin Signaling: Roles of NfKB and SREBP-1C and the Effects of Spexin. *Semin. Liver Dis.* **2016**, *36*, 360–372. [CrossRef]
32. Ring, A.; Le Lay, S.; Pohl, J.; Verkade, P.; Stremmel, W. Caveolin-1 is required for fatty acid translocase (FAT/CD36) localization and function at the plasma membrane of mouse embryonic fibroblasts. *Biochim. Biophys. Acta (BBA)-Mol. Cell Biol. Lipids* **2006**, *1761*, 416–423. [CrossRef]
33. Ameer, F.; Scandiuzzi, L.; Hasnain, S.; Kalbacher, H.; Zaidi, N. De novo lipogenesis in health and disease. *Metabolism* **2014**, *63*, 895–902. [CrossRef] [PubMed]
34. Yamaguchi, K.; Yang, L.; McCall, S.; Huang, J.; Yu, X.X.; Pandey, S.K.; Bhanot, S.; Monia, B.P.; Li, Y.-X.; Diehl, A.M. Inhibiting triglyceride synthesis improves hepatic steatosis but exacerbates liver damage and fibrosis in obese mice with nonalcoholic steatohepatitis. *Hepatology* **2007**, *45*, 1366–1374. [CrossRef] [PubMed]
35. Kawano, Y.; Cohen, D.E. Mechanisms of hepatic triglyceride accumulation in non-alcoholic fatty liver disease. *J. Gastroenterol.* **2013**, *48*, 434–441. [CrossRef] [PubMed]
36. Malaguarnera, M.; Rosa, M.D.; Nicoletti, F.; Malaguarnera, L. Molecular mechanisms involved in NAFLD progression. *J. Mol. Med.* **2009**, *87*, 679–695. [CrossRef] [PubMed]
37. Zheng, F.; Cai, Y. Concurrent exercise improves insulin resistance and nonalcoholic fatty liver disease by upregulating PPAR-γ and genes involved in the beta-oxidation of fatty acids in ApoE-KO mice fed a high-fat diet. *Lipids Health Dis.* **2019**, *18*, 6. [CrossRef] [PubMed]
38. Prins, G.H. *Metabolic-Associated Fatty Liver Disease: A Major Problem Miniaturised*; University of Groningen: Groningen, The Netherlands, 2022. [CrossRef]
39. Jiang, Z.Y.; Zhou, Y.; Zhou, L.; Li, S.W.; Wang, B.M. Identification of Key Genes and Immune Infiltrate in Nonalcoholic Steatohepatitis: A Bioinformatic Analysis. *Biomed. Res. Int.* **2021**, *2021*, 7561645. [CrossRef]
40. Söderberg, C.; Marmur, J.; Eckes, K.; Glaumann, H.; Sällberg, M.; Frelin, L.; Rosenberg, P.; Stål, P.E.R.; Hultcrantz, R. Microvesicular fat, inter cellular adhesion molecule-1 and regulatory T-lymphocytes are of importance for the inflammatory process in livers with non-alcoholic steatohepatitis. *APMIS* **2011**, *119*, 412–420. [CrossRef]
41. Mackay, C.R.; Imhof, B.A. Cell adhesion in the immune system. *Immunol. Today* **1993**, *14*, 99–102. [CrossRef]
42. Kumar, S.; Duan, Q.; Wu, R.; Harris, E.N.; Su, Q. Pathophysiological communication between hepatocytes and non-parenchymal cells in liver injury from NAFLD to liver fibrosis. *Adv. Drug Deliv. Rev.* **2021**, *176*, 113869. [CrossRef]
43. Fabregat, I.; Caballero-Díaz, D. Transforming Growth Factor-β-Induced Cell Plasticity in Liver Fibrosis and Hepatocarcinogenesis. *Front. Oncol.* **2018**, *8*, 357. [CrossRef]
44. Ahmed, H.; Umar, M.I.; Imran, S.; Javaid, F.; Syed, S.K.; Riaz, R.; Hassan, W. TGF-β1 signaling can worsen NAFLD with liver fibrosis backdrop. *Exp. Mol. Pathol.* **2022**, *124*, 104733. [CrossRef]
45. Fabregat, I.; Moreno-Càceres, J.; Sánchez, A.; Dooley, S.; Dewidar, B.; Giannelli, G.; Dijke, P.; IT-LIVER Consortium. TGF-β signalling and liver disease. *FEBS J.* **2016**, *283*, 2219–2232. [CrossRef] [PubMed]
46. Dewidar, B.; Meyer, C.; Dooley, S.; Meindl-Beinker, N. TGF-β in hepatic stellate cell activation and liver fibrogenesis—Updated 2019. *Cells* **2019**, *8*, 1419. [CrossRef] [PubMed]
47. Moylan, C.A.; Pang, H.; Dellinger, A.; Suzuki, A.; Garrett, M.E.; Guy, C.D.; Murphy, S.K.; Ashley-Koch, A.E.; Choi, S.S.; Michelotti, G.A.; et al. Hepatic gene expression profiles differentiate presymptomatic patients with mild versus severe nonalcoholic fatty liver disease. *Hepatology* **2014**, *59*, 471–482. [CrossRef] [PubMed]
48. Ghouri, Y.A.; Mian, I.; Rowe, J.H. Review of hepatocellular carcinoma: Epidemiology, etiology, and carcinogenesis. *J. Carcinog.* **2017**, *16*, 1. [CrossRef] [PubMed]
49. Baffy, G.; Brunt, E.M.; Caldwell, S.H. Hepatocellular carcinoma in non-alcoholic fatty liver disease: An emerging menace. *J. Hepatol.* **2012**, *56*, 1384–1391. [CrossRef]
50. Dongiovanni, P.; Romeo, S.; Valenti, L. Hepatocellular carcinoma in nonalcoholic fatty liver: Role of environmental and genetic factors. *World J. Gastroenterol.* **2014**, *20*, 12945–12955. [CrossRef]
51. Nishida, N. Role of Oncogenic Pathways on the Cancer Immunosuppressive Microenvironment and Its Clinical Implications in Hepatocellular Carcinoma. *Cancers* **2021**, *13*, 3666. [CrossRef]

nutrients

Review

Health Effects and Mechanisms of Inulin Action in Human Metabolism

Jaime Alonso-Allende [1,2], Fermín I. Milagro [1,2,3,4,*] and Paula Aranaz [2,3]

1 Department of Nutrition, Food Science and Physiology, Faculty of Pharmacy and Nutrition, University of Navarra, 31008 Pamplona, Spain; jalonsoallende@unav.es
2 Center for Nutrition Research, University of Navarra, 31008 Pamplona, Spain; paranaz@unav.es
3 Navarra Institute for Health Research (IdiSNA), 31009 Pamplona, Spain
4 Centro de Investigación Biomédica en Red de la Fisiopatología de la Obesidad y Nutrición (CIBEROBN), Instituto de Salud Carlos III, 28029 Madrid, Spain
* Correspondence: fmilagro@unav.es; Tel.: +34-948-425600 (ext. 806553)

Abstract: Inulin is a plant polysaccharide which, due to its chemical structure, is not digestible by human gut enzymes but by some bacteria of the human microbiota, acting as a prebiotic. Consequently, inulin consumption has been associated with changes in the composition of the intestinal microbiota related to an improvement of the metabolic state, counteracting different obesity-related disturbances. However, the specific mechanisms of action, including bacterial changes, are not exactly known. Here, a bibliographic review was carried out to study the main effects of inulin on human metabolic health, with a special focus on the mechanisms of action of this prebiotic. Inulin supplementation contributes to body weight and BMI control, reduces blood glucose levels, improves insulin sensitivity, and reduces inflammation markers, mainly through the selective favoring of short-chain fatty acid (SCFA)-producer species from the genera *Bifidobacterium* and *Anaerostipes*. These SCFAs have been shown to ameliorate glucose metabolism and decrease hepatic lipogenesis, reduce inflammation, modulate immune activity, and improve anthropometric parameters such as body weight or BMI. In conclusion, the studies collected suggest that inulin intake produces positive metabolic effects through the improvement of the intestinal microbiota and through the metabolites produced by its fermentation.

Keywords: inulin; SCFA; microbiota; obesity; bifidobacteria; pathway; health; insulin

1. Introduction

1.1. Inulin Structure and Properties

Inulin is a linear fructan consisting of fructosyl units linked by β (2→1) bonds, typically with a glucose moiety attached at the end through an α (1→2) linkage [1]. Structurally, it is a polysaccharide made up of D-fructofuranose units, rendering it resistant to hydrolysis by human gastrointestinal enzymes due to its β-configuration (Figure 1) [1]. The molecular formula GFn denotes the presence of a terminal glucose unit (G) and fructose units (F), with 'n' indicating the number of fructose units [2].

Depending on the degree of polymerization (DP), ranging from 2 to 60, inulin can be categorized into two main varieties: short-chain inulin, containing 2–10 fructose units, and long-chain inulin, containing 10–60 fructose units [2,3]. The DP determines inulin's properties (viscosity, solubility or even color) and it is influenced by several factors such as plant maturity, climate, and extraction techniques [1,3]. The DP also affects the organoleptic properties of inulin, the short-chain form being sweeter than the long-chain form, which is why it is often that these varieties are commonly used as a sucrose substitute and as a fat substitute, respectively [3]. In addition, its chemical structure, which contains fructose monomers with anomeric C2 in beta-configuration, makes it resistant to hydrolysis by digestive enzymes, making it a non-digestible carbohydrate for humans [2,4]. Consequently, inulin has been principally studied and used as a prebiotic, as it can be digested by some bacteria of the human microbiota and can have positive health effects, particularly in obesity-related metabolic diseases [2,3].

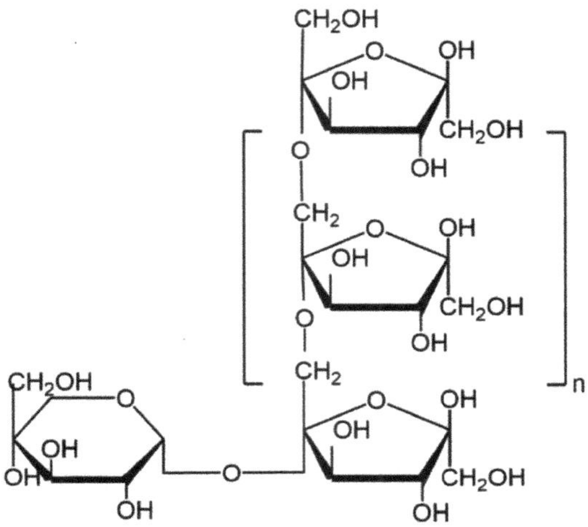

Figure 1. Haworth projection of inulin molecule chemical structure [3].

1.2. Sources of Inulin

Inulin is a polysaccharide that can be found in a wide variety of plant families, with *Liliaceae*, *Amaryllidaceae*, and *Asteraceae* being its main natural sources [2,5]. Inulin is stored in different regions of the plant, including bulbs, roots, and tubers, depending on the needs and physiology of each species [5]. In terms of species, chicory (*Cichorium intybus* L.) is a major natural source of inulin, containing approximately 20% inulin by wet weight and 80% by dry weight in the roots [6]. Chicory inulin presents the highest fructose/glucose ratio [7]. Jerusalem artichoke (*Helianthus tuberosus*), which belongs to the *Asteraceae* family, also contains a high amount of inulin (17–20% by wet weight) and stores it in its tubers [8]. Garlic (*Allium sativum* L.) stores inulin in its bulb and has similar content as chicory, around 75% by dry weight [3].

In addition, since inulin's main function is to store energy as carbohydrates, the time in the plant's life cycle at which inulin is extracted must be considered. This determines the degree of polymerization, since it is increased or reduced according to energetic and physiological needs. Roughly speaking, the degree of polymerization increases from spring to mid-autumn and declines with the onset of winter until the following spring. Therefore, even if two inulin samples are taken from the same plant, they can be very different if they are collected at two different stages of the life cycle.

1.3. Study Aim

The present article aims to review the effects of inulin on obesity and human microbiota by compiling the results obtained in the clinical trials available to date that have investigated the effect of inulin on both aspects of human health. Furthermore, this study tries to emphasize the mechanisms of action that justify its use in obesity and metabolic syndrome.

2. Materials and Methods

This article is based on the available literature in MEDLINE (PubMed), Scopus, and Cochrane Library to identify clinical trials related to the effect of the prebiotic inulin on human microbiota and obesity. The study encompasses all full-text publications that were available between January 2009 and 28 June 2024. The search strategy is shown in Figure 2.

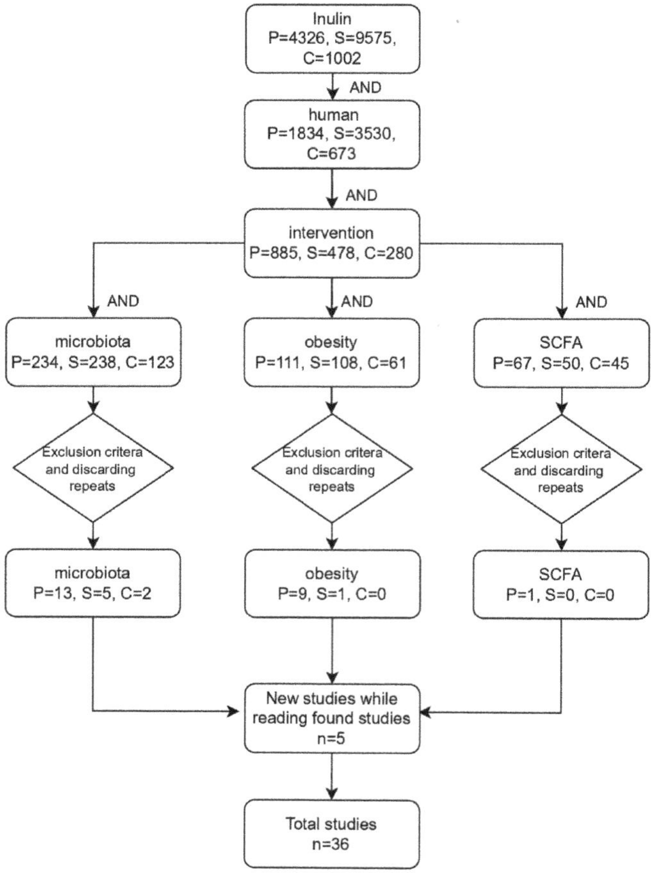

Figure 2. Flowchart depicting the search strategy. P: PubMed; S: Scopus; C: Cochrane Library.

Exclusion criteria were settled for the purpose of filtering the studies obtained, which included studies without human subjects, studies with the main topic of the study not related to the research, studies using symbiotics without specifying the inulin content, studies with a low number of human subjects (below 10) or short study duration, and those studies where only the abstract was available. For the analysis of the results, the following parameters were considered in each article: type of study, population, cohort, patient status, age, sample size, sex (M/F), inulin origin, dose, characterization, characterization control, duration (weeks), intervention group, control group, increasing effect, decreasing effect, no effect, inulin mechanism of action, and adverse effects.

3. Study Characteristics

The studies analyzed are listed in Table 1, which describes the general characteristics and the status of their participants.

The most common studies analyzed include the implementation of randomized, controlled trials (RCTs) and various combinations of blinding and control measures such as double-blinding, placebo control, and crossover designs. Conversely, characteristics such as single blinding, multicentric settings, and specific population focuses like obese individuals appear less frequently.

Table 1. Study designs and subjects' characteristics among the available clinical trials.

Study Type	Inulin Source	Intervention Group	Control Group	Population	Age	Sample Size (Male/Female)	Reference
Randomized, controlled trial	Beneo-Orafti HP, Kreglinger Europe, Antwerpen, Belgium	Inulin-propionate ester	Inulin and Cellulose	English adults with overweight and obesity	18–65	12 (3/9)	[9]
Clinical trial	Oligofructose-inulin (Synergy1; BENEO GmbH, Mannheim, Germany)	Oligofructose-inulin	Maltodextrin placebo	Canadian healthy children with overweight and obesity	7–12	38	[10]
Randomized, double-blind, placebo-controlled trial	Extracted inulin powder from Thai Jerusalem artichoke	Extracted inulin powder	Isocaloric maltodextrin	Thai children with obesity	7–15	165	[11]
Randomized, controlled, cross-over study	Not available	Inulin-propionate ester	Inulin and Cellulose	English adults with excess of weight	40–65	155	[12]
Double-blind, randomized, cross-over intervention study	Chicory-derived inulin (Orafti inulin)	Chicory-derived inulin	Maltodextrin	German healthy men and women having constipation	20–75	54 (33/11)	[13]
Randomized, placebo-controlled, double-blind, cross-over trial	50/50 mixture of oligofructose and inulin; Orafti® Synergy1, Beneo GmbH, Germany)	Inulin-type fructans	Maltodextrin	Norwegian patients with type 2 diabetes	63.1 (mean)	25 (10/15)	[14]
Randomized, double-blind, cross-over design	Chicory inulin (Frutafit supplied by Imperial Suiker-Unie, Sugar Land, TX, USA; produced by Sensus, Roosendaal, The Netherlands)	Low-fiber diet with chicory inulin	Low-fiber control diet	U.S. healthy human subjects	27–49	12 (12/0)	[15]
Four-arm parallel, double-blind, randomized, placebo-controlled trial	Orafti P95, DP 3–9, average DP 4; BENEO-Orafti	OF + maltodextrin/FL	Maltodextrin + FL	U.K. healthy adults	18–50	92 (30/62)	[16]
Multicenter, randomized, placebo-controlled trial performed in obese individuals	Extracted from chicory root, Cosucra, Pecq, Belgium	Inulin and inulin + PA	Maltodextrin and Maltodextrin + PA	Belgian obesity-related metabolic disorder adults	18–65	59	[17]

Table 1. *Cont.*

Study Type	Inulin Source	Intervention Group	Control Group	Population	Age	Sample Size (Male/Female)	Reference
Randomized, controlled trial	>85% fructo-oligosaccharides; average chain length: seven monomers; Sensus, Roosendaal, The Netherlands	Green tea + Inulin beverage	Placebo beverage	Taiwanese adults with excess of weight	20–50	30	[18]
Randomized, double-blind, parallel, placebo-controlled trial	Synergy 1, namely, inulin/oligofructose 50/50 mix, Orafti, Oreye, Belgium	Inulin + oligofructose 50/50	Maltodextrin	Belgian women with obesity	18–65	30 (0/30)	[19]
Placebo-controlled, double-blind, cross-over trial	Not available	Inulin snack	Control snack	Canadian healthy adults	18–65	48 (22/28)	[20]
Randomized, controlled, cross-over trial	BIOAGAVE agave inulin fiber; Ingredion Incorporated	Inulin-containing chews	control chews	U.S. healthy adults	18–65	29	[21]
Randomized, placebo-controlled trial	Orafti® Synergy 1, Beneo, Tienen, Belgium	Oligofructose-enriched inulin	Maltodextrin	Polish children with celiac disease	4–18	34 (13/21)	[22]
Randomized, placebo-controlled, double-blind, parallel intervention	Fibruline® Instant, Cosucra Group Warcoing, Warcoing, Belgium	Inulin + maltodextrin	Maltodextrin	Danish adults with obesity or excess of weight	18–60	86 (31/54)	[23]
Randomized, single-blind, multicentric, placebo-controlled trial	Extracted from chicory root, Cosucra, Belgium	Inulin	Maltodextrin	Belgian adults with obesity	18–65	110	[24]
Randomized, controlled clinical trial	Not available	Inulin	Maltodextrin	Female patients with type 2 diabetes mellitus	>18	49 (0/49)	[25]
Multicentric, single-blind, placebo-controlled trial	Extracted from chicory root, Cosucra, Belgium	Inulin	Maltodextrin	Belgian patients with obesity	>18	24	[26]
Cross-over randomized, controlled trial	Not available	Inulin	Glucose	Spanish patients with excess of weight	18–65	25 (12/13)	[27]

Table 1. *Cont.*

Study Type	Inulin Source	Intervention Group	Control Group	Population	Age	Sample Size (Male/Female)	Reference
Randomized, double-blind, placebo-controlled, cross-over study	Orafti® Synergy1–50:50 inulin to fructo-oligosaccharide mix; Beneo GmbH	Inulin-type fructan	Maltodextrin	New Zealand healthy adults	19–65	33 (13/20)	[28]
Double-blind, placebo-controlled study	FrutafitTEX!, Sensus, Roosendaal, The Netherlands	Chicory long-chain inulin	Glucose	Caucasian old patients	55–80	26 (18/8)	[29]
Interventional prospective controlled study	Not available	Low-protein diet and inulin	Low-protein diet	Italian patients with chronic kidney disease	18–80	41 (25/16)	[30]
Double-blind, randomized, placebo-controlled, cross-over study	Bayer BioScience GmbH, Hermannswerder, Potsdam, Germany	VLCI	Maltodextrin	UK healthy adults	20–42	31	[31]
Randomized, double-blind, placebo-controlled, cross-over study	Orafti Synergy, Beneo, Belgium	Inulin	Maltodextrin	U. S. individuals undergoing HD	55 (mean)	12 (6/6)	[32]
Randomized, double-blind, cross-over design	Fibruline Instant; Cosucra Group Warcoing	Inulin	Maltodextrin	Swiss healthy adult women	18–40	32 (0/32)	[33]
Randomized, controlled trial	Chicory root	Inulin	Baseline	U. S. healthy young adults	17–29	174	[34]
Randomized, triple-blind, controlled trial	Chicory root	Inulin	Baseline	Iranian type 2 diabetic women	20–65	49 (0/49)	[35]
Double-blind, placebo-controlled, intervention study	Chicory root	Inulin and oligofructose (50/50)	Maltodextrin	Belgian women with obesity	18–65	30 (0/30)	[36]
Randomized, triple-blind, controlled trial	Quantum High-Tech Biologicals Co. Ltd., Jiangmen, China	Inulin	Baseline	Chinese healthy adults	18–65	57 (22/35)	[37]
Randomized, controlled trial	Not available	Inulin	Baseline	Czechs Patients with type 2 diabetes	18–65	27	[38]
Randomized, controlled trial	Chicory root	Inulin and Arabinoxylan (50/50)	Maltodextrin	U.K. healthy adult men	19–55	20	[39]
Prospective single-arm study	Inulin Biosciences Company, Wuhan, China	Inulin	Baseline	Chinese adult patients with prediabetes	37–69	49 (16/39)	[40]

Table 1. *Cont.*

Study Type	Inulin Source	Intervention Group	Control Group	Population	Age	Sample Size (Male/Female)	Reference
Three-arm parallel, placebo-controlled, randomized, double-blind study	Jerusalem artichoke	Inulin + fruit juice	Fruit juice	U.K. healthy adults	18–55	66 (33/33)	[41]
Simple randomized intervention study	Frutafit® IQ, Roosendaal, The Netherlands	Inulin	Baseline	Kuwaiti adult women with obesity	18–65	12	[42]
RCT	Not available	Inulin + pomegranate juice	Pomegranate juice	Chinese adults with obesity	18–65	67 (33/34)	[43]
Randomized, controlled trial	Jerusalem artichoke	Inulin	Maltodextrin	Thai children with obesity	7–15	143	[44]

Among the countries where these studies have been conducted, the most common ones include Belgium, the United States, and the United Kingdom. Additionally, Canada, Spain, Germany, and Italy are mentioned multiple times, indicating their significant contribution to the body of research in this field. Comparatively, it must be emphasized that there are very few studies from Asia and none from Africa or South America. These findings are conclusive, as the geographical distribution of the studies coincides with the biodistribution of the main natural sources of inulin, such as chicory or Jerusalem artichoke. However, it is noteworthy that countries with higher numbers of studies also tend to have better socioeconomic conditions.

Moreover, there is a notable diversity in the origins of inulin utilized in the analyzed studies. The most common source of inulin in the analyzed studies is chicory, including other natural sources such as Jerusalem artichoke and agave, whereas some studies do not specify their inulin origin. Additionally, specific commercial brands are also observed, where Beneo Orafti stands out as the main supplier.

The intervention groups consisted of a variety of treatments involving inulin or an inulin-containing compound, such as inulin–propionate ester or combinations of inulin with other substances like maltodextrin or polyphenols [9,18,23]. In addition, the administration form varies widely, including powder, snack, chews, and beverages [18,20,21,27,41,43]. Notably, inulin alone is frequently used as an intervention across multiple studies, appearing multiple times as a treatment group. On the other hand, the control groups typically involve the administration of maltodextrin as placebo, as it is very similar to inulin in taste and flavor. However, glucose, cellulose, and even fruits are sometimes used as well [9,12,27,29,41,43]. Control groups also include baseline measurements or control diets without the intervention substance. Among the provided doses of inulin, the range varies from 3 g per day to 20 g per day.

Most of the analyzed studies include adults with overweight or obesity [9,12,18,19,23,24,27,42]. Additionally, healthy adults represent a prevalent group [15,16,20,21,28,31,33,34,37,39,41,43]. Specific medical conditions such as type 2 diabetes mellitus and obesity-related metabolic disorders also happen to appear, indicating a focus on investigating interventions in these populations [14,17,25]. Notably, there is a significant representation of studies involving children, particularly those with obesity-related conditions like obese children with elevated body mass index (BMI) or healthy children with overweight and obesity [10,11,44]. Furthermore, certain studies target specific groups such as patients with hypertrygliceridemic and hypercholesterolemic status, chronic kidney disease, or celiac disease [22,30]. Overall, these findings reflect a diverse range of patient populations targeted in clinical research, spanning from healthy adults to individuals with various medical conditions across different age groups.

Among the age ranges utilized in the analyzed studies, a broad spectrum is observed, with the most common range being adults aged 18 to 65 years old [19,23,24,27,28,37,38,41–43]. This range encompasses a significant portion of the research population, indicating a focus on adult participants across various studies. Additionally, certain studies targeted narrower age groups, such as children aged 7 to 15 or adults aged 40 to 80 [10–12,22,29]. For the total number of individuals studied per trial, the highest numbers of individuals studied in a single trial was 174, while the lowest was 12. Above this mean value, (approximately 58) there are 11 trials with participant numbers ranging from 59 to 174, indicating larger-scale studies. Conversely, below the mean value, there are 16 trials with participant numbers ranging from 12 to 49, suggesting smaller-scale studies or trials with more specific inclusion criteria. Regarding the male/female ratio, the highest male/female ratio is 33/11, while the lowest ratio is 0/30 [13,19]. Overall, while the differences may not be statistically significant, there is a slight tendency towards a higher proportion of men studied, being predominant on 16 of the analyzed trials.

4. Inulin Effect on Human Metabolic Health

The studies compiled on the role of inulin on metabolic health are listed in Table 2, which specifies the physiological effects and its activity modulating gut microbiota.

Table 2. Effects of inulin on metabolic disturbances and gut microbiota according to its origin, dose, and intervention duration.

Source	Effect on Gut Microbiota	Effect on Metabolic Disturbances and SCFAs	Dose (g/d)	Time (Weeks)	Reference
Chicory	**Increase**: *Bifidobacterium, Anaerostipes* **Decrease**: *Bilophila* **No effect**: *Akkermansia, Eubacterium, Faecalibacterium, Lactobacillus*	None	12	4	[13]
Chicory	**Increase**: Total anaerobes, *Lactobacillus* **Decrease**: None **No effect**: *Clostridium, Bifidobacterium, Enterobacteriaceae*	None	20	6	[15]
Chicory	**Increase**: *Bifidobacterium, Anaerostipes, B. angulatum* **Decrease**: *Clostridium sensu stricto* **No effect**: *B. adolescentis, B. bifidum*	**Increase**: Plasma AST levels **Decrease**: Body weight, BMI, liver stiffness, TC **No effect**: None	16	12	[17]
Chicory	**Increase**: *B. bifidum, B. longum B. adolescentis, Catenibacterium* **Decrease**: *Desulfovibrio, Roseburia*, **No effect**: None	**Increase**: Insulin HOMA-IR, HOMA-ISI **Decrease**: Body weight, BMI **No effect**: HDL-C, LDL-C, HOMA-IR, TGs, TC, SBP, DBP	16	12	[24]
Chicory	**Increase**: *Bifidobacterium, Anaerostipes, Catenibacterium* **Decrease**: *Actinomyces*, Erysipelotrichaceae, Lachnospiraceae, Enterobacteriaceae **No effect**: None	**Increase**: Linolenic acid **Decrease**: None **No effect**: SCFAs, body weight, BMI, fat mass, waist, SBP, DBP, TC, LDL-C, HDL-C, TG	16	12	[26]
Chicory	**Increase**: *B. angulatum, B. ruminantium, B. adolescentis* **Decrease**: None **No effect**: None	**Increase**: None **Decrease**: None **No effect**: SCFAs	8	9	[29]

Table 2. *Cont.*

Source	Effect on Gut Microbiota	Effect on Metabolic Disturbances and SCFAs	Dose (g/d)	Time (Weeks)	Reference
Chicory	**Increase**: None **Decrease**: None **No effect**: *Bifidobacterium*, *Faecalibacterium*	**Increase**: None **Decrease**: None **No effect**: Butyrate, propionate, acetate	10 or 15	13	[32]
Chicory	**Increase**: *Bifidobacterium* **Decrease**: None **No effect**: None	**Increase**: Lactate **Decrease**: Fecal pH **No effect**: Butyrate, propionate, fumarate, acetate, iron absorption	20	4	[33]
Chicory	**Increase**: *B. longum*, *B. adolescentis*, *Anaerostipes hadrus* **Decrease**: None **No effect**: None	**Increase**: Total SCFA **Decrease**: None **No effect**: None	20	4	[34]
Chicory	**Increase**: *A. hadrus*, *B. faecale*, *Bacteroides caccae* **Decrease**: *Ruminococcus faecis*, *Blautia obeum*, *Blautia faecis* **No effect**: None	**Increase**: Insulin sensitivity, IL-10 **Decrease**: Fasting insulin, IL-8 **No effect**: SCFAs, body weight, food intake	20	6	[9]
Jerusalem artichoke	None	**Increase**: FFMI **Decrease**: BMI-z, FMI, LDL-C **No effect**: TC, HDL-C, TGs, FPG, SBP	13	24	[11]
Agave	**Increase**: *B. adolescentis*, *B. breve*, *B. longum*, *B. pseudolongum* **Decrease**: *Desulfovibrio*, *Lachnobacterium* *, *Ruminococcus* * **No effect**: *B. animalis*, *B. bifidum*, *Akkermansia*, *Faecalibacterium*, *Coprococcus*	**Increase**: None **Decrease**: None **No effect**: Propionate, butyrate, acetate	5 or 7 *	12	[21]
Global artichoke	**Increase**: *Bifidobacterium*, lactobacilli–enterococci **Decrease**: *Bacteroides–Prevotella* **No effect**: *Escherichia coli*, *Eubacterium rectale–Clostridium coccoides group*, *Ruminococcus*	**Increase**: None **Decrease**: None **No effect**: SCFAs	10	6	[31]
Not available	None	**Increase**: HDL-C **Decrease**: Serum insulin, TC, TGs **No effect**:	19	24	[30]
Not available	None	**Increase**: None **Decrease**: FBS, HbA1c, fasting insulin, HOMA-IR, hs-CRP, TNF-α **No effect**: None	10	8	[25]
Not available	**Increase**: *Bifidobacterium*, *Cellulomonas*, *Nesterenkonia* *, *Brevibacterium* * **Decrease**: *Ruminococcus*, *Dorea* **No effect**: *Lachnospira*, *Oscillospira*	**Increase**: None **Decrease**: None **No effect**: SCFAs	3 or 7 *	12	[20]

Table 2. *Cont.*

Source	Effect on Gut Microbiota	Effect on Metabolic Disturbances and SCFAs	Dose (g/d)	Time (Weeks)	Reference
Not available	None	**Increase**: None **Decrease**: BMI, body weight, fasting glucose, HbA1c, **No effect**: Fasting insulin, HOMA-IR	10	8	[35]
Not available	**Increase**: *Bifidobacterium, Eubacterium,* **Decrease**: *Ruminococcus* **No effect**: None	**Increase**: Butyrate, propionate **Decrease**: None **No effect**: Acetate	10	16	[37]
Not available	**Increase**: *Bifidobacterium, Faecalibacterium, Akkermansia, Anaerostipes* **Decrease**: *Bacteroides* **No effect**: None	**Increase**: Insulin sensitivity, butyrate, propionate **Decrease**: None **No effect**: Acetate	10	12	[38]
Not available	**Increase**: *Bifidobacterium, Lactobacillus, Anaerostipes* **Decrease**: None **No effect**: None	**Increase**: TGs **Decrease**: HDL-c, LDL-c **No effect**: None	15	24	[40]
Not available	None	**Increase**: None **Decrease**: Body weight, BMI **No effect**: None	21	6	[42]
Jerusalem artichoke	**Increase**: *Bifidobacterium, Faecalibacterium,* **Decrease**: None **No effect**: None	**Increase**: None **Decrease**: None **No effect**: Acetate, butyrate, propionate	13	24	[44]

"None" means either that the study showed no outcomes or results were considered irrelevant. * Exclusive results obtained at 7 g/d dose. LDL-C: low-density lipoprotein cholesterol; HDL-C: high-density lipoprotein cholesterol; TC: total cholesterol; TG: triglycerides; FFMI: free fat mass index; FMI: fat mass index; SBP: systolic blood pressure; DBP: diastolic blood pressure; HOMA-IR: Homeostatic Model Assessment of Insulin Resistance; SCFA: short-chain fatty acids; TNF-α: tumor necrosis factor alpha. AST: aspartate aminotransferase; HOMA-ISI: homeostasis model assessment of insulin sensitivity index; BMI-z: body mass index z-score; FPG: fasting plasma glucose; hs-CRP: high-sensitivity c-reactive protein; HbA1c: hemoglobin A1c.

4.1. Physiological Effects of Inulin

All inulin physiological effects are detailed in Table 2. Several human studies have shown that inulin supplementation leads to a significant improvement in anthropometric parameters, particularly body weight and BMI. In fact, two independent clinical trials found that 16 g/d of inulin for 12 weeks significantly decreased both parameters in adults with obesity or metabolic disorder, even though 10 g/d for 8 weeks has been observed to achieve similar results [17,24,35]. However, two studies with similar dosages (21 g/d and 20 g/d) and identical intervention periods (6 weeks) showed contradictory results as they found beneficial or no effects on anthropometric parameters, respectively [9,42]. Connected to this, anthropometric parameters are closely linked to an individual's metabolic status, such as blood glucose levels or insulin sensitivity, which are key determinants of overall metabolic health. In relation to this, it has been found that inulin supplementation could reduce insulin levels in blood and improve insulin sensitivity [9,24]. For example, inulin consumption at 10 g/d for 8 weeks ultimately ends up lowering blood glucose levels, improving glucose overall metabolism [35]. In addition, it is noteworthy that inulin can achieve these effects within a wide range of dosages and intervention time periods [9,30,35,38]. Regarding other physiological parameters, inulin can positively affect the metabolic lipid profile. Thus, two studies have shown that inulin can reduce LDL-c, total cholesterol, and triglyceride blood levels, even at different doses [30]. However, the use of natural sources seems to be an important factor, since Jerusalem artichoke inulin and chicory inulin showed contradictory effects at similar doses. In addition, inulin is capable of increasing HDL-c serum levels when it is consumed for a long time, whereas short-time consumption has been seen not to be effective [24,30]. Interestingly, inulin supplementation improves inflammation status as it has been observed to decrease proinflammatory cytokines and biomarkers, such as TNF-α or calprotectin, as well as increase anti-inflammatory parameters such as IL-10 [9,25,26].

4.2. Link between Microbiota Modulation and Physiological Effects Produced by Inulin Supplementation

As discussed in the introduction, inulin can only be digested by some bacterial species present in the human microbiota, generating changes in its composition that eventually produce the physiological effects observed after inulin supplementation. Among them, most of the studies show that inulin significantly increases different *Bifidobacterium* species, regardless of dose or inulin source, highlighting *B. longum*, *B. adolescentis*, and *B. angulatum* [17,24,29,34]. Surprisingly, inulin was observed not to affect *Bifidobacterium* populations at 20 g/d dose for 6 weeks when following a low-fiber diet or when presenting a metabolic disorder status, suggesting that metabolic status may influence inulin prebiotic action, since 20 g/d dose for 4 weeks in healthy subjects has been found twice to significantly increase *Bifidobacterium* species [15,33,34]. In addition to this, inulin has been found to negatively impact *B. bifidum* at low doses but has an increasing effect at doses over 16 g/d [24]. Nevertheless, it may be possible that this outcome is not caused directly by inulin supplementation but due to competitive exclusion, as other Bifidobacteria species may be better at utilizing inulin as an energy source, which would explain why higher doses positively affect this species [15,32].

This marked bifidogenic effect has been related to different effects on either anthropometric or physiological parameters. Notably, *Bifidobacterium* increase has been associated with higher insulin sensitivity and improved insulin blood levels [17,38]. In addition, the revisions performed suggest that the dosage plays a crucial role in this bifidogenic effect of inulin, modifying its influence over anthropometric parameters. Thus, studies have shown that at high doses administered over a prolonged period, the increase in Bifidobacteria is associated with improvements in these values, especially body weight and BMI [17,24]. Conversely, at lower doses or shorter administration periods, an increase in these species is observed without a corresponding improvement in body measurements [20]. In relation to the lipid profile, the increase in Bifidobacteria following inulin supplementation presents highly contradictory data regarding total cholesterol levels. Two studies using the same source of inulin, at the same dosage, over the same duration and on the same geographical population yielded conflicting results: one study reported an increase in total cholesterol, while the other showed no significant changes [17,24]. These discrepancies highlight the need for further research to elucidate the effects of inulin on lipid profiles.

Although *Bifidobacterium* increase is the most common inulin effect on the intestine, it has also been reported that inulin positively affects the *Anaerostipes* genus independently of dosages or intervention periods. Increases in species of this genus have been related to beneficial physiological effects, including the improvement of glycemia and anthropometric parameters [9]. Specifically, *Anaerostipes* is related to a rapid improvement of inflammatory markers as 6 weeks inulin intervention showed increased anti-inflammatory and decreased proinflammatory markers [9]. On the other hand, inulin's modulation of the gut microbiota has other positive effects on health since by increasing their number so effectively and lowering the pH of the colon, they hinder and prevent the development of pathogenic or non-beneficial species. For example, inulin-induced microbiota modulation significantly reduces the population number of species belonging to the *Clostridium* genus, leading to reduced inflammation and enhanced overall digestive health and nutrient absorption.

Lastly, it should be clarified that, even though SCFAs increase theoretically, several studies have found no effects on SCFA levels. Nevertheless, as most of the studies analyzed them in fecal samples, it may be possible that enterocytes have consumed most of them, since SCFAs are the main source of energy for them, and therefore, samples containing a lower amount of SCFAs than actually produced were being analyzed.

5. Effect of Inulin Combinations with Other Compounds on Human Metabolic Health

The studies compiled on the role of inulin combinations on metabolic health are listed in Table 3, which specifies the physiological effects and its activity modulating gut microbiota.

Table 3. Effect of inulin combined with other compounds on human metabolism and microbiota according to its origin, dose, and time of intervention. "None" means either that the study showed no outcomes or results were considered irrelevant.

Combination	Effect on Gut Microbiota	Effect on Metabolic Disturbancesand SCFAs	Dose (g/d)	Time (Weeks)	Reference
Inulin–propionate ester	**Increase**: *Bacteroides uniformis, Bacteroides xylanisolvens* **Decrease**: *Blautia obeum, Eubacterium ruminantium, A. hadrus, B. faecale, Prevotella copri* **No effect**: *Bifidobacterium*	**Increase**: Insulin sensitivity, adipose tissue insulin resistance **Decrease**: Fasting insulin **No effect**: SCFAs, body weight, food intake	20	6	[9]
Inulin–propionate ester	None	**Increase**: PYY and GLP-1 **Decrease**: IHCL, Weight gain, intra-abdominal adipose tissue distribution **No effect**: None	10	24	[12]
Glucose + inulin	None	**Increase**: Acetate, propionate, and butyrate **Decrease**: None **No effect**: GLP-1, PYY, Grhelin	99	3	[27]
Catechins + inulin	None	**Increase**: None **Decrease**: Body weight, fat mass, BMI, blood pressure, glucose, **No effect**: Waist, hip, HDL-C, LDL-C, TGs, TC	534 mg catechins + 11.7 g inulin	3	[18]
Inulin + maltodextrin	**Increase**: *Parabacteroides, Bifidobacterium* **Decrease**: *Bilophila, Ruminococcus* **No effect**: None	**Increase**: None **Decrease**: Insulin, SBP, DBP, white blood cells **No effect**: BMI, body weight, fat mass, waist, hip, TC, HDL-C, LDL-C, TGs	20	12	[23]
Oligofructose-enriched inulin	**Increase**: *Bifidobacterium* **Decrease**: None **No effect**: None	**Increase**: Acetate, propionate, valerate **Decrease**: None **No effect**: Isovalerate, isobutyrate, butyrate	10	12	[22]
Oligofructose-enriched inulin	**Increase**: *Bifidobacterium,* **Decrease**: *Coprococcus, Dorea, Ruminococcus* **No effect**: None	**Increase**: None **Decrease**: None **No effect**: SCFAs	16	3	[28]

Table 3. *Cont.*

Combination	Effect on Gut Microbiota	Effect on Metabolic Disturbancesand SCFAs	Dose (g/d)	Time (Weeks)	Reference
Oligofructose-enriched inulin	**Increase**: *Bifidobacterium, Faecalibacterium prausnitzii* **Decrease**: *Bacteroides intestinalis, B vulgatus* **No effect**: None	**Increase**: None **Decrease**: None **No effect**: BMI, body weight, waist/hip ratio, HbA1c, fasting glycemia, insulinemia, TC, HDL-C or LDL-C, and TG	16	12	[36]
Oligofructose-enriched inulin	**Increase**: *B. adolescentis, B. longum, B. pseudocatenulatum* **Decrease**: None **No effect**: None	**Increase**: None **Decrease**: Acetate, butyrate **No effect**: Isobutyrate and isovalerate	16	12	[19]
Oligofructose-enriched inulin	**Increase**: *B. adolescentis, B. longum, Bacteriodes vulgatus, Faecalibacterium prausnitzii* **Decrease**: *Roseburia* sp., *Eubacterium eligens, B. bifidum, Anaerostipes butyraticus* **No effect**: *Actinomyces*	**Increase**: None **Decrease**: Body weight **No effect**: BMI, hist, waist, IL-6, HOMA-IR, insulin,	8	16	[10]
Inulin + Arabinoxylan	**Increase**: *Bifidobacterium, Propionibacterium* **Decrease**: None **No effect**: None	**Increase**: Acetate **Decrease**: None **No effect**: Butyrate, propionate	8	12	[39]
Inulin + fruit juice	**Increase**: *Bifidobacterium, Lactobacillus* **Decrease**: *Eubacterium* **No effect**: None	None	10	3	[41]
Inulin + pomegranate	**Increase**: *Bifidobacterium, Akkermansia,* **Decrease**: *Lachnospira, Klebsiella* **No effect**: None	**Increase**: None **Decrease**: None **No effect**: Body weight, BMI	10	3	[43]

"None" means either that the study showed no outcomes or results were considered irrelevant. GLP-1: Glucagon-like Peptide-1; PYY: Peptide YY; IHCL: Intrahepatic lipid content.

5.1. Effects of Inulin Combinations with Other Compounds on Human Metabolic Health

Different studies have evaluated the effect of supplementation of inulin combined with different compounds or prebiotics in order to find a synergistic effect that enhances the effects produced by inulin alone, which are summarized in Table 3.

Regarding anthropometric measures, inulin improves body weight levels when it is combined with either propionate ester, oligofructose, or green tea catechins [10,12,18]. Among them, regarding dose effect, inulin seems to act more synergistically with oligofructose, as a lower dose of inulin was required to achieve weight improvement when both were combined. Remarkably, catechins accelerate inulin's beneficial effects, as observed after just 3 weeks of intervention.

On the other hand, inulin combined with propionate ester has been observed to be time dependent, as beneficial effects on weight have been observed at long time intervention periods. However, inulin has not been found to decrease body weight when it is combined with glucose or maltodextrin, suggesting an antagonistic effect between them. Nevertheless, the study that analyzed the effects produced by the combination of inulin and glucose used a very high dose of the latter, which might be considered to affect the results. In relation to this, it has also been observed that inulin–propionate ester can increase GLP-1 and PYY, promoting satiety and reducing ghrelin levels, which decreases the sensation of hunger [12]. However, inulin is incapable of improving satiety hormone levels when combined with high doses of glucose, whereas it still can decrease ghrelin [27].

Regarding glucose metabolism, inulin has been shown to be capable of improving insulin parameters depending on which compound it is combined with. Interestingly, it seems that inulin and oligofructose act antagonistically as no effects on insulin or glucose levels have been observed. In addition, inulin with catechins has been observed to reduce glucose blood levels, but differences in insulin parameters have not been described [18]. Therefore, it may enhance insulin action or improve glucose metabolism by other mechanisms.

Interestingly, the studies collected show that the effect of inulin supplementation on the lipid profile is not maintained when administered in combination with other compounds. However, one of the reasons for this may be that, even if the studied population had obesity or metabolic disorder, they presented a normal lipid profile. Therefore, there would be no significant differences or improvements when presenting baseline levels within the normal health range. Consequently, further investigations are necessary to clarify these results.

5.2. *Link between Microbiota Modulation and Physiological Effects*

In the same way that inulin alone is capable of generating changes in the bacterial proliferation, its combination with other prebiotics modulates gut microbiota in a different way depending on the compound with which it is combined.

Inulin combined with propionate ester, surprisingly, has been observed not to affect bifidobacteria populations, but other genera such as *Bacteroides*, *Anaerostipes*, or *Blautia* were related to improved insulin metabolic parameters. However, no microbiota modulation has been studied to relate gut hormone regulation with inulin–propionate ester supplementation [9,12]. On the other hand, when inulin was combined with oligofructose or maltodextrin, changes in the bifidobacteria population have been found, in accordance with inulin's observed effects. Moreover, inulin combined with oligofructose has been also found to increase the population of *Faecalibacterium prausnitzii*, which is one of the most common beneficial colon bacteria [10,36]. However, although changes in the microbiota are maintained at different doses and intervention periods of oligofructose–inulin supplementation, the physiological effects are controversial at the anthropometric level, as both improvements in body weight and the absence of a significant effect have been observed [10,36].

Therefore, although inulin appears to have the capacity to act synergistically with other compounds, due to the limited number of studies utilizing inulin in combination with other treatments and even fewer studies examining the modulation of the microbiota alongside physiological effects in patients, it is challenging to correlate alterations in bacterial species with physiological outcomes. This underscores the necessity for further research in this area.

6. Effect of Inulin Intake on Other Diseases

The role of inulin in body weight management, glucose metabolism, and inflammatory markers have evidenced the potential use of this fiber to ameliorate obesity-related diseases, including type 2 diabetes and cardiovascular disease. However, due to its prebiotic and anti-inflammatory capacity, supplementation with this ingredient has been evaluated in diseases where the intestinal microbiota and the resulting inflammation play a role in their pathogenesis, such as kidney and inflammatory bowel diseases.

In this regard, the prebiotic activity of inulin supplementation has been investigated for the amelioration of chronic kidney disease (CKD) due to the gut dysbiosis and dysmetabolism that characterize patients with this pathology, which seems to contribute to the progression to CKD-related complications, including cardiovascular disease [45]. Thus, gut microbiota modulation through the administration of prebiotics such as inulin has been suggested as a potential therapeutic target for CKD due to its effect on the profile of circulating blood metabolites [46]. In this regard, Silvia Lai and colleagues observed that supplementation with inulin (19 g/day) with a low-protein diet (LPD) for 6 months was able to modulate gut microbiota, increasing the abundance of Bifidobacteriaceae, accompanied by a reduction of inflammatory circulating markers in patients with CKD [45]. Interestingly, inulin was able to reduce serum insulin and fasting glucose levels and to improve cholesterol metabolism in the participants of the study, suggesting a cardioprotective role of inulin in CKD [30].

The effect of inulin on intestinal health has also been evaluated in different models of intestinal disease, with variable results. Initially, the use of inulin or inulin-enriched products has been proposed to improve intestinal transit and reduce constipation. Thus, a meta-analysis comprising individuals with chronic constipation revealed a positive effect of inulin consumption on bowel function, improving the stool frequency, stool consistency, transit time, and hardness of stool, while no efficacy was shown for pain and bloating [47]. A similar finding was observed in another study, where supplementation with 12 g per day of inulin for 4 weeks improved bowel function in volunteers with chronic constipation [48]. However, the anti-inflammatory activity of inulin and its role contributing to intestinal function is not always observed in individuals with inflammatory bowel disease, including ulcerative colitis (UC) or Crohn's disease [49]. Studies using inulin as prebiotic for the treatment of chronic intestinal inflammation have shown benefit in animal models of colitis [50]. In the case of humans, the potential use of oligofructose-enriched inulin has been suggested to ameliorate symptoms in individuals with ulcerative colitis [51,52]. However, despite promising data on inulin in gut health, its usefulness in individuals with inflammatory bowel disease is controversial since some individuals report an increase in flatulence and bloating [49]. Therefore, the use inulin in patients with inflammatory bowel disease requires additional studies to demonstrate its convenience.

7. Mechanisms of Action

The mechanisms of action by which inulin induces beneficial effects in human health are not fully understood. In fact, only a few studies in humans have been designed to explain the underlying mechanisms, which has been a limitation for the present investigation. This is probably due to the fact that, in order to study these mechanisms, the techniques required are excessively invasive. However, we present the proposed mechanisms of action that have been studied in both animal models and in vitro studies that could explain the results observed in humans (Figure 3).

Although inulin primarily exerts its effects through gut microbiota modulation, as previously mentioned, its consumption can also directly influence human physiology. Thus, since it is not digestible, it delays gastric emptying, which increases satiety hormone secretion, such as GLP-1, ending up with a lower food and energy intake and therefore improving anthropometric parameters like body weight [35].

In addition, it also delays glucose absorption in the intestine, helping to regulate the increase in blood glucose levels after meals, favoring glycemic homeostasis [35]. Once inulin arrives in the colon, it is fermented by gut microbiota, principally by *Bifidobacterium* species, even though they represent a low number of the human microbiota, due to several genes which encode for different β-fructofuranosidases, which cleave through inulin structure and produce fermentable oligosaccharides or monosaccharides [53]. Obviously, the ability to utilize inulin varies among species and specific strains of this genus. Those metabolic products are then converted into acetate and lactate by bifidobacteria, which enables cross-feeding interactions between them and SCFA-producing species, such as *Anaerostipes* spp., leading to increased butyrate and propionate levels [53]. In vitro studies have shown that both propionate and butyrate increase GLP-1 and PYY secretion by L-cells via G-protein-coupled receptor 43 (GPR43) human receptor stimulation in a dose-dependent way, regulating food and energy intake [54]. Therefore, depending on the ability of inulin to modulate the gut microbiota, propionate and butyrate would be able to alter GLP-1 and PYY serum levels, which explains why consistent findings are not always found. In addition, hormone secretion regulation has been observed to also impact glucose metabolism, as higher levels of GLP-1 have been observed to improve insulin sensitivity.

Therefore, both inulin and inulin-derived SCFAs enhance each other's effects by similarly altering the secretion of these hormones [54].

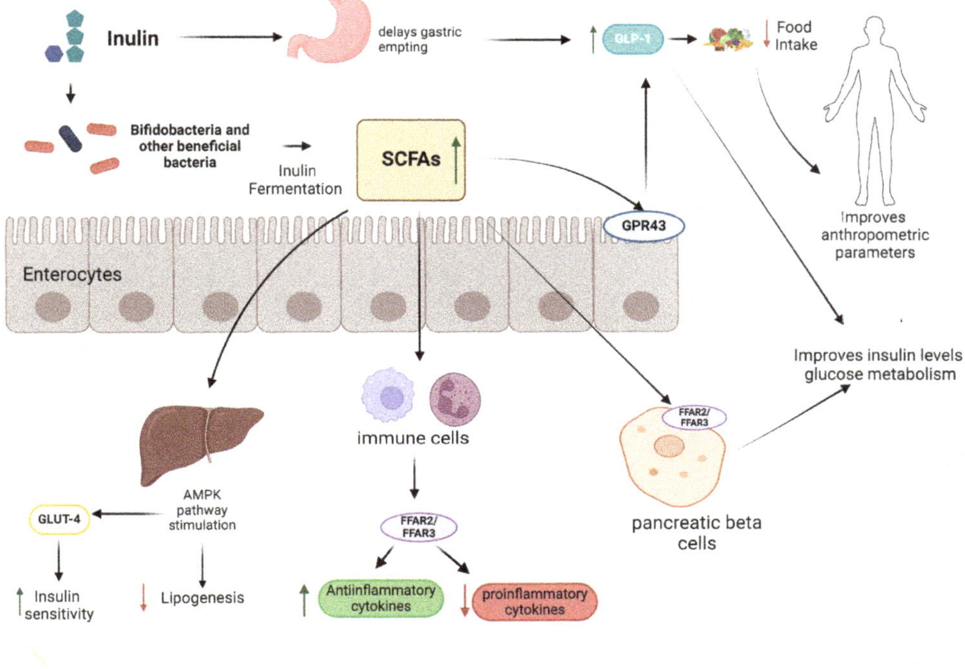

Figure 3. Overview of the primary mechanisms of action and physiological effects of inulin.

SCFAs are able to improve glucose metabolism by other metabolic pathways. In vivo models have found that propionate, butyrate, and especially acetate activate free fatty acid receptor 2 (FFAR2) and FFAR3 receptors in pancreatic beta cells, regulating insulin secretion and improving glucose homeostasis [55,56]. In addition, acetate can improve insulin sensitivity via glucose transporter 4 (GLUT4) by upregulating 5′-AMP-activated protein kinase signaling in liver tissues [57]. Regarding inflammation effects, SCFAs modulate cytokine secretion in a dose-dependent way. It has been reported that SCFAs activate immune cell receptor GPR43 by propionate and acetate or GPR41 by butyrate when they reach a certain concentration level [58,59]. Moreover, it has been suggested that, through the FFAR2 receptor, SCFAs can affect gene expression by inhibiting histone deacetylases and increasing histone acetylation, leading to an increased or decreased secretion of inflammatory cytokines [58]. Related to the lipid profile, acetate and butyrate are used as substrates for lipid formation. However, SCFAs are able to regulate lipogenesis in several ways. In vivo and in vitro studies have shown that activating the AMPK pathway increases the expression of peroxisome proliferator-activated receptor gamma coactivator 1-alpha (PGC-1α), which in turn increases the expression of the transcription factors peroxisome proliferator-activated receptor-α (PPAR-α) and PPAR-γ, which stimulate lipid oxidation [60]. In addition, the AMPK pathway also inhibits transcription factors associated with fat deposition and lipid synthesis.

8. Conclusions

According to the collected data, inulin demonstrates significant potential in improving human physiology by acting independently and through modulation of the gut microbiota, particularly species from the *Bifidobacterium* and *Anaerostipes* genera, leading to better anthropometric parameters, improved glucose metabolism, lower insulin levels, and beneficial effects on inflammation and immune function, ultimately improving metabolic states. These positive outcomes have been observed when inulin is used alone or in combination with other compounds, underscoring its versatility and efficacy in promoting metabolic health. However, further research is warranted to fully elucidate its therapeutic potential and optimize its clinical applications.

Author Contributions: Conceptualization, P.A. and F.I.M.; methodology, J.A.-A.; investigation, J.A.-A.; resources, J.A.-A.; data curation, J.A.-A. and P.A.; writing—original draft preparation, J.A.-A. and P.A.; writing—review and editing, F.I.M.; visualization, J.A.-A.; supervision, F.I.M. and P.A.; project administration, F.I.M. and P.A.; funding acquisition, F.I.M. and P.A. All authors have read and agreed to the published version of the manuscript.

Funding: This research was funded by Gobierno de Navarra (grant numbers 0011-1383-2022-000000 [PARABIOTICS] and 0011-1411-2021-000100 [ALISSEC]), MICINN (Gobierno de España; POSTBIOTICS project, PID2022-141766OB-I00), and CIBER (CB12/03/30002).

Conflicts of Interest: The authors declare that this study received funding from LABORATORIOS CINFA, S.A. and VISCOFAN, S.A. The funder was not involved in the study design, collection, analysis, interpretation of data, the writing of this article or the decision to submit it for publication.

References

1. Kays, S.J.; Nottingham, S.F. *Biology and Chemistry of Jerusalem Artichoke:* Helianthus tuberosus *L.*, 1st ed.; CRC Press: Boca Raton, FL, USA, 2007; pp. 52–96.
2. Shoaib, M.; Shehzad, A.; Omar, M.; Rakha, A.; Raza, H.; Sharif, H.R.; Shakeel, A.; Ansari, A.; Niazi, S. Inulin: Properties, health benefits and food applications. *Carbohydr. Polym.* **2016**, *147*, 444–454. [CrossRef]
3. Qin, Y.Q.; Wang, L.Y.; Yang, X.Y.; Xu, Y.J.; Fan, G.; Fan, Y.G.; Ren, J.N.; An, Q.; Li, X. Inulin: Properties and health benefits. *Food Funct.* **2023**, *14*, 2948–2968. [CrossRef]
4. Illippangama, A.U.; Jayasena, D.D.; Jo, C.; Mudannayake, D.C. Inulin as a functional ingredient and their applications in meat products. *Carbohydr. Polym.* **2022**, *275*, 118706. [CrossRef] [PubMed]
5. Ahmed, W.; Rashid, S. Functional and therapeutic potential of inulin: A comprehensive review. *Crit. Rev. Food Sci. Nutr.* **2019**, *59*, 1–13. [CrossRef]
6. Perović, J.; Tumbas Šaponjac, V.; Kojić, J.; Krulj, J.; Moreno, D.A.; García-Viguera, C.; Bodroža-Solarov, M.; Ilić, N. Chicory (*Cichorium intybus* L.) as a food ingredient—Nutritional composition, bioactivity, safety, and health claims: A review. *Food Chem.* **2021**, *336*, 127676. [CrossRef] [PubMed]
7. Gupta, N.; Jangid, A.K.; Pooja, D.; Kulhari, H. Inulin: A novel and stretchy polysaccharide tool for biomedical and nutritional applications. *Int. J. Biol. Macromol.* **2019**, *132*, 852–863. [CrossRef]
8. Gupta, D.; Chaturvedi, N. Prebiotic Potential of underutilized Jerusalem artichoke in Human Health: A Comprehensive Review. *Int. J. Environ. Agric. Biotech.* **2020**, *5*, 97–103. [CrossRef]
9. Chambers, E.S.; Byrne, C.S.; Morrison, D.J.; Murphy, K.G.; Preston, T.; Tedford, C.; Garcia-Perez, I.; Fountana, S.; Serrano-Contreras, J.I.; Holmes, E.; et al. Dietary supplementation with inulin-propionate ester or inulin improves insulin sensitivity in adults with overweight and obesity with distinct effects on the gut microbiota, plasma metabolome and systemic inflammatory responses: A randomised cross-over trial. *Gut* **2019**, *68*, 1430–1438. [PubMed]
10. Nicolucci, A.C.; Hume, M.P.; Martínez, I.; Mayengbam, S.; Walter, J.; Reimer, R.A. Prebiotics Reduce Body Fat and Alter Intestinal Microbiota in Children Who Are Overweight or with Obesity. *Gastroenterology* **2017**, *153*, 711–722. [CrossRef]
11. Visuthranukul, C.; Chamni, S.; Kwanbunbumpen, T.; Saengpanit, P.; Chongpison, Y.; Tepaamorndech, S.; Panichsillaphakit, E.; Uaariyapanichkul, J.; Nonpat, N.; Chomtho, S. Effects of inulin supplementation on body composition and metabolic outcomes in children with obesity. *Sci. Rep.* **2022**, *12*, 13014. [CrossRef]
12. Chambers, E.S.; Viardot, A.; Psichas, A.; Morrison, D.J.; Murphy, K.G.; Zac-Varghese, S.E.K.; MacDougall, K.; Preston, T.; Tedford, C.; Finlayson, G.S.; et al. Effects of targeted delivery of propionate to the human colon on appetite regulation, body weight maintenance and adiposity in overweight adults. *Gut* **2015**, *64*, 1744–1754. [CrossRef] [PubMed]
13. Vandeputte, D.; Falony, G.; Vieira-Silva, S.; Wang, J.; Sailer, M.; Theis, S.; Verbeke, K.; Raes, J. Prebiotic inulin-type fructans induce specific changes in the human gut microbiota. *Gut* **2017**, *66*, 1968–1974. [CrossRef]
14. Birkeland, E.; Gharagozlian, S.; Birkeland, K.I.; Valeur, J.; Måge, I.; Rud, I.; Aas, A.-M. Prebiotic effect of inulin-type fructans on faecal microbiota and short-chain fatty acids in type 2 diabetes: A randomised controlled trial. *Eur. J. Nutr.* **2020**, *59*, 3325–3338. [CrossRef]
15. Slavin, J.; Feirtag, J. Chicory inulin does not increase stool weight or speed up intestinal transit time in healthy male subjects. *Food Funct.* **2011**, *2*, 72–77. [CrossRef]

16. Jackson, P.P.; Wijeyesekera, A.; Williams, C.M.; Theis, S.; van Harsselaar, J.; Rastall, R.A. Inulin-type fructans and 2′fucosyllactose alter both microbial composition and appear to alleviate stress-induced mood state in a working population compared to placebo (maltodextrin): The EFFICAD Trial, a randomized, controlled trial. *Am. J. Clin. Nutr.* **2023**, *118*, 938–955. [CrossRef] [PubMed]
17. Rodriguez, J.; Neyrinck, A.M.; Van Kerckhoven, M.; Gianfrancesco, M.A.; Renguet, E.; Bertrand, L.; Cani, P.D.; Lanthier, N.; Cnop, M.; Paquot, N.; et al. Physical activity enhances the improvement of body mass index and metabolism by inulin: A multicenter randomized placebo-controlled trial performed in obese individuals. *BMC Med.* **2022**, *20*, 110. [CrossRef]
18. Yang, H.Y.; Yang, S.C.; Chao, J.C.J.; Chen, J.R. Beneficial effects of catechin-rich green tea and inulin on the body composition of overweight adults. *Br. J. Nutr.* **2012**, *107*, 749–754. [CrossRef]
19. Salazar, N.; Dewulf, E.M.; Neyrinck, A.M.; Bindels, L.B.; Cani, P.D.; Mahillon, J.; de Vos, W.M.; Thissen, J.-P.; Gueimonde, M.; de los Reyes-Gavilan, C.G.; et al. Inulin-type fructans modulate intestinal Bifidobacterium species populations and decrease fecal short-chain fatty acids in obese women. *Clin. Nutr.* **2015**, *34*, 501–507. [CrossRef]
20. Reimer, R.A.; Soto-Vaca, A.; Nicolucci, A.C.; Mayengbam, S.; Park, H.; Madsen, K.L.; Menon, R.; Vaughan, E.E. Effect of chicory inulin-type fructan-containing snack bars on the human gut microbiota in low dietary fiber consumers in a randomized crossover trial. *Am. J. Clin. Nutr.* **2020**, *111*, 1286–1296. [CrossRef]
21. Holscher, H.D.; Bauer, L.L.; Gourineni, V.; Pelkman, C.L.; Fahey, G.C.; Swanson, K.S. Agave inulin supplementation affects the fecal microbiota of healthy adults participating in a randomized, double-blind, placebo-controlled, crossover trial. *J. Nutr.* **2015**, *145*, 2025–2032. [CrossRef]
22. Drabińska, N.; Jarocka-Cyrta, E.; Markiewicz, L.H.; Krupa-Kozak, U. The effect of oligofructose-enriched inulin on faecal bacterial counts and microbiota-associated characteristics in celiac disease children following a gluten-free diet: Results of a randomized, placebo-controlled trial. *Nutrients* **2018**, *10*, 201. [CrossRef] [PubMed]
23. Hess, A.L.; Benítez-Páez, A.; Blædel, T.; Larsen, L.H.; Iglesias, J.R.; Madera, C.; Sanz, Y.; Larsen, T.M.; The MyNewGut Consortium. The effect of inulin and resistant maltodextrin on weight loss during energy restriction: A randomised, placebo-controlled, double-blinded intervention. *Eur. J. Nutr.* **2020**, *59*, 2507–2524. [CrossRef]
24. Hiel, S.; Gianfrancesco, M.A.; Rodriguez, J.; Portheault, D.; Leyrolle, Q.; Bindels, L.B.; da Silveria Cauduro, C.G.; Mulders, M.D.; Zamariola, G.; Azzi, A.-S.; et al. Link between gut microbiota and health outcomes in inulin -treated obese patients: Lessons from the Food4Gut multicenter randomized placebo-controlled trial. *Clin. Nutr.* **2020**, *39*, 3618–3628. [CrossRef] [PubMed]
25. Dehghan, P.; Gargari, B.P.; Jafar-Abadi, M.A.; Aliasgharzadeh, A. Inulin controls inflammation and metabolic endotoxemia in women with type 2 diabetes mellitus: A randomized-controlled clinical trial. *Int. J. Food Sci. Nutr.* **2014**, *65*, 117–123. [CrossRef]
26. Neyrinck, A.M.; Rodriguez, J.; Zhang, Z.; Seethaler, B.; Sánchez, C.R.; Roumain, M.; Hiel, S.; Bindels, L.B.; Cani, P.D.; Paquot, N.; et al. Prebiotic dietary fibre intervention improves fecal markers related to inflammation in obese patients: Results from the Food4Gut randomized placebo-controlled trial. *Eur. J. Nutr.* **2021**, *60*, 3159–3170. [CrossRef] [PubMed]
27. Rahat-Rozenbloom, S.; Fernandes, J.; Cheng, J.; Wolever, T.M.S. Acute increases in serum colonic short-chain fatty acids elicited by inulin do not increase GLP-1 or PYY responses but may reduce ghrelin in lean and overweight humans. *Eur. J. Clin. Nutr.* **2017**, *71*, 953–958. [CrossRef]
28. Healey, G.; Murphy, R.; Butts, C.; Brough, L.; Whelan, K.; Coad, J. Habitual dietary fibre intake influences gut microbiota response to an inulin-type fructan prebiotic: A randomised, double-blind, placebo-controlled, cross-over, human intervention study. *Br. J. Nutr.* **2018**, *119*, 176–189. [CrossRef]
29. Kiewiet, M.B.G.; Elderman, M.E.; El Aidy, S.; Burgerhof, J.G.M.; Visser, H.; Vaughan, E.E.; Faas, M.M.; de Vos, P. Flexibility of Gut Microbiota in Ageing Individuals during Dietary Fiber Long-Chain Inulin Intake. *Mol. Nutr. Food Res.* **2021**, *65*, 2000390. [CrossRef]
30. Lai, S.; Mazzaferro, S.; Muscaritoli, M.; Mastroluca, D.; Testorio, M.; Perrotta, A.; Esposito, Y.; Carta, M.; Campagna, L.; Di Grado, M.; et al. Prebiotic therapy with inulin associated with low protein diet in chronic kidney disease patients: Evaluation of nutritional, cardiovascular and psychocognitive parameters. *Toxins* **2020**, *12*, 381. [CrossRef]
31. Costabile, A.; Kolida, S.; Klinder, A.; Gietl, E.; Buerlein, M.; Frohberg, C.; Landschütze, V.; Gibson, G.R. A double-blind, placebo-controlled, cross-over study to establish the bifidogenic effect of a very-long-chain inulin extracted from globe artichoke (*Cynara scolymus*) in healthy human subjects. *Br. J. Nutr.* **2010**, *104*, 1007–1017. [CrossRef]
32. Biruete, A.; Cross, T.W.L.; Allen, J.M.; Kistler, B.M.; de Loor, H.; Evenepoel, P.; Fahey, G.C.; Bauer, L.; Swanson, K.S.; Wilund, K.R. Effect of Dietary Inulin Supplementation on the Gut Microbiota Composition and Derived Metabolites of Individuals Undergoing Hemodialysis: A Pilot Study. *J. Ren. Nutr.* **2021**, *31*, 512–522. [CrossRef]
33. Petry, N.; Egli, I.; Chassard, C.; Lacroix, C.; Hurrell, R. Inulin modifies the bifidobacteria population, fecal lactate concentration, and fecal pH but does not influence iron absorption in women with low iron status. *Am. J. Clin. Nutr.* **2012**, *96*, 325–331. [CrossRef]
34. Baxter, N.T.; Schmidt, A.W.; Venkataraman, A.; Kim, K.S.; Waldron, C.; Schmidt, T.M. Dynamics of Human Gut Microbiota and Short-Chain Fatty Acids in Response to Dietary Interventions with Three Fermentable Fibers. *Am. Soc. Microbiol. J.* **2019**, *10*, e02566-18. [CrossRef]
35. Gargari, B.P.; Dehghan, P.; Aliasgharzadeh, A.; Jafar-Abadi, M.A. Effects of high performance inulin supplementation on glycemic control and antioxidant status in women with type 2 diabetes. *Diabetes Metab. J.* **2013**, *37*, 140–148. [CrossRef]
36. Dewulf, E.M.; Cani, P.D.; Claus, S.P.; Fuentes, S.; Puylaert, P.G.; Neyrinck, A.M.; Bindels, L.B.; de Vos, W.M.; Gibson, G.R.; Thissen, J.-P.; et al. Insight into the prebiotic concept: Lessons from an exploratory, double blind intervention study with inulin-type fructans in obese women. *Gut* **2013**, *62*, 1112–1121. [CrossRef]

37. Yin, P.; Du, T.; Yi, S.; Zhang, C.; Yu, L.; Tian, F.; Chen, W.; Zhai, Q. Response differences of gut microbiota in oligofructose and inulin are determined by the initial gut Bacteroides/Bifidobacterium ratios. *Food Res. Int.* **2023**, *174*, 113598. [CrossRef]
38. Ďásková, N.; Modos, I.; Krbcová, M.; Kuzma, M.; Pelantová, H.; Hradecký, J.; Heczková, M.; Bratová, M.; Videňská, P.; Šplíchalová, P.; et al. Multi-omics signatures in new-onset diabetes predict metabolic response to dietary inulin: Findings from an observational study followed by an interventional trial. *Nutr. Diab.* **2023**, *13*, 7. [CrossRef]
39. Collins, S.M.; Gibson, G.R.; Stainton, G.N.; Bertocco, A.; Kennedy, O.B.; Walton, G.E.; Commane, D.M. Chronic consumption of a blend of inulin and arabinoxylan reduces energy intake in an ad libitum meal but does not influence perceptions of appetite and satiety: A randomised control-controlled crossover trial. *Eur. J. Nutr.* **2023**, *62*, 2205–2215. [CrossRef]
40. Wang, X.; Wang, T.; Zhang, Q.; Xu, L.; Xiao, X. Dietary Supplementation with Inulin Modulates the Gut Microbiota and Improves Insulin Sensitivity in Prediabetes. *Int. J. Endocrinol.* **2021**, *2021*, 5579369. [CrossRef]
41. Ramnani, P.; Gaudier, E.; Bingham, M.; Van Bruggen, P.; Tuohy, K.M.; Gibson, G.R. Prebiotic effect of fruit and vegetable shots containing Jerusalem artichoke inulin: A human intervention study. *Br. J. Nutr.* **2010**, *104*, 233–240. [CrossRef]
42. Salmean, Y.A. Using inulin fiber supplementation with MyPlate recommendations promotes greater weight loss in obese women. *Prog. Nutr.* **2019**, *21*, 81–85.
43. Hou, C.; Shi, H.; Xiao, J.; Song, X.; Luo, Z.; Ma, X.; Shi, L.; Wei, H.; Li, J. Pomegranate Juice Supplemented with Inulin Modulates Gut Microbiota and Promotes the Production of Microbiota-Associated Metabolites in Overweight/Obese Individuals: A Randomized, Double-Blind, Placebo-Controlled Trial. *J. Agric. Food Chem.* **2024**, *72*, 14663–14677. [CrossRef]
44. Visuthranukul, C.; Sriswasdi, S.; Tepaamorndech, S.; Chamni, S.; Leelahavanichkul, A.; Joyjinda, Y.; Aksornkitti, V.; Chomtho, S. Enhancing gut microbiota and microbial function with inulin supplementation in children with obesity. *Int. J. Obes.* **2024**, 1–9. [CrossRef]
45. Lai, S.; Molfino, A.; Testorio, M.; Perrotta, A.M.; Currado, A.; Pintus, G.; Pietrucci, D.; Unida, V.; La Rocca, D.; Biocca, S.; et al. Effect of low-protein diet and inulin on microbiota and clinical parameters in patients with chronic kidney disease. *Nutrients* **2019**, *11*, 3006. [CrossRef]
46. Sohn, M.B.; Gao, B.; Kendrick, C.; Srivastava, A.; Isakova, T.; Gassman, J.J.; Fried, L.F.; Wolf, M.; Cheung, A.K.; Raphael, K.L.; et al. Targeting Gut Microbiome with Prebiotic in Patients With CKD: The TarGut-CKD Study. *Kidney Int. Rep.* **2024**, *9*, 671–685. [CrossRef]
47. Yurrita, L.C.; Martín, I.S.M.; Ciudad-Cabañas, M.J.; Calle-Purón, M.E.; Cabria, M.H. Eficacia de la ingesta de inulina sobre los indicadores del estreñimiento crónico; un meta-análisis de ensayos clínicos aleatorizados controlados. *Nutr. Hosp.* **2014**, *30*, 244–252.
48. Micka, A.; Siepelmeyer, A.; Holz, A.; Theis, S.; Schön, C. Effect of consumption of chicory inulin on bowel function in healthy subjects with constipation: A randomized, double-blind, placebo-controlled trial. *Int. J. Food Sci. Nutr.* **2017**, *68*, 82–89. [CrossRef]
49. Limketkai, B.N.; Godoy-Brewer, G.; Shah, N.D.; Maas, L.; White, J.; Parian, A.M.; Mullin, G.E. Prebiotics for Induction and Maintenance of Remission in Inflammatory Bowel Disease: Systematic Review and Meta-Analysis. *Inflamm. Bowel Dis.* **2024**, izae115. [CrossRef]
50. Leenen, C.H.M.; Dieleman, L.A. Health Benefits and Claims-A Critical Review Inulin and Oligofructose in Chronic Inflammatory Bowel Disease 1–3. *J. Nutr.* **2007**, *137*, 2572S–2575S. [CrossRef]
51. Valcheva, R.; Koleva, P.; Martínez, I.; Walter, J.; Gänzle, M.G.; Dieleman, L.A. Inulin-type fructans improve active ulcerative colitis associated with microbiota changes and increased short-chain fatty acids levels. *Gut Microbes* **2019**, *10*, 334–357. [CrossRef]
52. Casellas, F.; Borruel, N.; Torrejón, A.; Varela, E.; Antolin, M.; Guarner, F.; Malagelada, J. Oral oligofructose-enriched inulin supplementation in acute ulcerative colitis is well tolerated and associated with lowered faecal calprotectin. *Aliment. Pharmacol. Ther.* **2007**, *25*, 1061–1067. [CrossRef]
53. Rivière, A.; Selak, M.; Lantin, D.; Leroy, F.; De Vuyst, L. Bifidobacteria and butyrate-producing colon bacteria: Importance and strategies for their stimulation in the human gut. *Front. Microbiol.* **2016**, *7*, 979. [CrossRef]
54. Angelini, G.; Russo, S.; Mingrone, G. Incretin hormones, obesity and gut microbiota. *Peptides* **2024**, *178*, 171216. [CrossRef]
55. Tang, C.; Ahmed, K.; Gille, A.; Lu, S.; Gröne, H.J.; Tunaru, S.; Offermanns, S. Loss of FFA2 and FFA3 increases insulin secretion and improves glucose tolerance in type 2 diabetes. *Nat. Med.* **2015**, *21*, 173–177. [CrossRef]
56. Kumar, J.; Rani, K.; Datt, C. Molecular link between dietary fibre, gut microbiota and health. *Mol. Biol. Rep.* **2020**, *47*, 6229–6237. [CrossRef]
57. Kim, Y.A.; Keogh, J.B.; Clifton, P.M. Probiotics, prebiotics, synbiotics and insulin sensitivity. *Nutr. Res. Rev.* **2018**, *31*, 35–51. [CrossRef]
58. Smith, P.M.; Howitt, M.R.; Panikov, N.; Michaud, M.; Gallini, C.A.; Bohlooly-Y, M.; Glickman, J.N.; Garrett, W.S. The microbial metabolites, short-chain fatty acids, regulate colonic T reg cell homeostasis. *Science* **2013**, *341*, 569–573. [CrossRef]
59. Watzl, B.; Girrbach, S.; Roller, M. Inulin, oligofructose and immunomodulation. *Br. J. Nutr.* **2005**, *93*, 49–55. [CrossRef]
60. He, J.; Zhang, P.; Shen, L.; Niu, L.; Tan, Y.; Chen, L.; Zhao, Y.; Bai, L.; Hao, X.; Li, X.; et al. Short-chain fatty acids and their association with signalling pathways in inflammation, glucose and lipid metabolism. *Int. J. Mol. Sci.* **2020**, *21*, 6356. [CrossRef]

nutrients

Review

Exploring the Therapeutic Potential of Royal Jelly in Metabolic Disorders and Gastrointestinal Diseases

Hesham R. El-Seedi [1,2,3,*], Suzy Salama [4], Aida A. Abd El-Wahed [5], Zhiming Guo [2], Alessandro Di Minno [6,7], Maria Daglia [6,7], Chuan Li [8], Xiao Guan [9,10], Daniele Giuseppe Buccato [6], Shaden A. M. Khalifa [11,*] and Kai Wang [12,*]

1 Pharmacognosy Group, Department of Pharmaceutical Biosciences, Uppsala University, Biomedical Centre, P.O. Box 591, SE-751 24 Uppsala, Sweden
2 School of Food and Biological Engineering, Jiangsu University, Zhenjiang 212013, China; guozhiming@ujs.edu.cn
3 International Joint Research Laboratory of Intelligent Agriculture and Agri-Products Processing, Jiangsu University, Zhenjiang 210024, China
4 Indigenous Knowledge and Heritage Center, Ghibaish College of Science and Technology, Ghibaish 51111, Sudan; s.salama999@hotmail.com
5 Department of Bee Research, Plant Protection Research Institute, Agricultural Research Centre, Giza 12627, Egypt; aidaabd.elwahed@arc.sci.eg
6 Department of Pharmacy, University of Napoli Federico II, Via D. Montesano 49, 80131 Naples, Italy; alessandro.diminno@unina.it (A.D.M.); maria.daglia@unipv.it (M.D.); d.buccato@studenti.unina.it (D.G.B.)
7 CEINGE-Biotecnologie Avanzate, Via Gaetano Salvatore 486, 80145 Naples, Italy
8 School of Food Science and Engineering, Hainan University, Haikou 570228, China; lichuan@hainanu.edu.cn
9 School of Health Science and Engineering, University of Shanghai for Science and Technology, Shanghai 200093, China; gnxo@163.com
10 National Grain Industry (Urban Grain and Oil Security) Technology Innovation Center, Shanghai 200093, China
11 Psychiatry and Neurology Department, Capio Saint Göran's Hospital, Sankt Göransplan 1, 112 19 Stockholm, Sweden
12 State Key Laboratory of Resource Insects, Institute of Apicultural Research, Chinese Academy of Agricultural Sciences, Beijing 100093, China
* Correspondence: hesham@kth.se or elseedi_99@yahoo.com (H.R.E.-S.); shaden.khalifa@regionstockholm.se (S.A.M.K.); wangkai@caas.cn (K.W.); Tel.: +46-700-434-343 (H.R.E.-S.)

Abstract: Metabolic disorders, encompassing diabetes mellitus, cardiovascular diseases, gastrointestinal disorders, etc., pose a substantial global health threat, with rising morbidity and mortality rates. Addressing these disorders is crucial, as conventional drugs often come with high costs and adverse effects. This review explores the potential of royal jelly (RJ), a natural bee product rich in bioactive components, as an alternative strategy for managing metabolic diseases. RJ exhibits diverse therapeutic properties, including antimicrobial, estrogen-like, anti-inflammatory, hypotensive, anticancer, and antioxidant effects. This review's focus is on investigating how RJ and its components impact conditions like diabetes mellitus, cardiovascular disease, and gastrointestinal illnesses. Evidence suggests that RJ serves as a complementary treatment for various health issues, notably demonstrating cholesterol- and glucose-lowering effects in diabetic rats. Specific RJ-derived metabolites, such as 10-hydroxy-2-decenoic acid (10-HDA), also known as the "Queen bee acid," show promise in reducing insulin resistance and hyperglycemia. Recent research highlights RJ's role in modulating immune responses, enhancing anti-inflammatory cytokines, and suppressing key inflammatory mediators. Despite these promising findings, further research is needed to comprehensively understand the mechanisms underlying RJ's therapeutic effects.

Keywords: royal jelly; diabetes mellitus; gastrointestinal diseases; cardiovascular diseases; bioactive compounds

1. Introduction

Obesity, insulin resistance, hypertension, and heart disease are groups of diseases categorized as metabolic disorders [1]. Metabolic disorders are caused by the disruption of regular metabolic functions triggered by oxidative stress and chronic inflammation. The characteristics of chronic inflammation include elevated levels of inflammatory mediators, such as chemokines and cytokines, which confirm the promotion of the development of metabolic disorders [2]. These chronic inflammations also contribute to the development of major chronic diseases such as non-alcoholic fatty liver disease and obesity. Type 2 diabetes and metabolic disorders now rank among the main risks to human health [3,4]. The significant rise in the prevalence of various metabolic diseases is related to aging, environmental factors, changes in lifestyle, and genetics [5]. Metabolic disorders have been estimated to affect 25% of the population globally [6].

Metformin and Glimepiride are the two synthetic drugs that are currently validated and available for the management of Type 2 diabetes mellitus (T2DM) [7,8]. Metformin is the first-line treatment and has been used for many decades to reduce blood sugar [9]. One of the alternate methods for treating metabolic disorders, and therefore reducing health risks, is utilizing natural product resources. Berberine, derived from the root of *Berberis vulgaris* L. and taken as a typical example, demonstrates significant potential in fighting T2DM [10].

Royal jelly (RJ) is a "rich source of nutrients" that nurse bees produce and feed to worker larvae and queen bees. RJ supplementation is beneficial for a variety of disorders, including diabetes [11,12], gastrointestinal diseases [13,14], and cardiovascular diseases [15,16]. The active ingredients of RJ, including its proteins, carbohydrates, and fats, as well as its minerals, amino acids, vitamins, enzymes, hormones, and polyphenols, are what provide its biological properties [17].

The current review intends to emphasize the protective properties of RJ and/or its components against metabolic disorders such as diabetes, gastrointestinal ailments, and cardiovascular diseases as a part of our ongoing project studying honeybee products [18–20].

2. Methodology

Sci-finder, PubMed, Google Scholar, Web of Science, ScienceDirect, Microsoft Academic Search, Core, and Scopus were all accessed to conduct a literature search. The following keywords were used to address the search terms: "royal jelly", "diabetes mellitus", "gastrointestinal diseases", and "cardiovascular diseases". The search included published studies, and therefore only articles in English were selected. Studies that explored the functions of RJ or its active components with the co-administration of prescription products were chosen. The search approach identified 310 specific studies, out of which 87 were disregarded due to the irrelevance of the study scope. Finally, the studies were further analyzed to offer an insightful overview of the field's progress. The authors, year of publication, RJ utilization, dosage form or percentage, category of investigation, main points, pathway of interaction, and techniques used were all data items.

3. Royal Jelly in Diabetes

Diabetes mellitus (DM) is a worldwide metabolic disorder. According to Saeedi et al., it affected 463 million individuals globally in 2019, with that figure expected to climb to 578 million by 2030 and 700 million by 2045 [21]. DM is anticipated to become more prevalent in the world, and thus health concerns are only liable to increase [22]. Diabetes is a prevalent health issue that affects both sexes equally and impairs sexual function (e.g., sexual disinclination, negative pregnancy outcomes, infertility, loss of penile erection, and diminished clitoral sensitivity). The male reproductive system is impacted by DM in a variety of anatomical and functional aspects, with reduced sperm parameters being an example of the secondary complications of diabetes [23]. Clinically, T2DM, which accounts for 90% of cases of diabetes, is characterized by hyperglycemia and insulin insufficiency caused by cell dysfunction and insulin resistance (IR) in target organs, including the liver, heart, skeletal muscle, and adipose tissue. Atherosclerosis, coronary heart disease, and kidney disease are all chronic consequences that are highly likely to develop in T2DM patients [24]. The two most common causes of morbidity in people with DM are infections and foot ulcers [25]. Up to 60% of non-traumatic lower-limb amputations in diabetics result from diabetic foot ulcers (DFU), the leading risk factor. Given the high expense of diabetic medical treatments, it is critical to explore alternative entities that can be cost-effective. These choices should not only manage blood glucose levels, but also reduce the probability of complications [26]. Thus, the race to find a sustainable and economically viable solution for diabetes remains ongoing. As demonstrated below, RJ treatments have shown therapeutic potential in both rodent and human diabetic models and are effective against hypercholesterolemia and diabetes, as mentioned in Figure 1 [12,27].

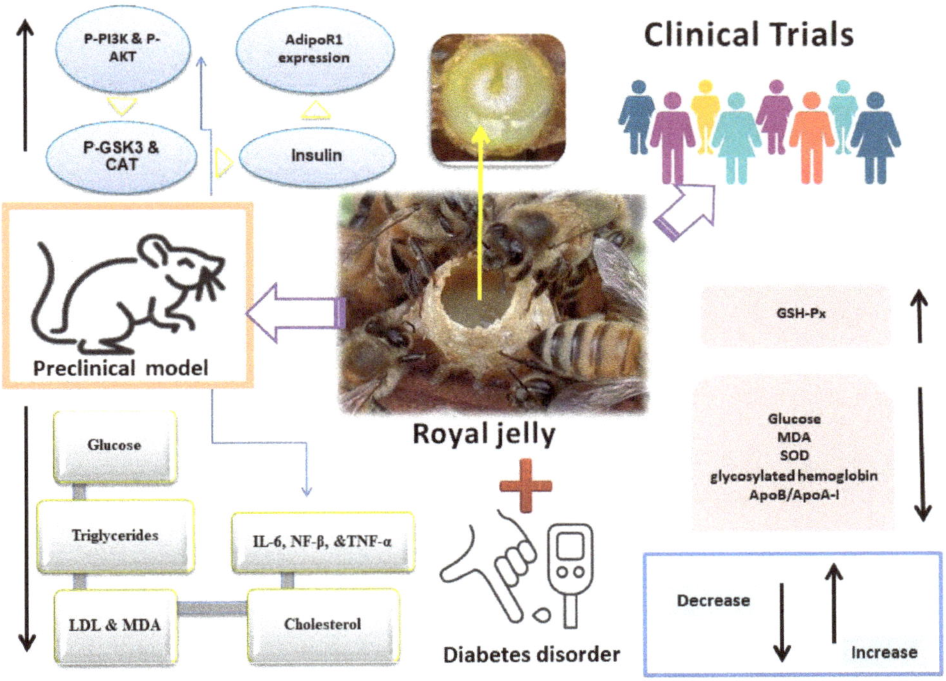

Figure 1. Mechanism of action of royal jelly in diabetes mellitus (DM), preclinical and clinical models [28–32]. LDL: low-density lipoprotein; SOD: superoxide dismutase, IL-6: interlukin-6; MDA: malondialdehyde; GSH-Px: glutathione peroxidase; ApoB/ApoA-I: apolipoprotein B/apolipoprotein A-I; AdipoR1: adiponectin receptor-1; CAT: catalase; p-GSK3β: glycogen synthase kinase 3β; p-AKT: phosphorylated Akt; PI3K: phosphoinositide 3-kinase; IL-1β: interleukin-1β; TNF-α: tumor necrosis factor-α; NF-κB: nuclear factor kappa-B; COX-2: cyclooxygenase-2.

3.1. Preclinical Studies

In preclinical studies, RJ was given orally to KK-Ay mice at a dose of 10 mg/kg body weight (BW). In obese/diabetic KK-Ay mice, RJ treatment improves hyperglycemia and partially lowers BW. RJ administration activates the expression of adiponectin (AdipoQ) and adiponectin receptor-1 (AdioR1), which then activate the expression of phosphorylated AMP-activated protein kinase (pAMPK). Additionally, RJ treatment that increases adiponectin receptor-1 (AdipoR1) expression also boosts Ppara and Pgc1a expression, which improves lipid utilization and causes a reduction in BW in KK-Ay mice [28]. Adult male Wistar rats were divided into four groups: diabetic, RJ, diabetic treated with RJ, and control. To induce diabetes, streptozotocin-induced diabetes (STZ)was administered intravenously at a dosage of 60 mg/kgBW. RJ was then administered via gavage at a dosage of 100 mg/kg BW for six weeks. Testicular weight, viability, sperm count, deformity, motility, chromatin quality, DNA integrity, testicular tissue malondialdehyde (MDA) levels, and serum testosterone were all enhanced via RJ in diabetic mice [33]. A total of 28 adult Wistar rats were randomized and divided into four groups: control, RJ, diabetic, and hyperglycemic treated with RJ. To induce diabetes, a single intraperitoneal injection of STZ at a dose of 50 mg/kg BW was used. The rats were then administered RJ (100 mg/kg BW) orally each day for a duration of six weeks. The treatment with RJ resulted in improved levels of catalase (CAT) and ferric reducing antioxidant power (FRAP) when compared to other groups [29]. RJ was given orally to the diabetic rats at a dosage of 100 mg/kg for 42 days following STZ. RJ improved the serum levels of aspartate aminotransferase (AST), alkaline phosphatase (ALP), high-density lipoprotein cholesterol (HDL-c), alanine aminotransferase (ALT), total protein (TP), fasting blood glucose (FBG) levels, insulin, and albumin. RJ dramatically lowered MDA levels in the liver and pancreatic tissues and simultaneously normalized the levels of CAT and FRAP [30]. In a similar experimental study, diabetes was induced in rats through an intraperitoneal injection of STZ (60 mg/kg BW). Then, RJ was administered via gavage for three days at doses of 100 and 200 mg/kg. This treatment notably decreased the levels of cholesterol, glucose, low-density lipoprotein (LDL), and triglycerides in the diabetic rats. Interestingly, the rats that received RJ treatment exhibited significantly elevated HDL levels compared to the untreated diabetic rats [11]. The 18 adult Wistar albino rats were grouped into three groups: control, STZ-induced diabetes, and STZ-induced diabetes plus RJ at a dosage of 400 mg/kg/day for a month. To induce diabetes, STZ (60 mg/kg) was injected intraperitoneally once. Both the RJ-treated and untreated diabetic rats exhibited lower body and testicular weights compared to the control group. Rats treated with STZ had significantly more degenerative alterations in their spermatogenesis and seminiferous tubules, according to the histological analysis. The RJ treatment group, on the other hand, revealed nearly normal morphology, in addition to more intense immunohistochemistry staining for Ki67-positive cells [34]. To induce diabetes, STZ was administered intravenously to rats once at a dosage of 75 mg/kg BW. The rats were categorized into four distinct groups: a healthy control group and three treatment groups. Three groups of rats; untreated diabetic group, 100 mg/kg/daily of metformin group, and one group received a honey–RJ (H–RJ) combination, containing 2% RJ and 98% honey. H–RJ was given daily to this rat group (100 mg/kg BW). The H–RJ treatment significantly lowered the levels of very low-density lipoprotein (VLDL) in the blood, compared to both the control therapy and metformin treatment. Rats with diabetes can effectively present with lower blood sugar when given H–RJ. This combination can also successfully lower triglycerides and VLDL-C lipids (TGs) [35]. Likewise, C57BL/6J mice were subjected to a high-fat diet (HFD) and administered a 5% RJ diet. Alloxan-induced diabetes in male Albino Wistar rats was discussed, as well as the hypoglycemic effect of composite formulations of *Moringa oleifera* seed oil extract and RJ. When compared to pure *M. oleifera* or pure RJ, the medication containing 20% RJ mixed with *M. oleifera* seed extract was found to be more effective in reducing blood sugar levels in treated mice [36]. Taken together, this dietary intervention with RJ effectively mitigated diet-induced obesity, hyperglycemia, and hepatic steatosis in mice by stimulating metabolic thermogenesis in brown adipose tissue (BAT) [37].

Fatty acid 10-hydroxy-2-decenoic acid (10H2DA) is a component found in RJ (Figure 2, Table 1). Female KK-Ay mice received 10H2DA orally at a dosage of 3 mg/kg BW via gavage for four weeks. It greatly reduced insulin resistance and hyperglycemia. In skeletal muscles, 10H2DA elevated the expression of the pAMPK protein; however, this expression was unrelated to elevated glucose transporter 4 (GLUT4) translocation. Adiponectin receptor mRNA expression was not improved by 10H2DA, and the liver's glycogen synthetase kinase-3β (GSK-3β) phosphorylation was not triggered by the insulin signaling cascade [38]. The main active component of RJ is 10-hydroxydecanoic acid (10-HDA) (Figure 2). Recent research findings indicate that 10-HDA may possess anti-T2DM properties. When administered orally at a dosage of 100 mg/kg BW, it stopped liver degeneration in the diabetic rats and boosted insulin levels while decreasing fasting blood glucose. Additionally, 10-HDA intervention improved lipid peroxidation, reduced liver NF-β nuclear translocation, reduced interlukin-6 (IL-6) and tumor necrosis factor (TNF-α) content, and elevated P-PI3K, phosphorylated Akt (p-AKT), and glycogen synthase kinase 3β (p-GSK3β(protein levels. It also enhanced glutathione peroxidase (GPx), superoxide dismutase (SOD), and CA activity in diabetic mouse livers. Through the PI3K/AKT/GSK3 signaling pathway, 10-HDA clearly exhibited hypoglycemic effects on diabetic mice [39].Growth factor deficiency and bacterial infection are two of the main factors causing non-healing wounds in diabetics [40]. 8-Bromoadenosine-3′, 5′-cyclic monophosphate (8Br-cAMP) and antimicrobial peptide Jelleine-1 (J-1) (Figure 2) were combined to form a hydrogel without the use of any other gelators or chemical crosslinkers. This hydrogel demonstrated remarkable antibacterial action in a wound model in diabetic rats infected with methicillin-resistant *Staphylococcus aureus* (MRSA) [41].

Table 1. Bioactive compounds identified from royal jelly as antidiabetic agents.

Identified Compounds	Dosage	Biological Activity (In Vitro/In Vivo)	References
Hesperetin	40 mg/kg body weight (BW) for 45 days	Reduces high blood sugar and lipid levels by enhancing insulin secretion (in vivo).	[42,43]
Naringenin	(25, 50, 100 mg/kg) for 4 weeks	The therapy significantly enhanced the control of blood glucose levels and also contributed to the recovery of BWin diabetic rats, in contrast to those that received a vehicle treatment (in vivo).	[43,44]
	50 mg/kg for 4 weeks	Adequate to mitigate the alterations in the lenses caused by diabetes-related oxidative stress (in vivo).	[45]
	50 mg/kg/day for 5 days	Significant decrease in blood glucose and triglyceride levels in diabetic rats (in vivo).	[46]
	100 mg/kg BW /day for 4 weeks	Restored the serum insulin and C-peptide levels, replenished liver glycogen, and reduced glucose-6-phosphatase and glycogen phosphorylase activity in the liver. Additionally, it improved the serum lipid profile and strengthened the liver's antioxidant defense system (in vivo).	[47]
Genistein	(20 and 40 mg/kg) for 8 weeks.	Improved glucose tolerance, blood glucose levels, insulin, glucagon, lipid profiles, and pro-inflammatory factors. It also improved liver function, reduced inflammation in the liver and colon, and positively altered gut microbiota composition (in vivo).	[43,48]
	25–200 mg/day	Improved hyperglycemia, glucose tolerance, and blood insulin levels, along with enhancing islet beta-cell proliferation, survival, and mass (in vivo).	[49]
	600 mg/kg for 4 weeks	Enhanced insulin sensitivity and increased expression of neurotrophic factors, such as nerve growth factor (NGF) and brain-derived neurotrophic factors (BDNF) (in vivo).	[50]
Formononetin	20 mg/kg for 28 days	Reduced serum glucose levels and increased serum insulin compared to the control group. It also decreased insulin resistance and reduced fasting glucose (C57BL/6 mice, in vivo).	[43,51]
	40 mg/kg/day for 16 weeks	Decreased insulin resistance and regulated hypoglycemia in male rats with diabetes (in vivo).	[52]

Table 1. *Cont.*

Identified Compounds	Dosage	Biological Activity (In Vitro/In Vivo)	References
Coumestrol	50 µM	Improved hepatic insulin resistance in primary at hepatocyte (in vivo).	[43,53]
Chrysin	100 mg/kg	It resulted in a reduction of fasting blood glucose and insulin levels in db/db mice when compared to the control group (in vivo).	[43,54]
	80 mg/kg BW for 10 days	Anti-diabetic effects via increasing insulin levels, reducing oxidative stress, and regulating the inflammatory pathway (in vivo).	[55]
10-Hydroxy-2-decenoic acid	100 mg per kg BW/Daily for 4 weeks.	Decreased fasting blood glucose and increased insulin levels in diabetic mice. Enhanced activity of crucial antioxidants in the livers of diabetic mice, such as superoxide dismutase, catalase, and glutathione peroxidase (in vivo).	[39,56]

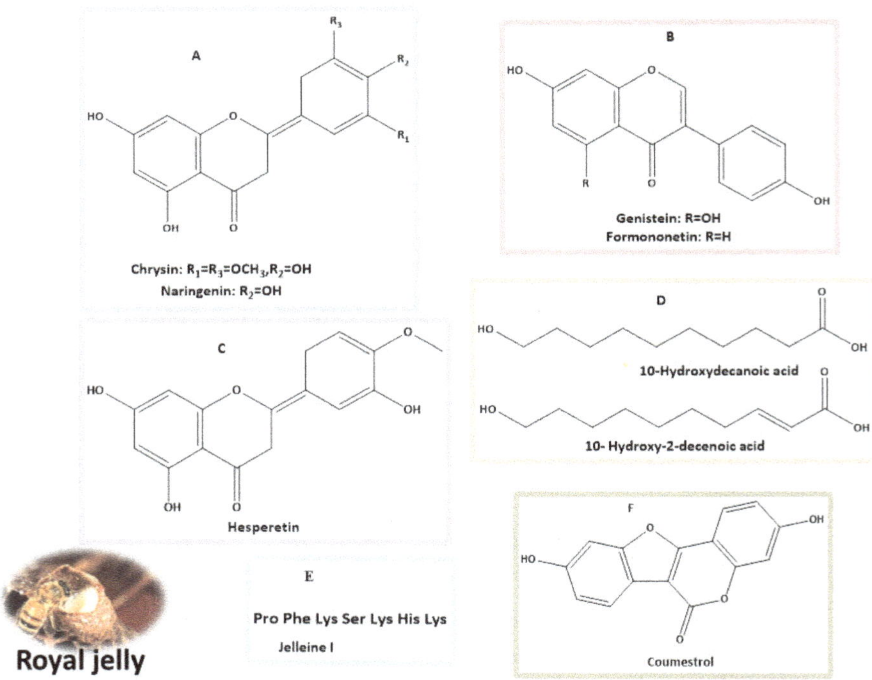

Figure 2. Major classes of natural products (**A–C**) flavonoids; (**D**) fatty acids; (**E**) peptide; and (**F**) coumestans identified in royal jelly with potential anti-diabetes properties.

3.2. Clinical Studies

A double-blind, placebo-controlled trial including 50 T2DM patients was carried out, where either 1000 mg of RJ or a placebo were administered to subjects three times per day for eight weeks. The groups were assigned to the RJ or placebo groups at intervals. Baseline characteristics and food intake between groups did not differ significantly. In the RJ group, the mean glucose level decreased (−9.4 mg/dL vs. 4 mg/dL), the mean ApoA-I concentration increased (34.4 mg/dL vs. −1.08 mg/dL), and there was a significant decrease in the mean apolipoprotein B (ApoB) /apolipoprotein A-I (ApoA-I), (0.008 vs. 0.13; $p < 0.044$, respectively) when comparing the RJ group to the placebo group [31].

For eight weeks, 50 female T2DM volunteers were divided into two groups and given either a daily dose of 1000 mg RJ (soft gel) or a placebo. In the RJ group, the average fasting blood sugar dropped significantly to 149.68 mg/dL after supplementation with RJ. The mean serum levels of glycosylated hemoglobin also significantly decreased to 7.05%, and the mean insulin concentration significantly decreased through RJ supplementation to 27.5 pmol/L. MDA levels declined, and GPx and erythrocyte superoxidase dismutase activity was dramatically elevated [32]. A daily dose of a 1000 mg soft gel of RJ or a placebo was administered to 50 female T2DM volunteers, divided into two respective groups, in a randomized clinical trial for 8 weeks. RJ supplementation reduced the daily total energy and carbohydrate intake as well as mean BW (72.45 vs. 71.00 kg) when compared to the control group [12]. Another randomized controlled trial comprised 46 T2DM patients aged 25–65 years with a hemoglobin A1c (HbA1c) of 6–8%. For eight weeks, the patients were randomized to take 1000 mg of RJ supplement or a placebo three times each day. In the RJ group, the insulin resistance index (HOMA-IR) decreased (1.98 vs. 3.13) while the serum total antioxidant capacity increased (907.63 vs. 765.69 mol/L) [57].

The effectiveness of topical RJ for treating diabetic foot ulcers has also been studied [58]. The trial design was randomized, controlled, and open-label, with a 12-week average follow-up time. After conservative debridement of necrotic tissue and irrigation with warm normal saline, 189 eligible patients with diabetic foot wounds from three outpatient clinics in Egypt were randomized to receive a local application of either RJ + Panthenol (PedyPhar® Ointment) or Panthenol ointment underdressing. The purpose of the research was to look at the use of PedyPhar® Ointment in the treatment of individuals suffering from limb-threatening diabetic foot infections [59]. At the end of the 12-week follow-up period, PedyPhar® revealed a greater degree (32.4%) of full healing of limb-threatening wounds in the target population, versus 12% in the Panthenol-treated (control) group [60].

Adiponectin is an adipokine released from adipose tissue that has a role in insulin sensitivity [61], and has been reported for its inhibitory effect on the glucogenesis process in the liver via down-expression of phosphoenolpyruvate carboxykinase and glucose-6-phosphatase genes, which play crucial enzymatic roles in glucose production by the liver [62]. Furthermore, research has shown that giving RJ to diabetic rats at a dosage of 10 mg/kg/day for a month controls hyperglycemia via the acceleration of adiponectin secretion [28]. The possible mechanism of RJ in diabetes was reported by Maleki et al. (2019), through the activation of the AMP-activated protein kinase pathway through increasing the production of adiponectin in skeletal muscle and liver cells [63]. Joshi et al. (2019) noted that the activation of the AMPK protein increases the cells' uptake of glucose, while reducing intracellular secretion of glucose. Therefore, the AMPK signaling pathway is the target in controlling diabetes [64].

4. Royal Jelly in Gastrointestinal Diseases

Gastrointestinal disorders are common illnesses, including irritable bowel syndrome, peptic ulcers, liver diseases, pancreatitis, gallstones, and Crohn's disease, and are often found in tropical regions [65]. Sperber et al. (2021) estimated that 40% of the people in 33 countries across six continents have functional gastrointestinal problems [66]. A strong link between gastrointestinal diseases and the diet of individuals in at-risk groups has been found. The dietary habits of people in the different areas contribute to the composition of their individual gut microbiota, and this is particularly noticeable when people move from urban to rural areas [67]. According to Rizello et al. (2019), a westernized diet rich in carbohydrates and animal proteins is a main contributor to the development and progression of chronic inflammatory bowel disease [68]. Due to involvement of the gastrointestinal tract in the absorption of nutrients, as well as its role in immune response, the risk of developing an inflammatory, autoimmune, chronic disease is inevitably increasing [69].

RJ is considered one of the most important super foods, having displayed much biological activity in preclinical and clinical studies [70]. As a honey bee product, it has been documented for its active potential against many disorders, including inflammation, liver disease, hypercholesterolemia, oxidative stress, and immune disease [71]. RJ contains many bioactive compounds such as proteins, vitamins, phenolics, and flavonoids. Additional pharmaceutical studies have revealed that the bioactive, major protein constituents of RJ (MRJPs) are considered the main therapeutic compounds of those tested [72,73].

4.1. Inflammatory Bowel Diseases

Ulcerative colitis and Crohn's disease are both chronic, inflammatory bowel illnesses. Inflammatory bowel disease describes a persistent, non-infectious inflammation with uncertain causes, affecting one or more locations in the digestive system. According to global estimations, the prevalence of inflammatory bowel disease accounted for approximately 7 million people in 2017 [74]. Hence, the exploration of natural product remedies would pave the way to natural and complementary tools for healing. For instance, the experimental induction of colon inflammation using 2,4,6-trinitrobenzene sulphonic acid was found to be significantly inhibited by the administration of RJ in mice at a dosage of 250 mg/kg/day for a week via the inhibition of pro-inflammatory cytokines, TNF-α, and interlukin-1β (IL-1β) along with the elevation of the anti-inflammatory cytokine interlukin-10 (IL-10) [75]. Another study revealed that daily administration of RJ (150 mg/kg) considerably ameliorated the damage caused by acetic acid in rats with induced colitis, manifesting as decreased lesion areas in the colon where the intestinal mast cells were also involved in inflammation [76]. Likewise, similar doses of RJ were found to reduce the proliferation of T-lymphocytes involved in the intestinal inflammation induced by acetic acid in rats [77]. According to a recent study, synergism between RJ and selenium exhibits significant anti-inflammatory activity in inflammatory bowel disease in mice and promotes intestinal health through the improvement of the gut microbiota [78].

The mechanism of RJ in treating inflammatory bowel syndrome has been reported by Guo et al. (2022) as shown on Figure 3. The investigation revealed that RJ boosted the activity of the anti-inflammatory cytokine IL-10 and the intracellular antioxidant enzyme GPx. Additionally, RJ decreased the number of CD3+, CD5+, CD8+, and CD45+ T-cells, the release of TNF-$\alpha$and the pro-inflammatory cytokines IL-1β, the nuclear factor Kappa-B (NF-κB), and cyclogenase-2 (COX-2) and tumor necrosis factor-induced injury in rats with colitis induced by 2,4,6-trinitrobenzene sulfonic acid [79].

4.2. Lactose Intolerance

Lactose intolerance is a gastrointestinal disorder that results from a lack of β-galactosidase, resulting in the maldigestion of lactose from milk and milk products. Patients with lactose intolerance present with symptoms such as pain in the abdomen, diarrhea, and flatulence, which appear after the intake of lactose-containing foods [80]. Recently, researchers have found that the synergism between RJ and probiotic yogurt has potent activity in treating lactose intolerance [81].

There is growing evidence that lactose intolerance symptoms can be treated with probiotic bacteria found in fermented and unfermented milk products [82]. The mechanism of RJ in reducing lactose intolerance relies on the activity of probiotics delivered via fermented milk products, which have been found to play an important role in health benefits, as reported by Hassan et al. (2022). The fermentation of milk with 1% RJ displayed the presence of abundant probiotics, namely *Lactobacillus helveticus*, which results in boosting the bioactive properties of fermented milk [83].

4.3. Chronic Diarrhea and Constipation

The symptoms of chronic constipation include uncomfortable defecation, marked by straining and difficulty along with extended time in stool passage [84]. Constipation in children is estimated to affect from 1% to 30% of the young generation worldwide [85]. Compared to standard antiviral medication, honey has been shown to reduce the incidence and duration of viral diarrhea [86]. As documented by Miyauchi-Wakuda et al. (2019), under in vitro circumstances, acetylcholine in RJ induced contractions of the smooth muscle of the mouse's ilium via the muscarinic acetylcholine receptor, which was independent of nicotinic acetylcholine activity. The intake of royal jelly does not result in severe symptoms like diarrhea in normal situations [14,87]. Further, the anti-diarrheal potency of RJ could be attributed to the antimicrobial activity of its peptide constituents, royalisin and royalactin [88,89]. Even though RJ has a high concentration of acetylcholine, only one oral dose of RJ was not enough to boost intestinal motility or alleviate constipation.

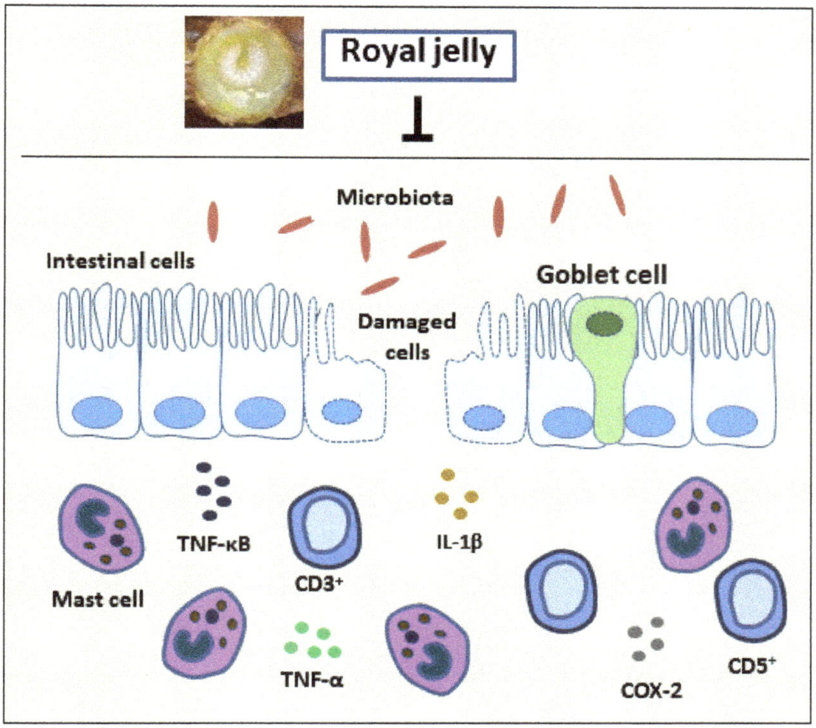

Figure 3. Treatment with royal jelly suppressed the rise of CD3+, CD5+, CD8+ and CD45+ T-cells, pro-inflammatory cytokines, IL-1β, TNF–α, and the expression of major inflammatory mediators (COX-2 and NF-κB) in the colon of rats with colitis. IL-1β: interleukin-1β; TNF-α: tumor necrosis factor-α; NF-κB: nuclear factor kappa-B; COX-2: cyclooxygenase-2.

4.4. Gastrointestinal Ulcer Disease

Gastric and intestinal ulcers induced by diclofenac (50 mg/kg) have been normalized using RJ at a dose of 150 mg/kg or 300 mg/kg via the increase of prostaglandin-2 (PGE-2) and COX-2 in the stomach tissues of mice, as well as reducing myeloperoxidase (MPO) and inducible nitric oxide synthase (iNOS) [71]. Another study revealed that acetic acid-induced peptic ulcers in rats could be treated significantly using a daily dose of RJ (200 mg/kg), in comparison to the commonly used anti-ulcer drug omeprazole (20 mg/kg), for 14 days of treatment [90]. Furthermore, when RJ was administered to rats at a dose of 250 mg/kg, it protected them from ulcers in the stomach caused by ethanol. This was relative to the ulcer-preventing medication, lansoprazole, given at a dose of 30 mg/kg. The mechanism of gastroprotection has been claimed to be due to the attenuation of pro-inflammatory cytokines, TNF-α, lipid peroxidation, and IL-1β in addition to the augmentation of the endogenous antioxidant enzyme SOD and CAT [13]. El-Naeem and Fareed (2022) reported the positive effects of 30-day administration of RJ (300 mg/kg) on ameliorating the gastric mucosal histopathological changes that were caused in rats by intra-peritoneal injection of 0.5 mg nicotine tartarate [91].

4.5. Liver Disease

4.5.1. Preclinical Studies

Synergistic, daily treatment of mice with the drug diclofenac (50 mg/kg) for seven days, following which RJ was given orally at dosages of 150/or 300 mg/kg for a month, was found to alleviate the hepato-renal toxicity of the drug through over-expression of PGE-2 and COX-2 in the animals' liver and stomach tissues [71]. The hepatic toxicity triggered by the immunosuppressive drug azathioprine was found to be altered by oral administration of RJ (200 mg/kg) in rats through the attenuation of the high levels of serum hepatic enzymes caused by an intra-peritoneal injection of 50 mg/kg dose of azathioprine [92]. According to a recent study, feeding diabetic rats 300 mg/kg of RJ for 16 weeks, development of non-alcoholic fatty liver disease (NAFLD) was reported. The study concluded that RJ has anti-inflammatory and antioxidant properties that protect against NAFLD, while also regulating the metabolism of fatty acids such as arachidonic acid and linoleic acid as well as the production of unsaturated fatty acids [93]. The study revealed that RJ treatment significantly raised the serum levels of adiponectin and concurrently raised the hepatic phosphorylation of 5′ adenosine monophosphate-activated protein kinase (AMPK). Where the hypolipidemic effect of RJ is mediated mainly by regulating AMPK, these effects have been noticed in rats that were fed a high-fat diet and subsequently developed NAFLD. RJ suggests a novel, independent mode of action by promoting fatty acid oxidation via activation of hepatic AMPK signaling and by suppressing cholesterol formation via sterol regulatory element-binding proteins SREBP1/2 without altering the production of adiponectin, the enzyme responsible for fatty acid oxidation, and lowering the synthesis of triglycerides and cholesterol [93,94].

4.5.2. Clinical Studies

A further investigation found that administration of 1 g/day of RJ for 30 days to patients with chronic hepatitis B could influence the immunological responses of patients by inhibiting the protein responsible for initiating inflammatory responses, NLRP1. Additionally, RJ up-regulated the functions that the inflammasome adaptor speck-like protein (ASC) performs in modulating immune responses [95]. Collectively, RJ was documented for its potent activity in treating gastrointestinal diseases as per in vitro and in vivo studies. The possible impact of RJ on various digestive tract illnesses is summarized in Figure 4.

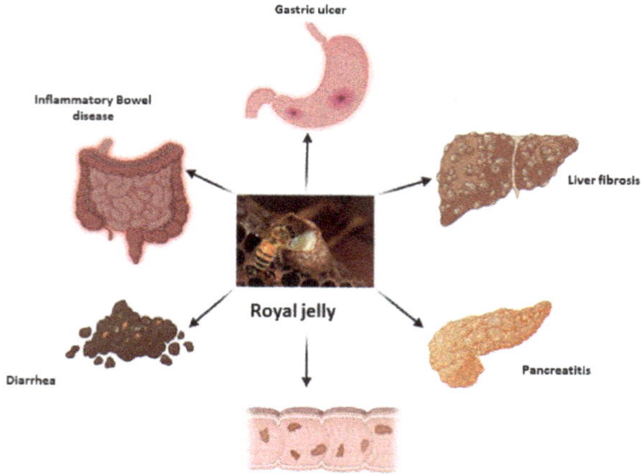

Figure 4. Potential activity of royal jelly in the most common gastrointestinal diseases.

5. Royal Jelly in Cardiovascular Disease

Among all disorders, cardiovascular conditions pose the greatest threat to human health and cause the greatest number of fatalities (more than 17 million/year) [96]. The latest mortality records in Europe have estimated there to be more than 3.5 million deaths per year due to cardiovascular diseases, ranking them first in the world for modern mortality causes [97]. Aslan et al. (2021) have reported that, in the rat model, the administration of RJ to drinking water at doses of 50 or 100 mg/kg for a month demonstrated cardioprotective activity against fluoride-induced heart injury through the down-expression of Bcl-2 protein and the enhanced expression of Bcl-2-associated X proteins (Bax) in the heart tissues and the caspase family (caspase-3, 6 and 9). Additionally, RJ revealed a significant reduction in the expression of cardiac glycogen synthase kinase-3 (Gsk-3) and Nf-κB proteins [98].

5.1. Antihypertensive Activity of Royal Jelly

The biggest risk factor for cardiovascular disease around the globe is hypertension, which is caused by a disturbance in the contractile or proliferative function of the vascular smooth muscle cells of blood vessels [99].

5.1.1. Preclinical Studies

Previous research has demonstrated that peptides isolated from RJ at doses of 1 g/kg considerably reduced the high blood pressure of hypertensive rats after 10 weeks of administration. The evidential effect was explained by the down-expression of angiotensin-1-converting enzyme, the main regulator of blood pressure [100]. The protein content of RJ represents half of its dry weight, while a major RJ protein, namely MRJP1-9, represents 80% of its total protein content [101]. A recent in vitro study revealed that incubation of aortic vascular smooth muscle cell lines from mice with MRJP1 showed a significant reduction in the cellular α-smooth muscle actin protein, the marker responsible for hypertension [102]. A similar in vivo study conducted on experimentally induced hypertension in Wistar rats using angiotensin-converting enzyme revealed that administration of RJ at a dosage of 15 mg/kg every day for four weeks to hypertensive rats suppressed their increases in blood pressure [103]. Another preclinical study on rats and rabbits showed that RJ exhibited a hypotensive impact via increased nitric oxide production. The study also demonstrated vasodilation effects through the suppression of the cyclic guanosine monophosphate pathway that mediates the contraction and relaxation of vascular smooth muscle cells [104]. Other studies suggested that a muscarinic receptor agonist, like acetylcholine, might be one of the vasodilators in RJ. In fact, it has already been noted that RJ contains more than 900 μg/g of acetylcholine-like substances. Additionally, acetylcholine stimulates the release of endothelium-derived relaxing factors (EDRFs), including nitric oxide (NO), prostacyclin (PGI2), and endothelium-derived hyperpolarizing factor (EDHF), by the vascular endothelial cells. They are the primary mediators of the vasorelaxant effects that are endothelium dependent [104,105].

5.1.2. Clinical Studies

At the clinical level, a randomized, placebo-controlled study demonstrated that a daily intake of RJ tablets (690 mg) for four weeks significantly improved the vascular endothelial activity of the participants' blood vessels, suggesting that RJ may exert anti-atherogenic activity [106]. Another clinical study reported that the treatment of renal failure patients suffering from cardiovascular disease with a daily dosage of RJ (3600 mg) for a year considerably attenuated the progression of atherosclerosis in hemodialysis patients [15]. Figure 5 illustrates the mechanism of the antihypertensive activity of RJ.

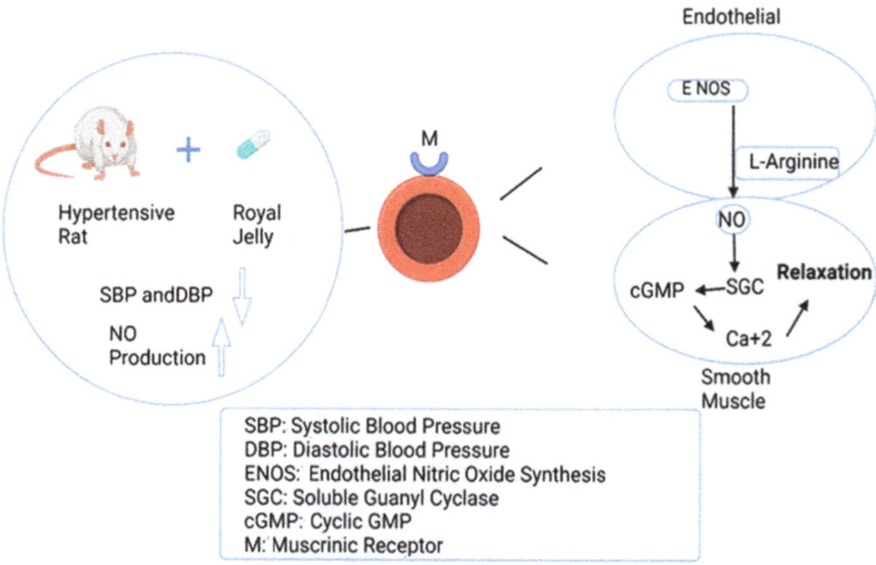

Figure 5. The mechanism of the antihypertensive activity of royal jelly.

5.2. Hypo-Cholesterolemic Activity of Royal Jelly

Cholesterol plays an important role in several physiological activities inside the body; nevertheless, increased levels in serum cause serious health problems, including cardiovascular disorders [107]. Previous studies reported that providing a diet containing 5% RJ to mice for seven days markedly reduced cholesterol levels in the blood through down-expression of squalene epoxidase, which is essential for the biosynthesis of cholesterol [108]. Animal experimentation proved that MRJP1 has anti-cholesterolemic potential through its interaction with bile acids that increase cholesterol catabolism in the liver and excretion of cholesterol in feces, compared to a β-sitosterol drug [109].

In a placebo-controlled trial, daily use of RJ capsules containing 350 mg RJ for three months was found to significantly alter low-density lipoprotein and total cholesterol levels [110]. The study's findings led the authors to the conclusion that RJ consumption significantly increased the levels of dehydroepiandrosterone sulphate (DHEA-S). By significantly raising the concentration of DHEA-S over the course of three months, nine RJ capsules taken daily might significantly reduce the levels of LDL-c and total cholesterol (TC) in the serum. In addition, the investigations revealed that DHEA-S can affect the activity of glycerol 3-phosphate dehydrogenase and glucose 6-phosphate dehydrogenase, which in turn can stop the creation of NADPH and hence prevent the biosynthesis of fatty acids, phospholipids, and cholesterol. The hypo-cholesterolemic potential of RJ has been confirmed by a meta-analysis study that supported the notion that RJ reduces total cholesterol levels while also increasing high-density lipoproteins and accordingly regulating the lipid profile [16]. According to Balan et al. (2020), atherosclerosis and cardiovascular diseases related to post-menopausal symptoms in women were found to be inhibited with the regular intake of RJ [111].

6. Conclusions and Future Prospects

The beneficial impact of RJ on diabetes, gastrointestinal ailments, and cardiovascular disease is well-documented in the research articles reviewed. These findings, derived from both preclinical and clinical studies, highlight the potential of RJ as a promising intervention in the field of metabolic health. Furthermore, RJ may enhance the treatment of these disorders by mitigating the adverse effects linked with the drugs used for managing diabetes, gastrointestinal, and cardiovascular conditions. However, despite these promising findings, the specific mechanisms through which RJ exerts its therapeutic effects remain a subject of ongoing research. This underscores the need for further scientific investigations to validate the therapeutic efficacy of RJ and to fully understand its role in disease management. As we move forward, it is crucial to continue exploring the potential of RJ in treating metabolic disorders, with a focus on elucidating its mechanisms of action and potential applications in personalized medicine.

Author Contributions: Conceptualization, H.R.E.-S. and S.A.M.K.; methodology, S.S., A.A.A.E.-W., H.R.E.-S. and S.A.M.K.; software, A.A.A.E.-W. and S.A.M.K.; writing of the original draft, S.S., A.A.A.E.-W. and S.A.M.K.; revision and editing of the final version, S.A.M.K., Z.G., A.D.M., M.D., C.L., X.G., D.G.B., H.R.E.-S. and K.W.; supervision, H.R.E.-S. and K.W.; project administration, H.R.E.-S. All authors have read and agreed to the published version of the manuscript.

Funding: This research received no external funding.

Acknowledgments: H.R. El-Seedi thanks Jiangsu University for the Adjunct Professor fellowship.

Conflicts of Interest: The authors declare that the research was conducted in the absence of any commercial or financial relationships that could be construed as a potential conflict of interest.

References

1. Han, T.S.; Lean, M.E.J. Metabolic Syndrome. *Medicine* **2011**, *39*, 24–31. [CrossRef]
2. Robbins, G.R.; Wen, H.; Ting, J.P.Y. Inflammasomes and Metabolic Disorders: Old Genes in Modern Diseases. *Mol. Cell* **2014**, *54*, 297–308. [CrossRef] [PubMed]
3. Chait, A.; den Hartigh, L.J. Adipose Tissue Distribution, Inflammation and Its Metabolic Consequences, Including Diabetes and Cardiovascular Disease. *Front. Cardiovasc. Med.* **2020**, *7*, 181–191. [CrossRef] [PubMed]
4. Rinaldi, L.; Pafundi, P.C.; Galiero, R.; Caturano, A.; Morone, M.V.; Silvestri, C.; Giordano, M.; Salvatore, T.; Sasso, F.C. Mechanisms of Non-Alcoholic Fatty Liver Disease in the Metabolic Syndrome. A Narrative Review. *Antioxidants* **2021**, *10*, 270. [CrossRef] [PubMed]
5. Heindel, J.J.; Blumberg, B.; Cave, M.; Machtinger, R.; Mantovani, A.; Mendez, M.A.; Nadal, A.; Palanza, P.; Panzica, G.; Sargis, R.; et al. Metabolism Disrupting Chemicals and Metabolic Disorders. *Reprod. Toxicol.* **2017**, *68*, 3–33. [CrossRef] [PubMed]
6. Korivi, M.; Liu, B.R. New Strategies from Natural Materials to Fight against Diet-Induced Metabolic Disorders (Part–II). *Curr. Pharm. Des.* **2021**, *27*, 761–762. [CrossRef] [PubMed]
7. Burant, C.F.; Viswanathan, P.; Marcinak, J.; Cao, C.; Vakilynejad, M.; Xie, B.; Leifke, E. TAK-875 versus Placebo or Glimepiride in Type 2 Diabetes Mellitus: A Phase 2, Randomised, Double-Blind, Placebo-Controlled Trial. *Lancet* **2012**, *379*, 1403–1411. [CrossRef] [PubMed]
8. Tabatabaei-Malazy, O.; Larijani, B.; Abdollahi, M. Targeting Metabolic Disorders by Natural Products. *J. Diabetes Metab. Disord.* **2015**, *14*, 57–77. [CrossRef] [PubMed]
9. Wang, H.; Zhu, C.; Ying, Y.; Luo, L.; Huang, D.; Luo, Z. Metformin and Berberine, Two Versatile Drugs in Treatment of Common Metabolic Diseases. *Oncotarget* **2018**, *9*, 10135–10146. [CrossRef]
10. Zhao, J.D.; Yu, C.J.; Li, J.Y.; Wang, S.H.; Yang, D.; Guo, C.L.; Du, X.; Zhang, W.J.; Cheng, R.D.; Diao, X.C.; et al. Effect of Berberine on Hyperglycaemia and Gut Microbiota Composition in Type 2 Diabetic Goto-Kakizaki Rats. *World J. Gastroenterol.* **2021**, *27*, 708–724. [CrossRef]
11. Asgari, M.; Asle-Rousta, M.; Sofiabadi, M. Effect of Royal Jelly on Blood Glucose and Lipids in Streptozotocin Induced Type 1 Diabetic Rats. *Arak Med. Univ. J.* **2017**, *20*, 48–56.
12. Pourmoradian, S.; Mahdavi, R.; Mobasseri, M.; Faramarzi, E.; Mobasseri, M. Effects of Royal Jelly Supplementation on Body Weight and Dietary Intake in Type 2 Diabetic Females. *Health Promot. Perspect.* **2012**, *2*, 231–235. [CrossRef]
13. Duran, Y.; Karaboğa, İ.; Polat, F.R.; Polat, E.; Erboğa, Z.F.; Ovalı, M.A.; Öztopuz, R.Ö.; Çelikkol, A.; Yılmaz, A. Royal Jelly Attenuates Gastric Mucosal Injury in a Rat Ethanol-Induced Gastric Injury Model. *Mol. Biol. Rep.* **2020**, *47*, 8867–8879. [CrossRef]
14. Miyauchi-Wakuda, S.; Kagota, S.; Maruyama-Fumoto, K.; Wakuda, H.; Yamada, S.; Shinozuka, K. Effect of Royal Jelly on Mouse Isolated Ileum and Gastrointestinal Motility. *J. Med. Food* **2019**, *22*, 789–796. [CrossRef]
15. Ohba, K.; Miyata, Y.; Shinzato, T.; Funakoshi, S.; Maeda, K.; Matsuo, T.; Mitsunari, K.; Mochizuki, Y.; Nishino, T.; Sakai, H. Effect of Oral Intake of Royal Jelly on Endothelium Function in Hemodialysis Patients: Study Protocol for Multicenter, Double-Blind, Randomized Control Trial. *Trials* **2021**, *22*, 950–957. [CrossRef]
16. Hadi, A.; Najafgholizadeh, A.; Aydenlu, E.S.; Shafiei, Z.; Pirivand, F.; Golpour, S.; Pourmasoumi, M. Royal Jelly Is an Effective and Relatively Safe Alternative Approach to Blood Lipid Modulation: A Meta-Analysis. *J. Funct. Foods* **2018**, *41*, 202–209. [CrossRef]
17. Bagameri, L.; Baci, G.-M.; Dezmirean, D.S. Royal Jelly as a Nutraceutical Natural Product with a Focus on Its Antibacterial Activity. *Pharmaceutics* **2022**, *14*, 1142. [CrossRef]

18. Khalifa, S.A.M.; Elashal, M.; Kieliszek, M.; Ghazala, N.E.; Farag, M.A.; Saeed, A.; Sabir, J.S.M.; Battino, M.; Xiao, J.; Zou, X.; et al. Recent Insights into Chemical and Pharmacological Studies of Bee Bread. *Trends Food Sci. Technol.* **2020**, *97*, 300–316. [CrossRef]
19. El-Seedi, H.; El-Wahed, A.A.; Yosri, N.; Musharraf, S.G.; Chen, L.; Moustafa, M.; Zou, X.; Al-Mousawi, S.; Guo, Z.; Khatib, A.; et al. Antimicrobial Properties of Apis Mellifera's Bee Venom. *Toxins* **2020**, *12*, 451. [CrossRef] [PubMed]
20. El-Wahed, A.A.A.; Farag, M.A.; Eraqi, W.A.; Mersal, G.A.M.; Zhao, C.; Khalifa, S.A.M.; El-Seedi, H.R. Unravelling the Beehive Air Volatiles Profile as Analysed via Solid-Phase Microextraction (SPME) and Chemometrics. *J. King Saud Univ. Sci.* **2021**, *33*, 101449–101456. [CrossRef]
21. Saeedi, P.; Petersohn, I.; Salpea, P.; Malanda, B.; Karuranga, S.; Unwin, N.; Colagiuri, S.; Guariguata, L.; Motala, A.A.; Ogurtsova, K.; et al. Global and Regional Diabetes Prevalence Estimates for 2019 and Projections for 2030 and 2045: Results from the International Diabetes Federation Diabetes Atlas, 9th Edition. *Diabetes Res. Clin. Pract.* **2019**, *157*, 107843. [CrossRef]
22. Heron, M. Deaths: Leading Causes for 2016. *Natl. Vital Stat. Rep.* **2018**, *67*, 1–77.
23. Van Cauwenberghe, J.; Enzlin, P.; Nefs, G.; Ruige, J.; Hendrieckx, C.; De Block, C.; Pouwer, F. Prevalence of and Risk Factors for Sexual Dysfunctions in Adults with Type 1 or Type 2 Diabetes: Results from Diabetes MILES–Flanders. *Diabet. Med.* **2022**, *39*, e14676. [CrossRef]
24. Bin Rakhis, S.A.; AlDuwayhis, N.M.; Aleid, N.; AlBarrak, A.N.; Aloraini, A.A. Glycemic Control for Type 2 Diabetes Mellitus Patients: A Systematic Review. *Cureus* **2022**, *14*, 6–13. [CrossRef] [PubMed]
25. Chamberlain, R.C.; Fleetwood, K.; Wild, S.H.; Colhoun, H.M.; Lindsay, R.S.; Petrie, J.R.; McCrimmon, R.J.; Gibb, F.; Philip, S.; Sattar, N.; et al. Foot Ulcer and Risk of LowerLimb Amputation or Death in People with Diabetes: A National Population-Based Retrospective Cohort Study. *Diabetes Care* **2022**, *45*, 83–91. [CrossRef]
26. McEwan, P.; Morgan, A.R.; Boyce, R.; Bergenheim, K.; Gause-Nilsson, I.A.M.; Bhatt, D.L.; Leiter, L.A.; Johansson, P.A.; Mosenzon, O.; Cahn, A.; et al. The Cost-Effectiveness of Dapagliflozin in Treating High-Risk Patients with Type 2 Diabetes Mellitus: An Economic Evaluation Using Data from the DECLARE-TIMI 58 Trial. *Diabetes Obes. Metab.* **2021**, *23*, 1020–1029. [CrossRef] [PubMed]
27. Switanek, M.; Crailsheim, K.; Truhetz, H.; Brodschneider, R. Modelling Seasonal Effects of Temperature and Precipitation on Honey Bee Winter Mortality in a Temperate Climate. *Sci. Total Environ.* **2017**, *579*, 1581–1587. [CrossRef] [PubMed]
28. Yoshida, M.; Hayashi, K.; Watadani, R.; Okano, Y.; Tanimura, K.; Kotoh, J.; Sasaki, D.; Matsumoto, K.; Maeda, A. Royal Jelly Improves Hyperglycemia in Obese/Diabetic KK-Ay Mice. *J. Vet. Med. Sci.* **2017**, *79*, 299–307. [CrossRef] [PubMed]
29. Ghanbari, E.; Nejati, V.; Khazaei, M. Antioxidant and Protective Effects of Royal Jelly on Histopathological Changes in Testis of Diabetic Rats. *Int. J. Reprod. Biomed.* **2016**, *14*, 511–518. [CrossRef]
30. Ghanbari, E.; Nejati, V.; Khazaei, M. Improvement in Serum Biochemical Alterations and Oxidative Stress of Liver and Pancreas Following Use of Royal Jelly in Streptozotocin-Induced Diabetic Rat. *Cell J.* **2016**, *18*, 362–370. [PubMed]
31. Khoshpey, B.; Djazayeri, S.; Amiri, F.; Malek, M.; Hosseini, A.F.; Hosseini, S.; Shidfar, S.; Shidfar, F. Effect of Royal Jelly Intake on Serum Glucose, Apolipoprotein A-I (ApoA-I), Apolipoprotein B (ApoB) and ApoB/ApoA-I Ratios in Patients with Type 2 Diabetes: A Randomized, Double-Blind Clinical Trial Study. *Can. J. Diabetes* **2016**, *40*, 324–328. [CrossRef]
32. Pourmoradian, S.; Mahdavi, R.; Mobasseri, M.; Faramarzi, E.; Mobasseri, M. Effects of Royal Jelly Supplementation on Glycemic Control and Oxidative Stress Factors in Type 2 Diabetic Female: A Randomized Clinical Trial. *Chin. J. Integr. Med.* **2014**, *20*, 347–352. [CrossRef]
33. Ghanbari, E.; Nejati, V.; Najafi, G.; Khazaei, M.; Babaei, M. Study on the Effect of Royal Jelly on Reproductive Parameters in Streptozotocin-Induced Diabetic Rats. *Int. J. Fertil. Steril.* **2015**, *9*, 113–120.
34. Karaca, T.; Demirtaş, S.; Karaboğa, İ.; Ayvaz, S. Protective Effects of Royal Jelly against Testicular Damage in Streptozotocin-Induced Diabetic Rats. *Turk. J. Med. Sci.* **2015**, *45*, 27–32. [CrossRef]
35. Nohair, S.F. Al Antidiabetic Efficacy of a Honey-Royal Jelly Mixture: Biochemical Study in Rats. *Int. J. Health Sci.* **2021**, *15*, 4–9.
36. Macadangdang, R.R., Jr.; Orodio, A.L.T.; Calabio, A.G.T.; Miciano, K.B.V.; Coleen, A.L.T. Hypoglycemic Effect of Moringa Oliefera Lam. Seed Oil Extract and Royal Jelly Composite Mixture in Alloxan-Induced Diabetes in Male Albino Wistar Rats. *Asian, J. Biol. Life Sci.* **2021**, *10*, 159–164. [CrossRef]
37. Yoneshiro, T.; Kaede, R.; Nagaya, K.; Aoyama, J.; Saito, M.; Okamatsu-Ogura, Y.; Kimura, K.; Terao, A. Royal Jelly Ameliorates Diet-Induced Obesity and Glucose Intolerance by Promoting Brown Adipose Tissue Thermogenesis in Mice. *Obes. Res. Clin. Pract.* **2016**, *10*, 127–137. [CrossRef] [PubMed]
38. Watadani, R.; Kotoh, J.; Sasaki, D.; Someya, A.; Matsumoto, K.; Maeda, A. 10-Hydroxy-2-Decenoic Acid, a Natural Product, Improves Hyperglycemia and Insulin Resistance in Obese/Diabetic KK-Ay Mice, but Does Not Prevent Obesity. *J. Vet. Med. Sci.* **2017**, *79*, 1596–1602. [CrossRef]
39. Hu, X.; Liu, Z.; Lu, Y.; Chi, X.; Han, K.; Wang, H.; Wang, Y.; Ma, L.; Xu, B. Glucose Metabolism Enhancement by 10-Hydroxy-2-Decenoic Acid via the PI3K/AKT Signaling Pathway in High-Fat-Diet/Streptozotocin Induced Type 2 Diabetic Mice. *Food Funct.* **2022**, *13*, 9931–9946. [CrossRef]
40. You, S.; Xiang, Y.; Qi, X.; Mao, R.; Cai, E.; Lan, Y.; Lu, H.; Shen, J.; Deng, H. Harnessing a Biopolymer Hydrogel Reinforced by Copper / Tannic Acid Nanosheets for Treating Bacteria-Infected Diabetic Wounds. *Mater. Today Adv.* **2022**, *15*, 100271–100286. [CrossRef]
41. Zhou, J.; Wang, Z.; Yang, C.; Zhang, H.; Yan, W.; Wang, K. A Carrier-Free, Dual-Functional Hydrogel Constructed of Antimicrobial Peptide Jelleine-1 and 8Br-CAMP for MRSA Infected Diabetic Wound Healing. *Acta Biomater. J.* **2022**, *151*, 223–234. [CrossRef]

42. Jayaraman, R.; Subramani, S.; Sheik Abdullah, S.H.; Udaiyar, M. Antihyperglycemic Effect of Hesperetin, a Citrus Flavonoid, Extenuates Hyperglycemia and Exploring the Potential Role in Antioxidant and Antihyperlidemic in Streptozotocin-Induced Diabetic Rats. *Biomed. Pharmacother.* **2018**, *97*, 98–106. [CrossRef]
43. López-Gutiérrez, N.; Aguilera-Luiz, M.D.M.; Romero-González, R.; Vidal, J.L.M.; Garrido Frenich, A. Fast Analysis of Polyphenols in Royal Jelly Products Using Automated TurboFlow^{TM}-Liquid Chromatography-Orbitrap High Resolution Mass Spectrometry. *J. Chromatogr. B Anal. Technol. Biomed. Life Sci.* **2014**, *973*, 17–28. [CrossRef]
44. Singh, P.; Bansal, S.; Kuhad, A.; Kumar, A.; Chopra, K. Naringenin Ameliorates Diabetic Neuropathic Pain by Modulation of Oxidative-Nitrosative Stress, Cytokines and MMP-9 Levels. *Food Funct.* **2020**, *11*, 4548–4560. [CrossRef] [PubMed]
45. Wojnar, W.; Zych, M.; Kaczmarczyk-Sedlak, I. Antioxidative Effect of Flavonoid Naringenin in the Lenses of Type 1 Diabetic Rats. *Biomed. Pharmacother.* **2018**, *108*, 974–984. [CrossRef] [PubMed]
46. Ortiz-Andrade, R.R.; Sánchez-Salgado, J.C.; Navarrete-Vázquez, G.; Webster, S.P.; Binnie, M.; García-Jiménez, S.; León-Rivera, I.; Cigarroa-Vázquez, P.; Villalobos-Molina, R.; Estrada-Soto, S. Antidiabetic and Toxicological Evaluations of Naringenin in Normoglycaemic and NIDDM Rat Models and Its Implications on Extra-Pancreatic Glucose Regulation. *Diabetes Obes. Metab.* **2008**, *10*, 1097–1104. [CrossRef]
47. Kumar, B.; Gupta, S.K.; Srinivasan, B.P.; Nag, T.C.; Srivastava, S.; Saxena, R. Hesperetin Ameliorates Hyperglycemia Induced Retinal Vasculopathy via Anti-Angiogenic Effects in Experimental Diabetic Rats. *Vascul. Pharmacol.* **2012**, *57*, 201–207. [CrossRef] [PubMed]
48. Yang, R.; Jia, Q.; Mehmood, S.; Ma, S.; Liu, X. Genistein Ameliorates Inflammation and Insulin Resistance through Mediation of Gut Microbiota Composition in Type 2 Diabetic Mice. *Eur. J. Nutr.* **2021**, *60*, 2155–2168. [CrossRef]
49. Fu, Z.; Zhang, W.; Zhen, W.; Lum, H.; Nadler, J.; Bassaganya-Riera, J.; Jia, Z.; Wang, Y.; Misra, H.; Liu, D. Genistein Induces Pancreatic β-Cell Proliferation through Activation of Multiple Signaling Pathways and Prevents Insulin-Deficient Diabetes in Mice. *Endocrinology* **2010**, *151*, 3026–3037. [CrossRef]
50. Li, R.Z.; Ding, X.W.; Geetha, T.; Al-Nakkash, L.; Broderick, T.L.; Babu, J.R. Beneficial Effect of Genistein on Diabetes-Induced Brain Damage in the OB/OB Mouse Model. *Drug Des. Devel. Ther.* **2020**, *14*, 3325–3336. [CrossRef]
51. Qnais, E.; Alqudah, A.; Wedyan, M.; Gammoh, O.; Alkhateeb, H.; Alnoaimi, M. Formononetin Suppresses Hyperglycaemia through Activation of GLUT4-AMPK Pathway. *Pharmacia* **2023**, *70*, 527–536. [CrossRef]
52. Oza, M.J.; Kulkarni, Y.A. Formononetin Ameliorates Diabetic Neuropathy by Increasing Expression of SIRT1 and NGF. *Chem. Biodivers.* **2020**, *17*, e2000162. [CrossRef] [PubMed]
53. Zywno, H.; Bzdega, W.; Kolakowski, A.; Kurzyna, P.; Harasim-Symbor, E.; Sztolsztener, K.; Chabowski, A.; Konstantynowicz-Nowicka, K. The Influence of Coumestrol on Sphingolipid Signaling Pathway and Insulin Resistance Development in Primary Rat Hepatocytes. *Biomolecules* **2021**, *11*, 268. [CrossRef] [PubMed]
54. Kim, S.M.; Imm, J.Y. The Effect of Chrysin-Loaded Phytosomes on Insulin Resistance and Blood Sugar Control in Type 2 Diabetic Db/Db Mice. *Molecules* **2020**, *25*, 5503. [CrossRef] [PubMed]
55. Salama, A.; Asaad, G.; Shaheen, A. Chrysin Ameliorates STZ-Induced Diabetes in Rats: Possible Impact of Modulation of TLR4/NF-Kβ Pathway. *Res. Pharm. Sci.* **2022**, *17*, 1–11. [CrossRef]
56. Honda, Y.; Araki, Y.; Hata, T.; Ichihara, K.; Ito, M.; Tanaka, M.; Honda, S. 10-Hydroxy-2-Decenoic Acid, the Major Lipid Component of Royal Jelly, Extends the Lifespan of Caenorhabditis Elegans through Dietary Restriction and Target of Rapamycin Signaling. *J. Aging Res.* **2015**, *2015*, 425261. [CrossRef]
57. Shidfar, F.; Jazayeri, S.; Mousavi, S.N.; Malek, M.; Hosseini, A.F.; Khoshpey, B. Does Supplementation with Royal Jelly Improve Oxidative Stress and Insulin Resistance in Type 2 Diabetic Patients? *Iran. J. Public Health* **2015**, *44*, 797–803.
58. Siavash, M.; Shokri, S.; Haghighi, S.; Mohammadi, M.; Shahtalebi, M.A.; Farajzadehgan, Z. The Efficacy of Topical Royal Jelly on Diabetic Foot Ulcers Healing: A Case Series. *J. Res. Med. Sci.* **2011**, *16*, 904–909.
59. Boukraâ, L.; Meslem, A.; Benhanifia, M.; Hammoudi, S.M. Synergistic Effect of Starch and Royal Jelly against Staphylococcus Aureus and Escherichia Coli. *J. Altern. Complement. Med.* **2009**, *15*, 755–757. [CrossRef]
60. Yakoot, M.; Abdelatif, M.; Helmy, S. Efficacy of a New Local Limb Salvage Treatment for Limb-Threatening Diabetic Foot Wounds–A Randomized Controlled Study. *Diabetes Metab. Syndr. Obes. Targets Ther.* **2019**, *12*, 1659–1665. [CrossRef]
61. Aljafary, M.A.; Al-suhaimi, E.A. Adiponectin System (Rescue Hormone): The Missing Link between Metabolic and Cardiovascular Diseases. *Pharmaceutics* **2022**, *14*, 1430. [CrossRef]
62. Geagea, A.G.; Mallat, S.; Matar, C.F.; Zerbe, R.; Filfili, E.; Francis, M.; Haidar, H.; Jurjus, A. Adiponectin and Inflammation in Health and Disease: An Update. *Open Med. J.* **2018**, *5*, 20–32. [CrossRef]
63. Maleki, V.; Jafari-Vayghan, H.; Saleh-Ghadimi, S.; Adibian, M.; Kheirouri, S.; Alizadeh, M. Effects of Royal Jelly on Metabolic Variables in Diabetes Mellitus: A Systematic Review. *Complement. Ther. Med.* **2019**, *43*, 20–27. [CrossRef] [PubMed]
64. Joshi, T.; Singh, A.K.; Haratipour, P.; Sah, A.N.; Pandey, A.K.; Naseri, R.; Juyal, V.; Farzaei, M.H. Targeting AMPK Signaling Pathway by Natural Products for Treatment of Diabetes Mellitus and Its Complications. *J. Cell. Physiol.* **2019**, *234*, 17212–17231. [CrossRef] [PubMed]
65. Ananthakrishnan, A.N.; Xavier, R.J. Gastrointestinal Diseases. In *Hunter's Tropical Medicine and Emerging Infectious Diseases*; Elsevier: Amsterdam, The Netherlands, 2020; pp. 16–26.

66. Sperber, A.D.; Bangdiwala, S.I.; Drossman, D.A.; Ghoshal, U.C.; Simren, M.; Tack, J.; Whitehead, W.E.; Dumitrascu, D.L.; Fang, X.; Fukudo, S.; et al. Worldwide Prevalence and Burden of Functional Gastrointestinal Disorders, Results of Rome Foundation Global Study. *Gastroenterology* **2021**, *160*, 99–114.e3. [CrossRef] [PubMed]
67. De Filippo, C.; Di Paola, M.; Ramazzotti, M.; Albanese, D.; Pieraccini, G.; Banci, E.; Miglietta, F.; Cavalieri, D.; Lionetti, P. Diet, Environments, and Gut Microbiota. A Preliminary Investigation in Children Living in Rural and Urban Burkina Faso and Italy. *Front. Microbiol.* **2017**, *8*, 1979–1983. [CrossRef] [PubMed]
68. Rizzello, F.; Spisni, E.; Giovanardi, E.; Imbesi, V.; Salice, M.; Alvisi, P.; Valerii, M.C.; Gionchetti, P. Implications of the Westernized Diet in the Onset and Progression of IBD. *Nutrients* **2019**, *11*, 1033. [CrossRef]
69. Vona, R.; Pallotta, L.; Cappelletti, M.; Severi, C.; Matarrese, P. The Impact of Oxidative Stress in Human Pathology: Focus on Gastrointestinal Disorders. *Antioxidants* **2021**, *10*, 201. [CrossRef]
70. Proestos, C. Superfoods: Recent Data on Their Role in the Prevention of Diseases. *Curr. Res. Nutr. Food Sci.* **2018**, *6*, 576–593. [CrossRef]
71. Mostafa, R.E.; El-Marasy, S.A.; Abdel Jaleel, G.A.; Bakeer, R.M. Protective Effect of Royal Jelly against Diclofenac-Induced Hepato-Renal Damage and Gastrointestinal Ulcerations in Rats. *Heliyon* **2020**, *6*, e03330. [CrossRef]
72. Ahmad, S.; Campos, M.G.; Fratini, F.; Altaye, S.Z.; Li, J. New Insights into the Biological and Pharmaceutical Properties of Royal Jelly. *Int. J. Mol. Sci.* **2020**, *21*, 382. [CrossRef]
73. Li, S.; Tao, L.; Yu, X.; Zheng, H.; Wu, J.H.F. Royal Jelly Proteins and Their Derived Peptides: Preparation, Properties, and Biological Activities. *J. Agric. Food Chem.* **2021**, *69*, 14415–14427. [CrossRef]
74. Jairath, V.; Feagan, B.G. Global Burden of Inflammatory Bowel Disease. *Lancet Gastroenterol. Hepatol.* **2020**, *5*, 2–3. [CrossRef]
75. Karaca, T.; Uz, Y.H.; Demirtas, S.; Karaboga, I.; Can, G. Protective Effect of Royal Jelly in 2,4,6 Trinitrobenzene Sulfonic Acid-Induced Colitis in Rats. *Iran. J. Basic Med. Sci.* **2015**, *18*, 370–379.
76. Karaca, T.; Bayiroglu, F.; Yoruk, M.; Kaya, M.S.; Uslu, S.; Comba, B.; Mis, L. Effect of Royal Jelly on Experimental Colitis Induced by Acetic Acid and Alteration of Mast Cell Distribution in the Colon of Rats. *Eur. J. Histochem.* **2010**, *54*, 158–161. [CrossRef]
77. Karaca, T.; Şimşek, N.; Uslu, S.; Kalkan, Y.; Can, I.; Kara, A.; Yörük, M. The Effect of Royal Jelly on CD3+, CD5+, CD45+ T-Cell and CD68+ Cell Distribution in the Colon of Rats with Acetic Acid-Induced Colitis. *Allergol. Immunopathol.* **2012**, *40*, 357–361. [CrossRef]
78. Chi, X.; Liu, Z.; Wang, H.; Wang, Y.; Xu, B.; Wei, W. Regulation of a New Type of Selenium-Rich Royal Jelly on Gut Microbiota Profile in Mice. *Biol. Trace Elem. Res.* **2022**, *200*, 1763–1775. [CrossRef] [PubMed]
79. Guo, J.; Ma, B.; Wang, Z.; Chen, Y.; Tian, W.; Dong, Y. Royal Jelly Protected against Dextran-Sulfate-Sodium-Induced Colitis by Improving the Colonic Mucosal Barrier and Gut Microbiota. *Nutrients* **2022**, *14*, 2069. [CrossRef] [PubMed]
80. Li, J.; Zhang, W.; Wang, C.; Yu, Q.; Dai, R.; Pei, X. Lactococcus Lactis Expressing Food-Grade β-Galactosidase Alleviates Lactose Intolerance Symptoms in Post-Weaning Balb/c Mice. *Appl. Microbiol. Biotechnol.* **2012**, *96*, 1499–1506. [CrossRef] [PubMed]
81. Kavas, N. Functional Probiotic Yoghurt Production with Royal Jelly Fortification and Determination of Some Properties. *Int. J. Gastron. Food Sci.* **2022**, *28*, 100519. [CrossRef]
82. Oak, S.J.; Jha, R. The Effects of Probiotics in Lactose Intolerance: A Systematic Review–The Effects of Probiotics in Lactose Intolerance: A Systematic Review. *Crit. Rev. Food Sci. Nutr.* **2018**, *8398*, 1425977. [CrossRef]
83. Hassan, A.A.M.; Elenany, Y.E.; Nassrallah, A.; Cheng, W.; Abd El-Maksoud, A.A. Royal Jelly Improves the Physicochemical Properties and Biological Activities of Fermented Milk with Enhanced Probiotic Viability. *LWT* **2022**, *155*, 112912. [CrossRef]
84. Vazhacharickal, J.P. A Review on Health Benefits and Biological Action of Honey, Propolis and Royal Jelly. *J. Med. Plants Stud.* **2021**, *9*, 1–13.
85. Ali, M.W.; Sabir, A.M.; Gadour, M.O. Gum Arabic in Treatment of Functional Constipation in Children in Sudan. *Sudan JMS* **2012**, *8*, 73–76.
86. Postali, E.; Peroukidou, P.; Giaouris, E.; Papachristoforou, A. Investigating Possible Synergism in the Antioxidant and Antibacterial Actions of Honey and Propolis from the Greek Island of Samothrace through Their Combined Application. *Foods* **2022**, *11*, 2041. [CrossRef]
87. Guo, J.; Wang, Z.; Chen, Y.; Cao, J.; Tian, W.; Ma, B.; Dong, Y. Active Components and Biological Functions of Royal Jelly. *J. Funct. Foods* **2021**, *82*, 104514. [CrossRef]
88. Bílikova, K.; Huang, S.-C.; Lin, I.-P.; Šimuth, J.; Peng, C.-C. Structure and Antimicrobial Activity Relationship of Royalisin, an Antimicrobial Peptide from Royal Jelly of Apis Mellifera. *Peptides* **2015**, *86*, 190–196. [CrossRef]
89. Chantawannakul, P. From Entomophagy to Entomotherapy. *Front. Biosci.* **2020**, *25*, 179–200. [CrossRef]
90. Sofiabadi, M.; Samiee-Rad, F. Royal Jelly Accelerates Healing of Acetate Induced Gastric Ulcers in Male Rats. *Gastroenterol. Hepatol. Bed Bench* **2020**, *13*, 18–22. [CrossRef]
91. Abd El-Naeem, A. Effect of Nicotine on the Structure of Gastric Mucosa of Adult Male Albino Rats and the Possible Protective Effect of Royal Jelly (Light and Scanning Electron Microscopic Study). *Egypt. J. Histol.* **2021**, *45*, 404–415. [CrossRef]
92. Ahmed, W.M.S.; Khalaf, A.A.; Moselhy, W.A.; Safwat, G.M. Royal Jelly Attenuates Azathioprine Induced Toxicity in Rats. *Environ. Toxicol. Pharmacol.* **2014**, *37*, 431–437. [CrossRef] [PubMed]
93. Felemban, A.H.; Alshammari, G.M.; Yagoub, A.E.G.A.; Al-Harbi, L.N.; Alhussain, M.H.; Yahya, M.A. Activation of AMPK Entails the Protective Effect of Royal Jelly against High-Fat-Diet-Induced Hyperglycemia, Hyperlipidemia, and Non-Alcoholic Fatty Liver Disease in Rats. *Nutrients* **2023**, *15*, 1471. [CrossRef] [PubMed]

94. Fang, C.; Pan, J.; Qu, N.; Lei, Y.; Han, J.; Zhang, J.; Han, D. The AMPK Pathway in Fatty Liver Disease. *Front. Physiol.* **2022**, *13*, 970292. [CrossRef]
95. Bahaaldin-Beygi, M.; Kariminik, A.; Arababadi, M.K. Royal Jelly Significantly Alters Inflammasome Pathways in Patients with Chronic Hepatitis B. *Indian J. Exp. Biol.* **2022**, *60*, 875–879.
96. Ezzati, M.; Obermeyer, Z.; Tzoulaki, I.; Mayosi, B.M.; Elliott, P.; Leon, D.A. Contributions of Risk Factors and Medical Care to Cardiovascular Mortality Trends. *Nat. Rev. Cardiol.* **2015**, *12*, 508–530. [CrossRef] [PubMed]
97. Townsend, N.; Kazakiewicz, D.; Lucy Wright, F.; Timmis, A.; Huculeci, R.; Torbica, A.; Gale, C.P.; Achenbach, S.; Weidinger, F.; Vardas, P. Epidemiology of Cardiovascular Disease in Europe. *Nat. Rev. Cardiol.* **2022**, *19*, 133–143. [CrossRef]
98. Aslan, A.; Beyaz, S.; Gok, O.; Can, M.I.; Parlak, G.; Ozercan, I.H.; Gundogdu, R. Royal Jelly Abrogates Flouride-Induced Oxidative Damage in Rat Heart Tissue by Activating of the Nrf-2/NF-KB and Bcl-2/Bax Pathway. *Toxicol. Mech. Methods* **2021**, *31*, 644–654. [CrossRef]
99. Fan, P.; Han, B.; Feng, M.; Fang, Y.; Zhang, L.; Hu, H.; Hao, Y.; Qi, Y.; Zhang, X.; Li, J. Functional and Proteomic Investigations Reveal Major Royal Jelly Protein 1 Associated with Anti-Hypertension Activity in Mouse Vascular Smooth Muscle Cells. *Sci. Rep.* **2016**, *6*, 30230–30241. [CrossRef]
100. Matsui, T.; Yukiyoshi, A.; Doi, S.; Sugimoto, H.; Yamada, H.; Matsumoto, K. Gastrointestinal Enzyme Production of Bioactive Peptides from Royal Jelly Protein and Their Antihypertensive Ability in SHR. *J. Nutr. Biochem.* **2002**, *13*, 80–86. [CrossRef]
101. Lin, N.; Chen, S.; Zhang, H.; Li, J.; Fu, L. Quantification of Major Royal Jelly Protein 1 in Fresh Royal Jelly by Ultraperformance Liquid Chromatography-Tandem Mass Spectrometry. *J. Agric. Food Chem.* **2018**, *66*, 1270–1278. [CrossRef]
102. Feng, M.; Fang, Y.; Han, B.; Xu, X.; Fan, P.; Hao, Y.; Qi, Y.; Hu, H.; Huo, X.; Meng, L.; et al. In-Depth N-Glycosylation Reveals Species-Specific Modifications and Functions of the Royal Jelly Protein from Western (*Apis mellifera*) and Eastern Honeybees (*Apis cerana*). *J. Proteome Res.* **2015**, *14*, 5327–5340. [CrossRef]
103. Escamilla, K.I.A.; Ordóñez, Y.B.M.; Sandoval-Peraza, V.M.; Fernández, J.J.A.; Ancona, D.A.B. Anti-Hypertensive Activity in Vitro and in Vivo on Royal Jelly Produced by Different Diets. *Emir. J. Food Agric.* **2022**, *34*, 9–15. [CrossRef]
104. Pan, Y.; Rong, Y.; You, M.; Ma, Q.; Chen, M.; Hu, F. Royal Jelly Causes Hypotension and Vasodilation Induced by Increasing Nitric Oxide Production. *Food Sci. Nutr.* **2019**, *7*, 1361–1370. [CrossRef]
105. Wei, W.; Wei, M.; Kang, X.; Deng, H.; Lu, Z. A Novel Method Developed for Acetylcholine Detection in Royal Jelly by Using Capillary Electrophoresis Coupled with Electrogenerated Chemiluminescence Based on a Simple Reaction. *Electrophoresis* **2009**, *30*, 1949–1952. [CrossRef]
106. Fujisue, K.; Yamamoto, E.; Sueta, D.; Arima, Y.; Hirakawa, K.; Tabata, N.; Ishii, M.; Ito, M.; Yamanaga, K.; Hanatani, S.; et al. A Randomized, Double-Blind Comparison Study of Royal Jelly to Augment Vascular Endothelial Function in Healthy Volunteers. *J. Atheroscler. Thromb.* **2022**, *29*, 1285–1294. [CrossRef] [PubMed]
107. Roman, W.A.; Piato, A.L.; Conterato, G.M.M.; Wildner, S.M.; Marcon, M.; Mocelin, R.; Emanuelli, M.P.; Emanuelli, T.; Nepel, A.; Barison, A.; et al. Hypolipidemic Effects of Solidago Chilensis Hydroalcoholic Extract and Its Major Isolated Constituent Quercetrin in Cholesterol-Fed Rats. *Pharm. Biol.* **2015**, *53*, 1488–1495. [CrossRef] [PubMed]
108. Kamakura, M.; Moriyama, T.; Sakaki, T. Changes in Hepatic Gene Expression Associated with the Hypocholesterolaemic Activity of Royal Jelly. *J. Pharm. Pharmacol.* **2010**, *58*, 1683–1689. [CrossRef]
109. Kashima, Y.; Kanematsu, S.; Asai, S.; Kusada, M.; Watanabe, S.; Kawashima, T.; Nakamura, T.; Shimada, M.; Goto, T.; Nagaoka, S. Identification of a Novel Hypocholesterolemic Protein, Major Royal Jelly Protein 1, Derived from Royal Jelly. *PLoS ONE* **2014**, *9*, e105073–e105084. [CrossRef] [PubMed]
110. Chiu, H.F.; Chen, B.K.; Lu, Y.Y.; Han, Y.C.; Shen, Y.C.; Venkatakrishnan, K.; Golovinskaia, O.; Wang, C.K. Hypocholesterolemic Efficacy of Royal Jelly in Healthy Mild Hypercholesterolemic Adults. *Pharm. Biol.* **2017**, *55*, 497–502. [CrossRef] [PubMed]
111. Bălan, A.; Moga, M.A.; Dima, L.; Toma, S.; Neculau, A.E.; Anastasiu, C.V. Royal Jelly–A Traditional and Natural Remedy for Postmenopausal Symptoms and Aging-Related Pathologies. *Molecules* **2020**, *25*, 3291. [CrossRef]

nutrients

Review

The Role of Probiotics in Skin Health and Related Gut–Skin Axis: A Review

Ting Gao, Xiaoyu Wang, Yixuan Li and Fazheng Ren *

Key Laboratory of Precision Nutrition and Food Quality, Key Laboratory of Functional Dairy, Ministry of Education, Beijing Laboratory of Food Quality and Safety, Department of Nutrition and Health, China Agricultural University, Beijing 100083, China; gaotinginging@163.com (T.G.); xy.wang@cau.edu.cn (X.W.); liyixuan@cau.edu.cn (Y.L.)

* Correspondence: fazhengren1962@163.com

Abstract: Aging skin, wrinkles, pigmentation, and dryness are problems that plague people, and researchers are working to solve them. Recent studies have shown that intestinal microbiota homeostasis can influence skin health, demonstrating the existence of a gut–skin axis. Recently, improving skin health through probiotic interventions has been proposed, and micro-ecological skin care is becoming a popular concept. By regulating skin health and gut–skin axis interactions, probiotics can be used as potential management tools to suppress and improve skin diseases in multiple ways, including decreasing oxidative stress, suppressing inflammatory responses, and keeping immune effects. The purpose of this paper is to provide a comprehensive review of the application and mechanisms of probiotic-mediated gut microbiota homeostasis in skin care and to offer a theoretical basis for the application of probiotics in skin care.

Keywords: skin; probiotics; intestinal microbiota; gut–skin axis

1. Introduction

Skin accounts for about 15 percent of the total body weight of adults, with an average surface area of 1.5–2 m^2 [1]. One of the main functions of the skin is its use as a mechanical barrier to disease-causing microorganisms and harmful substances; in fact, it could be viewed as one of the host's vital defenses against infections, as well as the innate and adaptive immune system [2]. Other important features include inhibition of transcutaneous water loss (TEWL), thermoregulation, structural support, and vitamin synthesis, all of which assist in maintaining a healthy host [2–4].

The quest for beauty never ends. It is often hard to know whether skin issues, including skin pigmentation, skin wrinkles, skin aging, and skin dehydration, occur due to external elements or internal changes. Skin issues have various causes, and investigators are continuously researching safe and efficient skin treatment products to address skin issues. Today, various cosmetic products contain chemicals, including titanium dioxide, which are more or less toxic and may be harmful to an individual's health [5]. There are also various researchers who use raw materials extracted from herbal medicines as important components in skin treatment products, while they show certain results due to the complexity of herbal ingredients, their influences sometimes fail to meet expectations and their quality still needs to be improved [6,7]. Thus, there is an urgent need to explore safe and efficient ingredients for skin treatment products that can effectively address skin issues. Recently, researchers have suggested that probiotics can be used as an efficient ingredient in cosmetics to address the above-mentioned skin issues to a better effect. In addition, experimental research have indicated that probiotics have no or less toxic influences on hosts and can be better used in the development of skin treatments.

Recently, we have become increasingly aware of the powerful effect of probiotics in many fields, and the effects of probiotics in skin care have been increasingly studied and argued by researchers. Probiotics can act through a variety of mechanisms (Figure 1).

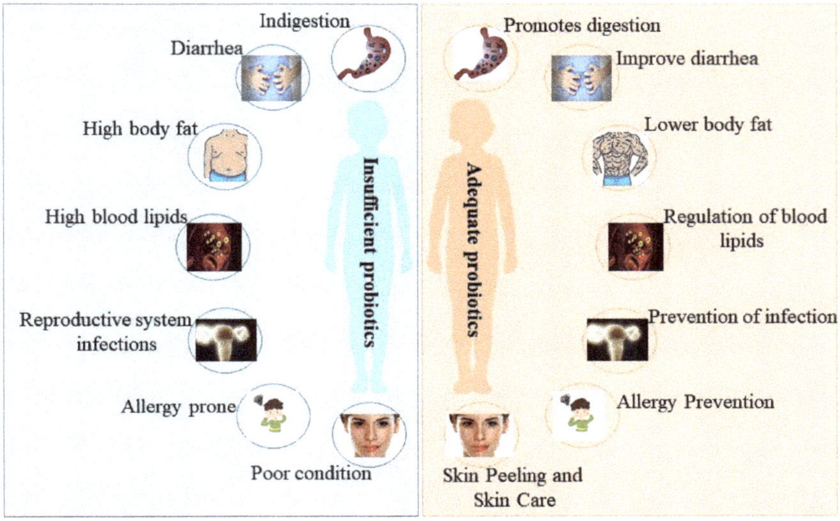

Figure 1. The beneficial effects of probiotics on the organism. When the abundance of probiotics in the organism is insufficient, the organism suffers from the following issues: indigestion, diarrhea, high body fat, high blood lipids, reproductive system infections, allergy prone, and poor skin condition (**left** picture). When the abundance of probiotics in the organism is sufficient, the organism behaves as follows: promotes digestion, improves diarrhea, lowers body fat, regulates blood lipids, prevents reproductive system infections, prevents allergies, and skin peeling and skin care (**right** picture).

This paper reviews the applications of probiotics in skin care, such as skin whitening, skin moisturizing, skin anti-aging, skin anti-wrinkle, and body odor removal, and their mechanisms, which offers a theory basis for further applications of probiotics in skin care in the future.

Although the applications of topical biologic therapies date back to 1912, when topical use of *Lactobacillus bulgaricus* ameliorated skin problems, including acne and seborrhea, the skin care industry has recently seen a proliferation of topical preparations containing microorganisms [8]. Table 1 lists skin care products containing probiotics sold worldwide.

Table 1. Skin care products with added probiotics and their effects.

Product	Probiotics	Efficacy
Okana	*Bacillus* bacterial ferment extract	Helps skin retain its firmness and elasticity and keeps it feeling smooth and plump.
Amperna	Unique probiotic complex	Soothes irritated skin and calms redness. Tested on eczema, dermatitis, perioral dermatitis, rosacea, and acne-prone skin.
Cream	1. *Lactobacillus acidophilus* 2. *Lactobacillus rhamnosus*	Anti-photoaging
Elissah Bio P2 Laviol Skin Care	16 types and 35 strains of bacteria including 14 *Bifidobacterium* and *Lactobacilli*.	Strengthens the skin's barrier against environmental threats and reduces the factors that trigger skin sensitivities, redness, and irritation.
Probiotic Skin Cream Melvory	lactobacilli probiotic (*Lactobacillus* ferment filtrate)	Cleans away the bad bacteria on the skin. For acne-prone or teenage skin.

Table 1. *Cont.*

Product	Probiotics	Efficacy
Andalou Brightening Probiotic + C Renewal Cream	*Bacillus coagulans*	Skin-friendly vegan probiotic microflora enzymatically supports dermal vitality, targeting over-exposed surface cells for a lighter, tighter, brighter looking appearance, and a luminous complexion.
Biossance Squalane + Probiotic Gel	*Lactococcus* ferment lysate	Helps restore the skin's balance and renew the skin barrier
Neogen Dermalogy Probiotics Double Action	The patented complex of Bifida ferment lysate, *Lactobacillus*, and *Streptococcus thermophilus* ferment	Protects the skin barrier
Elemis Dynamic Resurfacing Facial Pads	*Lactococcus* ferment lysate	Stimulates skin-cell renewal and reinforce the skin barrier
Manyo Factory Bifida Complex Ampoule	Bifida ferment lysate, Bifida ferment filtrate, *Lactobacillus* ferment lysate, and *Lactococcus* ferment lysate	Encourages self-repair of skin, hydrates, replenishes moisture and prevents aging
LaFlore Probiotic Serum Concentrate	*Lactococcus* ferment lysate and live kefir Probiotics (*Hansenula / Kloeckera / Lactobacillus / Lactococcus / Leuconostoc / Pediococcus / Saccharomyces*)	Helps calm and smooth fine lines and wrinkles and boosts the skin's natural defense system.
Elizabeth Arden Superstart Probiotic Boost Skin Renewal Biocellulose Mask	*Lactococcus* ferment lysate; inactivated strains of *Lactobacillus casei* and *Lactobacillus acidophilus*	Optimizes skin's microflora and natural defense. Moisturizes and smoothens skin
Dot and Key 72 h hydrating gel and Probiotics	*Saccharomyces* black tea ferment, *Lactobacillus*	Provides hours long moisturization and restores microbiome balance

However, there are many problems with topical probiotics that have not yet been solved. External products cannot be manufactured under sterile status and, therefore, do not require sterility testing. These manufacturers usually include antiseptics to regulate the microorganism's growth. These antiseptics may influence the viability of probiotic strains and also inadvertently change the microbiota of the receptor [8]. Topical formulations containing probiotics have not yet moved outside the personal care manufacture category; because they assess a high load of colony-forming units, such formulations have difficulty passing the US Food and Drug Administration's (USFDA) modulatory provisions for microbiota load. Antiseptic effect testing is a vital barrier to measuring these applications. According to the United States Pharmacopeia (USP), topical probiotic preparations for the treatment of acne were tested for microbiota counts. Studies have shown that topical products do not contain "undesirable" amounts of live microorganisms and, therefore, according to the USP [9], do not require less than 1000 colony-forming units (CFU). Meanwhile, since the stratum corneum maintains the skin's strict natural and protective barrier function, it regulates the absorption of effective substances into the deeper layers of the skin, thus also limiting the choice of treatments [10]. The preparation requirements for topical applications, including live microorganisms, are significantly different from those for products containing only smaller molecules resulting from the need to maintain microbial stabilization. Key factors that are required for microbial control are pH and osmolarity contents, as well as temperature and humidity levels of the storage environment [11].

Topical probiotics have been used to maintain skin health since the beginning of the 20th century, and the last decade has seen a dramatic rise in commercially available topical probiotics [12]. With the increasing popularity of these topical products and the dearth of clinical trials or efficacy studies to establish their clinical efficiency, we are gradually focusing on internal probiotics in treating skin disorders. Since internal probiotics first enter the intestinal tract and inevitably interfere with skin conditions by affecting intestinal homeostasis, this article elaborates on the relationship between probiotics, the intestine and skin, in an attempt to provide potential solutions and clinical value for finding appropriate skin interventions.

2. The Different Effects of Probiotics on the Skin (Figure 2)

2.1. Skin Whiting

Recently, there has been an increasing interest in skin lightening, and the focus of brightening products is to decrease melanin content and suppress overproduction pigmentation [13]. Melanin is photoprotective and protects the skin from ultraviolet (UV) radiation, but the overexpression of pigmentation can affect skin tone and even induce various skin disorders, including freckles and melasma [14–16]. The process of melanin production involves various enzymes and chemical catalytic reactions [17,18]. There are three main enzymes associated with melanogenesis, containing tyrosinase, tyrosinase-related protein 1 (TYRP-1), and tyrosinase-related protein 2 (TYRP-2), with tyrosinase being the indispensable primary enzyme [19]. A lot of whitening cosmetics can precisely suppress the tyrosinase activity, thus reducing melanin content and achieving a brightening effect. Recently, probiotics have been increasingly used in brightening products, which is tightly associated with their great antagonistic influence on tyrosinase (Figure 2).

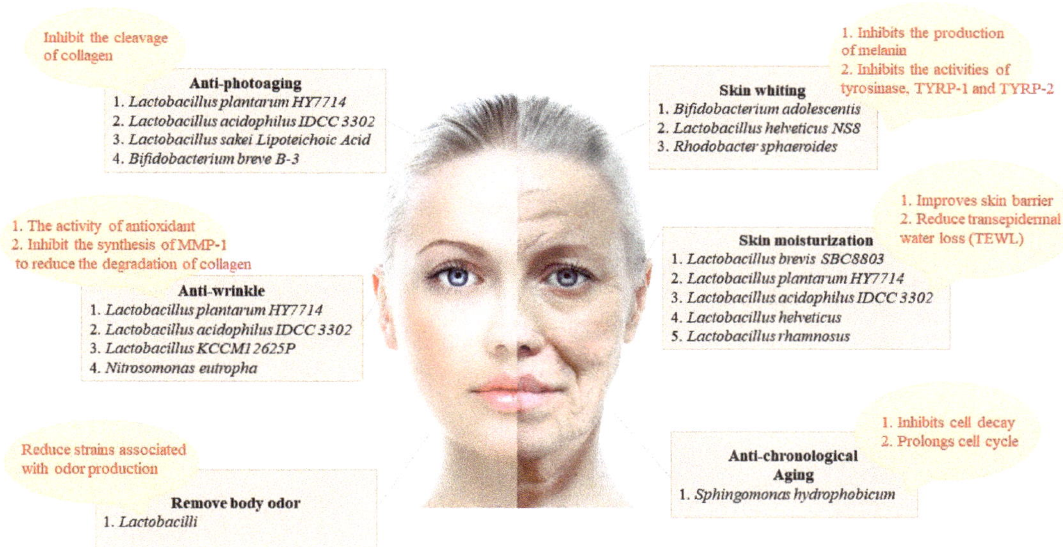

Figure 2. The skin improvement effect of probiotics and its related mechanism. The skin improving effects of probiotics include: anti-photoaging (inhibit the cleavage of collagen), skin whiting (inhibits the production of melanin and inhibits the activities of tyrosinase, TYRP-1 and TYRP-2), anti-wrinkle (the activity of antioxidant and inhibition of the synthesis of matrix metalloproteinase-1 (MMP-1) to reduce the degradation of collagen), skin moisturization (improves skin barrier and reduces TEWL), body odor removal (reduce strains associated with odor production), and anti-chronological aging (inhibits cell decay and prolongs cell cycle).

A study suggested that the antagonistic influence of *Bifidobacterium adolescentis* culture filtrate on mushroom tyrosinase and tyrosinase activity strengthened with increasing content, thereby reducing the melanin levels in B16F10 cells [20]. According to these studies, *Bifidobacterium adolescentis* culture filtrate could modulate tyrosinase activity via its antioxidant effect, thereby decreasing melanin content and achieving a whitening purpose. Moreover, they also uncovered that lactic acid in *Lactobacillus* can suppress melanin synthesis directly by down-regulating tyrosinase activity and also regulate melanin synthesis by affecting tyrosinase expression or tyrosinase, tyrosine 1 and tyrp-2 to exhibit a brightening effect [21].

Probiotics can decrease melanin content not only by regulating tyrosinase activity but also by other means to achieve a whitening effect. Jingjing Rong et al. indicated the brightening effect of *Lactobacillus helveticus* NS8 fermented milk supernatant (NS8-FS) [22]. Data demonstrated that NS8-FS decreased melanin levels in B16F10 cells by suppressing the activity of tyrosinase and proteins associated with tyrosinase expression. Furthermore, a UV radiation-induced pigmentation model was established in guinea pigs. Masson-Fontana staining and tyrosinase staining tests confirmed that NS8-FS improved skin pigmentation. The potential mechanism by which NS8-FS improved skin pigmentation is the modulation of the Nrf2 activity, which promotes melanogenesis in melanocytes under UV-mediated oxidative stress [23]. Liu et al. explored the inhibitory effect of *Rhodobacter spheroides* (Lysogen™) on melanin synthesis [24]. Research demonstrated that melanin content in B16F10 cells up-regulated after α-MSH supplementation and down-regulated in a dose-dependent manner after Lysogen™ treatment.

2.2. Skin Moisturization

There are various reasons for dry skin, such as seasonal alterations, skin barrier damage, and disorderly flaking [25]. Skin hydration is important for health and beauty, and it exerts vital effects on maintaining proper body activity and beauty. Thus, we are continuously searching for molecules that are beneficial for maintain skin moisture. Probiotics can decrease TEWL and improve skin dryness, which can be used to modulate dry skin. Moreover, probiotics can also decrease skin water-loss by modulating skin–barrier function and are good skin moisturizers [26].

A study confirmed that the oral supplementation of *Lactobacillus plantarum* HY7714 increased ceramide levels by elevating serine palmitoyltransferase (SPT) mRNA expression and decreasing ceramidase mRNA expression [27]. Ceramides exert an important effect on keeping the structural support of the epidermal barrier and the epidermal hydration [28, 29]. Elevated ceramide contents lead to lower TEWL values and up-regulated hydration. Studies used ELISA to detect hyaluronic acid (HA) content and found that the use of acidophilic lactic acid IDCC 3302 exerts a beneficial effect on skin hydration [30,31]. In conclusion, the treatment of *Lactobacillus acidophilus* IDCC 3302 resulted in improved skin dryness and decreased TWEL, thereby up-regulating skin hydration. Hidoko BABA et al. indicated that the administration of *Lactobacillus helveticus*-fermented milk whey (LHMW) led to an obvious reduction in TWEL of intact skin and an increase in skin water content, proving that LHMW milk has a moisturizing effect and is used in cosmetics [32].

2.3. Skin Barrier Integrity

Damage to the skin barrier function can adversely affect the skin by disrupting the moisture balance on the skin surface [33]. Ye-On Jung et al. demonstrated experimentally that *Lactobacillus rhamnosus* (LR) can effectively improve the skin barrier and can be regarded as a moisturizing skin care product [34]. They used immunofluorescence staining to identify up-regulated expression levels of Claudin-1 and Occludin, two tightly bound molecules and indicated that the stratum corneum of tissues treated with LR lysate was tighter and more organized. Moreover, qPCR results suggested elevated expression levels of loricrin and filaggrin which exert a vital effect on the restoration of skin barrier function [35]. Furthermore, the strengthened skin barrier function was further suggested by reducing sodium dodecyl sulfate (SLS)-induced cytotoxicity and decreasing skin permeability.

2.4. Anti-Aging

Chronological aging and photoaging are two primary forms of skin aging [36]. Chronological aging is mainly influenced by internal elements, whereas photoaging is mainly influenced by external elements [37]. These influences are different but have similar regulatory mechanisms, but probiotics have a beneficial effect on both forms of skin aging.

2.4.1. Anti-Chronological Aging

Chronological aging is mainly associated with genetic elements and is a regular physiological process in the human body. As we age, the body ages, and so does the skin, which is characterized by thinning and dryness [38]. Probiotics achieve anti-aging mainly by suppressing cell decay and prolonging the cell cycle. Sandie Gervason et al. indicated that the exclusion of *Sphingomonas hydrophobicum* (SH) could suppress the production of proteins associated with aging, such as P16 and P21, using an immunohistological experiment. P16 and P21 are cell cycle antagonists, which suppress the cell cycle and result in cell aging [39,40]. The production level of P16 and P21 in the experimental group were obviously down-regulated versus the control group without SH extraction. SH extraction was also demonstrated to suppress the SA-β-galactosidase level, which is linked to aging, to improve cell senescence. Moreover, the level of fibrillin-1 and Versican was up-regulated after SH extraction supplementation. Previous research indicated that fibrillin-1 participates in the production of elastic skin fibers [41] and that an up-regulated Versican level can inhibit the apoptosis response of fibroblasts [42], both of which can slow down cell senescence. In the case of SH extraction, it can be used as an anti-aging skin-care product.

2.4.2. Anti-Photoaging

Photoaging is primarily influenced by external environmental elements, such as UV radiation and toxins. These external elements will induce injury to the skin, causing it to lose elasticity, lose moisture, thicken, and become rough and sluggish [43]. Probiotics have a significant influence on the treatment of photoaging, which is primarily achieved by suppressing collagen division.

Research indicated that patients taking *Lactobacillus plantarum* HY7714 had decreased epidermal moisture loss, decreased wrinkle depth, and ameliorated skin gloss and elasticity [44]. Research demonstrated that tyndallized *Lactobacillus acidophilus* IDCC 3302 could restore the reduction in collagen expression after UV irradiation via Western blot analysis [45]. Meanwhile, it was shown that tyndallized *Lactobacillus acidophilus* IDCC 3302 could obviously decrease the contents of MMP-1, MMP-2, and MMP-9 in HaCaT, which were up-regulated due to exposure to UV rays, primarily by suppressing the MAPK signaling pathway. Moreover, tyndallized *Lactobacillus acidophilus* IDCC 3302 can improve the inflammation response by reducing the levels of proinflammatory cytokines, including IL-1β, IL-8, and TNF-α. The above data showed that tyndallized *Lactobacillus acidophilus* IDCC 3302 can inhibit photoaging and ameliorate inflammation responses induced by UV irradiation. You et al. suggested that *Lactobacillus sakei* Lipoteichoic Acid (sLTA) could suppress the phosphorylation of MAPK and further block the MMP-1 synthesis when hosts are exposed to UV rays [46].

2.5. Anti-Wrinkle

Wrinkles are induced by atrophy of the skin and repeated contractions of facial muscles underneath [47]. The use of probiotics has been proven to regulate facial wrinkles. The antioxidant activity of probiotics is closely associated with their anti-wrinkle properties. Moreover, MMP-1 synthesis activates the degradation of collagen produced by fibroblasts, resulting in wrinkles on the surface of human skin [48]. Probiotics could suppress MMP-1 synthesis and decrease collagen degradation, resulting in anti-wrinkle properties.

Researchers found that tyndallized *Lactobacillus* KCCM12625P (AL) can suppress the MMP-1 synthesis, thus preventing wrinkle formation. AL can effectively inhibit the formation of facial wrinkles and act as an anti-wrinkle mainly through the above two aspects.

Hyun Mee Kim et al. found that *Lactobacillus plantarum* HY7714 had a powerful blocking function on UV-induced MMP-1 according to Western blot [49]. Moreover, *Lactobacillus plantarum* HY7714 inhibited the MMP-1 expression and the MMP-2 and MMP-9 activity, which effectively improved the area and depth of wrinkles and exerted a vital effect on wrinkle elimination. A study indicated that tyndallized *Lactobacillus acidophilus* IDCC 3302 could effectively decrease MMP-1, MMP-2, and MMP-9 contents in UV-exposed HaCaT cells measured by ELISA. Therefore, tyndallized Lactobacillus acidophilus IDCC 3302 could decrease wrinkles by suppressing MMPs.

3. Presentation of the Gut–Skin Axis

The microbiota of the gut is similar to that of the skin. Various studies have linked inflammatory skin status to intestinal microbiota disorder. The intestinal microbiota affects the body's immunological function. The immune system defends itself from external pathogenic bacteria. Once the intestinal microbiota is imbalanced, the changed intestinal microbiota may lead to autoimmune and inflammation status not only in the intestines but also in remote organs, including the skin [50]. Various research prove the notion that disorders in the intestinal microbiota could contribute to skin diseases, including acne [51,52], atopic dermatitis [53], psoriasis [54], and rosacea [55]. The immune system appears to mediate the connection between the skin and the intestine. Microbial interactions with the host immune system are important for maintaining skin homeostasis [56]. Thus, a balance between skin and intestines is a logical way to cure many skin disorders. Probiotics exert a crucial effect in improving the microbiota and are a vital therapeutic modality in the cure of inflammatory skin disorders [50].

The gut–skin axis suggests a relationship in which the immune properties of the gut microbiota can also influence skin health. The positive modulation of the skin or intestinal microbiota via oral probiotics is regarded as a potential clinical approach to prevent photoaging of the skin. Oral probiotics are a kind of live microbiota that modify the intestinal microbe and can have direct photoprotective influences on special skin cells via regulating immune responses and inflammation factors. In addition, they can increase the serum contents of short-chain fatty acids (SCFAs), which induce a range of immune and inflammatory responses. Oral probiotics have been investigated as a means to directly improve the intestinal microbiota to suppress and cure skin photoaging. In addition, oral probiotics have been used in the treatment of other skin diseases, including atopic dermatitis, acne, rosacea, and psoriasis, by regulating the skin microbiota and gut–skin axis [57–61]. This section discusses the effect of oral probiotics in photoaging and their associated mechanisms.

3.1. Improvement of Intestinal Homeostasis by Probiotics

3.1.1. Enhancement of Barrier Function

Probiotics ameliorate gut barrier dysfunction through a variety of potential mechanisms. These mechanisms include upregulation of the expression of mucin proteins, including mucin-type glycoproteins (MUC)1, MUC2, and MUC3, which in turn restrict the movement of bacteria in the mucus, and upregulation of the secretion and expression of antimicrobial peptides and tight junction proteins, including α-defensins, β-defensins and occlusion bands (ZO), to prevent cell proliferation [62].

3.1.2. Suppression of Pathogens

Probiotics compete with pathogenic bacteria or commensals to combine with mucins or epithelial cells and prevent the overgrowth of potentially pathogenic bacteria. Moreover, probiotics offer anti-microbial ingredients, including anti-microbial peptides, SCFA, and bacteriocins, which are associated with inhibiting or killing pathogenic microorganisms. In addition, SCFA, including butyrate, for example, help regulate the expression of occludin and ZO, both of which are associated with the improvement of the epithelial barrier integrity [63].

Probiotics also up-regulate the production of IgA in the host gastrointestinal (GI) tract. Secretory IgA defends the intestinal epithelium from colonization and/or invasion via ligation of pathogenic or commensal antigens, induces reverse transcriptional transport of antigens to dendritic cells (DCs), and reduces pro-inflammatory factors [64].

3.2. The Pathway of Probiotic-Mediated Intestinal Microbiota Regulating Skin Status (Figure 3)

3.2.1. Immunologic Pathway

In terms of immunity to *Staphylococcus aureus*, the most popular bacterial strain influencing atopic dermatitis (AD), an association was indicated between the intestines and skin. *Staphylococcus aureus* is the most common pathogenic bacteria in the skin of AD patients. By contrast, a new birth cohort research showed that colonization with *Staphylococcus aureus* strains exerted a vital effect on preventing the development of AD in infancy, as early responses to *Staphylococcus aureus*, similar to other skin strains, promoted the maturation of the infant's immune system. *Staphylococcus aureus* strains on the mucosa can play a protective role through immune stimulation [65]. This research supports the speculation that the intestines and skin control the immune environment through the gut microbiota (Figure 3).

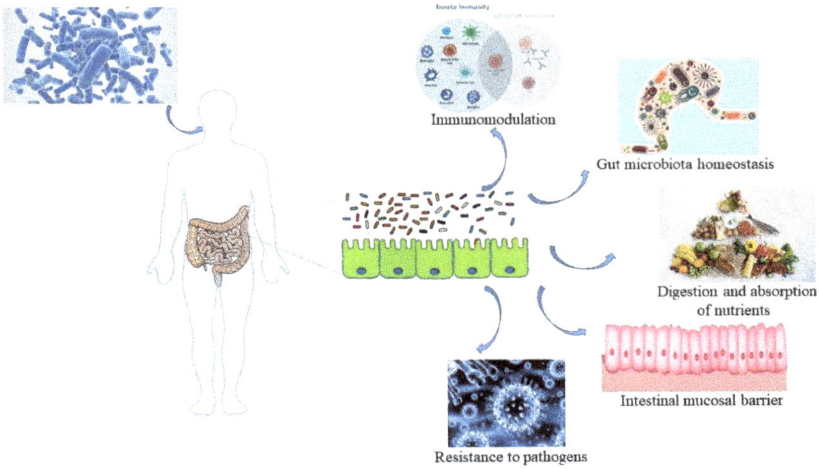

Figure 3. Oral probiotics mediate the beneficial effects of intestinal homeostasis on the organism. After the host ingests probiotics through the oral route, the probiotics enter the intestinal tract and play a role in improving intestinal homeostasis, mainly manifested as: immunomodulation, gut microbiota homeostasis, digestion and absorption of nutrients, and intestinal mucosal barrier.

Specific intestinal microbiota and their metabolites, including retinoic acid and polysaccharides A separated from *Bacteroides fragilis*, *Facecalibacterium prausnitzii*, and bacterial species attributing to *Clostridium* groups IV and XI, facilitate the accumulation of Tregs and lymphocytes that activate anti-inflammatory responses. In addition, some SCFAs, particularly butyrate, can modulate immune cell activation and apoptosis [66].

The gut microbiome has been studied as an important contributing factor to the immunologic pathway of skin disorders via probiotics. Orally administered probiotics could interact with gastrointestinal mucosa and gut-associated lymphoid tissue (GALT), where more than 70% of immune cells are located [67]. Probiotics interact with epithelial cells, mucosal DC, and macrophages in diverse ways. Depending on the probiotic strain, they can either induce immune activation signaling by producing IL-12, IL-18, and TNF-α or trigger tolerance signaling by stimulating anti-inflammatory cytokines, such as IL-10 and TGF-β [68,69]. In the IL-10/TGF-β-enriched cytokine milieu, DCs and macrophages can enhance the generation of the induced Treg cells that play key roles in maintaining peripheral immune tolerance by balancing the ratio of effector and Treg cells. Apart from probiotics, alterations in the gut microbiome might affect the development of host immune cell function through differences in gut microbiome genes, particularly in infants with AD [70].

3.2.2. Metabolite Pathway

Metabolites originating from intestinal microbiota, including SCFAs, or further supplemented by oral administration, also resolve the link between the intestines and the skin through the microbiota. SCFAs produced by intestinal microbiota including *Akkermansia muciniphila* exert an important effect on the pathology and etiology of AD, which could account for its correlation with the cutaneous immune system. A study showed that linoleic acid and 10-hydroxy-cis-12-octadecenoic acid alleviated AD diseases and controlled the intestinal microbiota in mice [71]. Further, three differential subgroups of the neonatal intestinal microbiota (NGM1–3) and its metabolites have been shown to play a role in early allergic sensitization [72]. Among these three subgroups, NGM3 is associated with multiple allergic sensitizations and was found to be relatively low in content in *Bifidobacterium, Ackermannia* and *Fasciola* [65]. For instance, 12,13-dihydroxy-9Zoctadecenoic acid (12,13-DiHome), a metabolite with inflammatory actions on in vitro immune control, was abundant in NGM3. In addition, 12,13-DiHome was up-regulated in the protective layer of casein, the white waxy coating on newborn host skin. These results may suggest the presence of a metabolite pathway in the gut–skin axis [65].

3.2.3. Neuroendocrine Pathway

Similarly to the skin, the lining of the GI is in direct contact with the outside environment, including food and microbes. One of the important roles of the skin and intestine is to suppress the entry of any pathogenic bacteria, and the microorganisms on both organs can be beneficial for the removal of these pathogenic bacteria through immune function, so it is essential to establish a balanced intestinal microbiota in both tissues and keep the appropriate balance for good health. In addition, these two microbiomes can interact with each other via neuroendocrine signaling. This influence can occur through two pathways: direct and indirect [65]. Tryptophan production by gut microbes leading to itchy skin in AD patients is an example of a direct signal. Conversely, γ-aminobutyric acid produced from *Lactobacillus* and *Bifidobacterium* in the intestines inhibits skin itching [65,73].

By indirect route, gut microbes control the content of cytokines such as IL-10 and IFN-γ in the blood, which can result in unusual alterations in brain function, leading to anxiety and stress [65]. Cortisol is a representative stress hormone in hosts that can change the permeability and barrier function of the intestinal epithelium by altering the composition of the intestinal microbe [74]. Cortisol can also alter the contents of circulating neuroendocrine molecules, including tryptamine, trimethylamine, and 5-hydroxytryptamine, further enhancing the skin barrier and immunological function [75].

4. Probiotics-Mediate Intestinal Microbiota to Improve Skin Disorders (Figure 4)

4.1. Acne

Acne patients have a special skin microbe. Available treatments for acne have various challenges because it injures the mechanical barrier of the skin, thereby drying it out and stimulating it. Research examining the skin–gut axis relationship in acne have shown that treatment with probiotics can improve the immune response beyond the intestines and extend it to the skin [61]. There is increasing evidence that topical probiotics also modulate the skin's mechanical barrier and generate a secondary up-regulation in antimicrobial peptides. For instance, the lactic acid bacterium *Streptococcus thermophiles* promoted ceramide synthesis when used as a cream, lasting one week in vitro and in vivo [76–78]. Ceramides can confine water to the skin, and certain ceramide sphingolipids, including sphingomyelin, have antibacterial activity against *Cutibacterium acnes*, further restoring acne. Through the generation of ceramides, probiotics are used to enhance the skin's mechanical barrier, which is helpful for acne-affected skin, as ceramides soothe irritated skin [79].

Thus, probiotics may be used to enhance protective barriers, suppress acne-causing bacteria, decrease pustules, and offer relief from skin irritation in acne patients (Figure 4).

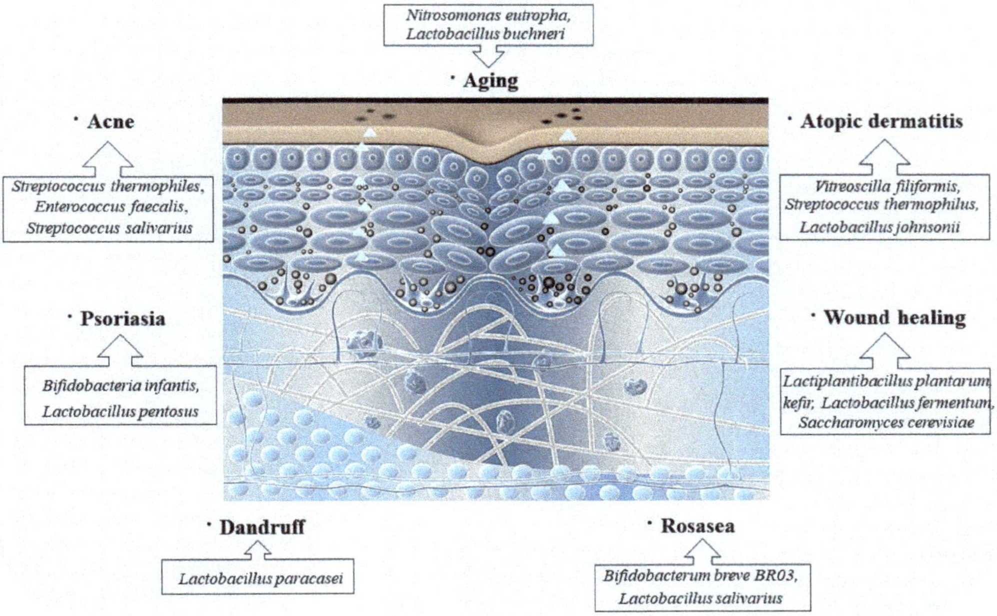

Figure 4. Probiotics can treat skin diseases. Different probiotics can treat different skin diseases, for example, *Nitrosomonas eutropha* and *Lactobacillus buchneri* can improve skin aging; *Streptococcus thermophiles*, *Enterococcus faecalis* and *Streptococcus salivarius* can improve acne; *Vitreoscilla filiformis*, *Streptococcus thermophilus* and *Lactobacillus johnsonii* can improve atopic dermatitis, *Bifidobacteria infantis* and *Lactobacillus pentosus* can improve psoriasia; *Lactiplantibacillus plantarum* kefir, *Lactobacillus fermentum* and *Saccharomyces cerevisiae* can improve wound healing; *Lactobacillus paracasei* can improve dandruff; *Bifidobacterum breve* BR03 and *Lactobacillus salivarius* can improve rosasea.

4.2. Atopic Dermatitis

AD is mainly caused by a decrease in microbial diversity; as mentioned above, the main microorganism in AD patients is *Staphylococcus aureus*. Various research suggest that oral probiotics may be used as a better option for its treatment [80]. One research showed that *Streptococcus thermophilus* obviously decreased eczema linked to AD and also decreased the severity of symptoms [78]. Another research suggested the validity of the lactic acid bacterium *Streptococcus thermophilus* on the stratum corneum by increasing ceramide levels in the skin [77].

In a randomized, double-blind experiment with patients with atopic dermatitis, researchers compared the use of emollients produced by *Lactobaciilus* with the use of regular emollients. Emollients containing Lactobacillus suppressed the extensions of *Staphylococcus aureus*, offered a mechanical barrier, and restored symptoms in patients with AD [81]. An experiment exploring the influence of lotions, including the heat-treated probiotic strain *Lactobacillus johnsonii* NCC on *Staphylococcus aureus* colonization, exhibited a useful effect in clinical symptoms in patients with atopic dermatitis.

Likewise, other experiments studying the treatment of *Roseomonas mucosa* via supplementation demonstrated an obvious decrease in disease severity, topical steroid needs, and *Staphylococcus aureus burden.* No adverse reactions or complications were reported in this trial [82]. Most of the trials conducted so far have shown that probiotics have a positive effect on patients with atopic dermatitis.

4.3. *Psoriasis*

Psoriasis is an autoimmune chronic skin disorder that is usually treated with topical emollients and oral immunosuppressants. Few research have included topical probiotics as a treatment for psoriasis. While research has shown that changes in the skin microbiota may help control psoriasis symptoms, oral probiotics have demonstrated therapeutic effects in clinical symptoms in some individuals. However, studies on the validity of oral probiotics in patients with psoriasis are needed to clinically demonstrate the advantages of oral probiotics [60].

4.4. *Seborrheic Dermatitis*

Yeast overgrowth on the scalp and decreased diversity of microbiota leads to dandruff and seborrheic dermatitis. Much research has been conducted to assess the treatment of probiotics in this context. A study of 60 patients exhibited a decrease in erythema, desquamation, and pruritus after the topical application of filamentous Staphylococci [78,83]. Another study revealed that *Vitreoscilla filiformis* lysate induced Treg activity through IL-10 production by dendritic cells [84]. Dandruff, seborrheic dermatitis, and scalp-associated disorders showed beneficial effects after oral supplementation of *Lactobacillus paracasei*. More research is necessary on the local efficacy of probiotics in treating this disease [85].

4.5. *Rosacea*

Overexpression of TLR2 receptors-induced rosacea arises, resulting in an inflammatory response and changed skin microbiota [86,87]. In addition to doxycycline as an antibiotic, oral probiotics are used to treat scalp rosacea; however, the use of topical probiotics for rosacea treatment has not been explored [88].

5. Probiotics That Regulate Skin Physiology (Figure 5)

5.1. *Nitrobacter*

Nitrobacter is a nitrifying bacterium that generates nitrate, a molecule that may have positive influences on the host, contained on the skin. Research has shown that dietary nitrates depletion, such as green leafy vegetables, has positive effects, such as increased blood flow to exercising skeletal muscles, decreased exercise oxygen demand, increased exercise tolerance in patients with peripheral arterial disorders, and lower blood pressure. Research has shown that these influences of nitrate depletion are largely the result of NOS-independent increases in NO synthesis and have been indicated to increase cutaneous reflex vasodilation through NOS-independent mechanisms in healthy hosts [89]. Antifungal activity of *Nitrobacter* spp. has been found to defend the skin from dermatophytes [90] and *Staphylococcus aureus* [91], a microbiota that can induce many dermatological infections. The ability of *Nitrobacter* to generate nitrate could also offer nitrate to the skin, which has been proven to conserve the skin's progenitor cells from UV injury [92,93].

5.2. *Lactobacillus*

Lactobacillus is the most abundant and diverse genus of lactic acid bacteria [94]. *Lactobacillus* strains have anti-inflammation activity on host keratinocytes and have specific irritation effects on the growth of *Staphylococcus epidermidis* in vitro [95]. *Lactobacillus* also suppresses substance P-caused skin inflammatory responses and promotes the restoration of skin barrier function [96]. A clinical experiment suggested that a six-week oral *Lactobacillus johnsonii* intervention regimen promoted the restoration of skin immune function in UV-caused immunosuppression [97]. In a randomized, double-blind, placebo-controlled experiment program with 20 adults, Fabroccini et al. found that a liquid intervention, including *Lactobacillus rhamnosus*, normalized skin expression of insulin signaling-related genes and improved adult acne [98]. The study showed that tyndalized *Lactobacillus acidophilus*, an oral probiotic, suppressed wrinkle formation resulting from UV irradiation in mice [99].

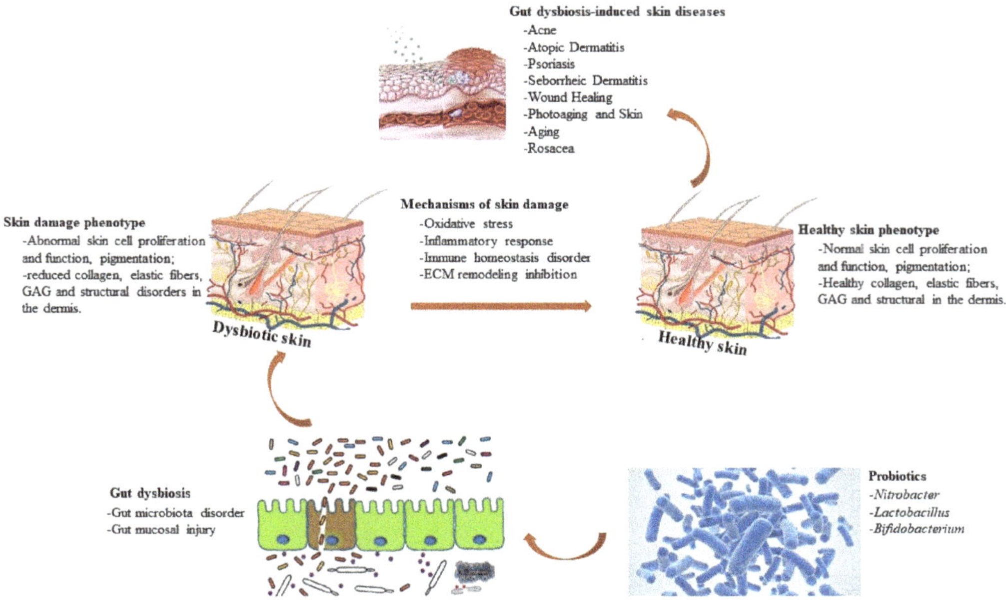

Figure 5. The mechanism of probiotics to improve skin diseases. Probiotics, including *Nitrobacter*, *Lactobacillus* and *Bifidobacterium*, can restore intestinal homeostasis by improving intestinal microbiota disorders and repairing intestinal mucosal damage, and then treat skin damage phenotype, including abnormal skin cell proliferation and function, pigmentation, reduced collagen, elastic fibers, glycosaminoglycan (GAG), and structural disorders in the dermis by inhibiting oxidative stress, inflammation response, immune homeostasis, and extracellular matrix (ECM) remodeling inhibition, ultimately treating skin diseases (acne, atopic dermatitis, psoriasis, seborrheic dermatitis, wound healing, photoaging and aging skin, and rosacea).

These beneficial influences were due to the reduction in MMPs. Furthermore, Park and Bae indicated in an in vitro study that combined the fermentation of *Lactobacillus* and *Bifidobacterium* to ferment A. koreanum extract inhibited the senescence phenotype of skin fibroblasts [100]. The senescence status is driven by UV- or hydrogen peroxide, while the protection of senescence is partially mediated by the inference of MMP-1 (Figure 5).

5.3. *Bifidobacterium*

Results showed that *Bifidobacterium breve* B-3, as an oral supplementation, obviously inhibited TEWL, skin dryness, changes in epidermal thickening, and improved injury to tight junction structures and basement membranes under excess UV irradiation in mice. *Bifidobacterium breve* B-3 supplementation also inhibited UV-caused production of IL-1β in the skin [101].

Bifidobacterium and *Lactobacillus* as lyophilized powders in capsules suppress atopic sensitization to general food allergens and decrease the prevalence of atopic eczema in early childhood [102]. In adult patients with atopic dermatitis, *Bifidobacterium bifidum* as an oral supplementation has antipruritic influences linked to up-regulated levels of antipruritic and analgesic metabolite acetonide [103]. In a double-blind, placebo-controlled, randomized trial, ingestion of fermented milk, including *Bifidobacterium breve* and galactooligosaccharides, inhibited reduced levels of stratum corneum hydration, inhibited increased histone l-like protease activity, and decreased serum and urinary phenol contents in healthy adult female volunteers [104].

6. Potential Mechanisms of Probiotic-Mediated Regulation of Skin Conditions by the Gut–Skin Axis

6.1. Oxidative Stress Level Decreases

The pathological physiology of skin photoaging is tightly related to ROS-caused injury, containing activation of MAPK and NF-κB loop, decrease in MMPs level, and collagen content, resulting in skin photoaging. A study suggested that the partial fermentation of *Agastache rugosa*-fermented extract (ARE-F) with a probiotic *Lactobacillus* promoted UV-caused concentrations of total glutathione and superoxide dismutase activity while down-regulating UV-caused ROS and MMPs levels in UV-treated human keratinocytes cells (HaCaT) keratinocytes [105]. A study indicated that partial *Lactobacillus acidophilus IDCC 3302* defended against UV-caused photodamage in the skin epidermis by promoting the antioxidant capacity of the skin, hydrating cytokines, and inhibiting MMPs synthesis via suppression of MAPK loop [45].

Another study suggested that *Lactobacillus acidophilus* KCCM12625 had great antioxidant functions and obviously decreased up-regulated ROS contents in vitro after UV irradiation and improved photodamage induced by oxidative injury [22]. A study suggested that the oral supplementation of *Bifidobacterium breve* Yakult suppressed ROS content and improved UV-caused skin mechanical barrier injury and oxidative stress in vivo [106]. Research showed that a plant extract fermented with *Lactobacillus buchneri* improved the influence of oxidative injury in UV-caused skin photoaging in vitro by up-regulating the type I procollagen content, suppressing elastase synthesis, and up-regulating the level of UV-caused MMPs on HaCaT keratinocytes and dermal fibroblasts [107]. A study suggested that *Limosilactobacillus fermentum* XJC60 could enhance mitochondrial capabilities, decrease ROS content in UV-damaged skin cells, and, therefore, keep the skin status [108]. In addition, new research has indicated antioxidant effect as the primary pathway by which *Lacticaseibacillus rhamnosus* GG (ATCC 53103, LGG) [109] and *Lacticaseibacillus casei* strain Shirota improve skin photoaging [110].

6.2. Inflammatory Response Suppression

Up-regulated skin inflammation elements lead to the dysfunction of barrier function, TEWL, incremental permeability of the epidermis, and rapid skin photoaging. Research suggested that oral supplementation of *Bifidobacterium breve* B-3 was effective in decreasing UV-caused IL-1β content in the skin in UV-irradiated mice. As a result, TEWL, skin dryness, and epidermal thickening were inhibited [111,112]. *Lactobacillus acidophilus* IDCC3302, in addition to its antioxidant effects, suppressed MAPK signaling pathway-mediated production of pro-inflammatory factors and decreased UV-radiation-induced skin inflammation [45]. To improve skin photoaging, a study indicated that the oral application of *Lactobacillus reuteri* DSM 17938 showed an anti-inflammation effect which resisted UV-induced IL-6 and IL-8 [113].

Keshari et al. suggested that butyrate from a new production of probiotic *Staphylococcus epidermidis* could reduce UV-caused pro-inflammation factor IL-6 factors using SCFA receptors [114]. A study showed that oral oligosaccharides regulated UV-induced inflammatory immune responses and reduced TEWL and sunburn erythema, therefore inhibiting skin photoaging [115].

6.3. Immune Homeostasis Maintaining

Many specified probiotics, such as *Lactobacillus paracasei*, regulate immune response to inhibit pathogens [116]. In addition, it inhibits unwanted immunological effects to maintain immune homeostasis against chronic inflammation diseases. It may be due to the modulation of the Tregs number by probiotics. Treg exerts a vital effect on the immunosuppression caused by skin photoaging. *Lactobacillus johnsonii* suppressed UV-caused decrease in epidermal Langerhans cell density and promoted the restoration of skin immune homeostasis after UV-caused immunosuppression. Moreover, probiotics exert different effects under different immune statuses. Under a physiological status, probiotics decrease cytotoxic T cell attack on the skin, up-regulate induction of functional damage to CD8$^+$ T cells, and cause the activation of quiescent dendritic cells and the activation and function of all regulatory T cell subsets. A study performed three clinical experiments to analyze the influences of dietary supplements (DS), including *Lactobacillus johnsonii* and nutritional carotenoids, on early UV-caused skin injury [117]. These results suggest that the ingestion of degenerative vertebral slippage has useful influences on the long-term and repetitive influences of UV exposure and is more specific to photoaging. Data showed that oral interventions, including *Bifidobacterium longum* and galacto-oligosaccharides, defend the skin from UV-caused photoaging, resulting in their anti-inflammation and anti-oxidative effects [118]. In addition, they up-regulated serum levels of SCFAs and acetates, which have been shown to up-regulate and activate histone acetylation-dependent skin resident treg.

6.4. ECM Remodeling Suppression

ROS contents were up-regulateed after UV exposure, leading to increased MMPs contents, skin collagen protein and elastin protein degradation, and rough, dry and sagging skin. Probiotics can not only directly decrease ROS content but also indirectly regulate MMP levels in the skin, decreasing collagen and elastin protein degradation following UV exposure [119]. Oral *Lactobacillus acidophilus* KCCM12625 could decrease the mRNA level of MMPs after skin photoaging by damaging the AP-1 loop of skin while up-regulating procollagen level and down-regulating collagen protein losses in the dermis [22]. A study indicated that oral *Lactobacillus plantarum HY7714* decreased the overproduction of MMP-13 content and the activities of MMP-2 and MMP-9 in UV-induced cell injuries by suppressing the activation of the JNK/AP-1 loop [49]. Research suggested that oral *Lactobacillus sakei* could suppress theAP-1 expression by inhibiting the MAPK loop to up-regulate collagen in the dermis and improve skin photoaging [46]. Another study demonstrated that extracellular *Lactobacilli* exopolysaccharides (LEPS) can decrease the MMPs level and increase TIMPs [120]. Data showed that LEPS of B9-1 from *Lactobacillus casei* can strengthen the anti-collagenase and anti-elastase functions of the skin and efficiently decrease the breakdown of collagen after UV radiation. A study demonstrated that localized extracts originating *Lactobacillus brucei* fermented plants in kimchi can significantly suppress UV-induced elastase activity and expression of MMPs and enhance the synthesis of type I procollagen. Negari et al. showed that the metabolites from the topical probiotic *Staphylococcus epidermidis* of Cetearyl isononanoate (CIN) as a potential carbon source could restore damaged collagen and cause the synthesis of collagen through phosphorylated extracellular signal-regulated kinase (p-ERK) activation, thereby inhibiting skin photoaging [121].

7. Concluding Remarks

In recent years, significant advances have been made in understanding the composition of skin probiotics and how dysbiosis affects skin health. Topical and internal probiotics in the form of various dermatological formulations are an important part of the treatment of skin conditions. While topical probiotics' functions and protective nature maintain the skin's homeostasis, their shortcomings and limitations result in inflammatory skin conditions that are difficult to completely cure via topical probiotics. Several clinical trials are being carried out in order to study the efficacy as well as the adverse effects of internal probiotics formulations for the treatment of conditions such as atopic dermatitis, acne, psoriasis, wound healing, and many other skin problems. We hope that this review forms a contribution to promoting enhanced research activities in the field of internal probiotics as a novel therapeutic approach for the treatment of skin disorders.

Author Contributions: T.G.: Conceptualization, Methodology, Formal analysis, Writing—original draft, Visualization; Y.L.: Resources, Supervision, Project administration, Funding acquisition; X.W.: Methodology; F.R.: Conceptualization, Formal analysis, Validation, Writing—review and editing, Project administration, Funding acquisition, Supervision. All authors have read and agreed to the published version of the manuscript.

Funding: This work was supported by the 111 project of the Education Ministry of China (No. B18053) and the National Natural Science Foundation (32130081).

Institutional Review Board Statement: Not applicable.

Informed Consent Statement: Not applicable.

Data Availability Statement: Data will be made available on request.

Conflicts of Interest: The authors declare that they have no known competing financial interest or personal relationships that could have appeared to influence the work reported in this paper.

Abbreviations

TEWL: transcutaneous water loss; MMP-1: matrix metalloproteinase-1; UV: ultraviolet; TYRP-1: tyrosinase-related protein 1; TYRP-2: tyrosinase-related protein 2; SPT: serine palmitoyltransferase; HA: hyaluronic acid; SLS: sodium dodecyl sulfate; SH: *Sphingomonas hydrophobicum*; sLTA: *Lactobacillus sakei* Lipoteichoic Acid; AL: *Lactobacillus* KCCM12625P; SCFAs: short-chain fatty acids; MUC: mucin-type glycoproteins; TJ: tight junction; GI: gastrointestinal; DCs: dendritic cells; AD: matitis; NGM1–3: neonatal intestinal microbiota; GAG: glycosaminoglycan; ECM: extracellular matrix; *ARE-F: Agastache rugosa*-fermented extract; HaCaT: human keratinocytes cells; DS: dietary supplements; *LEPS: Lactobacilli* exopolysaccharides; CIN: Cetearyl isononanoate; CFU: colony forming units; GALT: gut-associated lymphoid tissue.

References

1. Sender, R.; Fuchs, S.; Milo, R. Revised estimates for the number of human and bacteria cells in the body. *PLoS Biol.* **2016**, *14*, e1002533. [CrossRef] [PubMed]
2. Grice, E.A.; Segre, J.A. The skin microbiome. *Nat. Rev. Microbiol.* **2011**, *9*, 244–253. [CrossRef]
3. Oh, J.; Byrd, A.L.; Park, M.; Kong, H.H.; Segre, J.A. Temporal stability of the human skin microbiome. *Cell* **2016**, *165*, 854–866. [CrossRef]
4. Schommer, N.N.; Gallo, R.L. Structure and function of the human skin microbiome. *Trends Microbiol.* **2013**, *21*, 660–668. [CrossRef]
5. Dréno, B.; Alexis, A.; Chuberre, B.; Marinovich, M. Safety of titanium dioxide nanoparticles in cosmetics. *J. Eur. Acad. Dermatol. Venereol.* **2019**, *33*, 34–46. [CrossRef]
6. Hu, Y.; Zeng, H.; Huang, J.; Jiang, L.; Chen, J.; Zeng, Q. Traditional Asian aerbs in skin whitening: The current development and limitations. *Front. Pharmacol.* **2020**, *11*, 982. [CrossRef]
7. Yu, J.; Ma, X.; Wang, X.; Cui, X.; Ding, K.; Wang, S.; Han, C. Application and mechanism of probiotics in skin care: A review. *J. Cosmet. Dermatol.* **2022**, *21*, 886–894. [CrossRef] [PubMed]
8. Puebla-Barragan, S.; Reid, G. Probiotics in cosmetic and personal care products: Trends and challenges. *Molecules* **2021**, *26*, 1249. [CrossRef]
9. Osborne, D.; Tan, P.; Varma, Y.; Carbol, J. Formulating topical products containing live microorganisms as the active ingredient. *Pharm. Technol.* **2018**, *42*, 32–36.
10. Marto, J.; Ascenso, A.; Simoes, S.; Almeida, A.J.; Ribeiro, H.M. Pickering emulsions: Challenges and opportunities in topical delivery. *Expert. Opin. Drug Deliv.* **2016**, *13*, 1093–1107. [CrossRef]
11. Sreeja, V.; Prajapati, J.B. Probiotic formulations: Application and status as pharmaceuticals-a review. *Probiotics Antimicrob. Proteins* **2013**, *5*, 81–91. [CrossRef]
12. Lee, G.R.; Maarouf, M.; Hendricks, A.J.; Lee, D.E.; Shi, V.Y. Topical probiotics: The unknowns behind their rising popularity. *Dermatol. Online J.* **2019**, *25*, 15–21. [CrossRef]
13. Pillaiyar, T.; Manickam, M.; Jung, S.H. Recent development of signaling pathways inhibitors of melanogenesis. *Cell. Signal.* **2017**, *40*, 99–115. [CrossRef] [PubMed]
14. Brenner, M.; Hearing, V.J. The protective role of melanin against UV damage in human skin. *Photochem. Photobiol.* **2008**, *84*, 539–549. [CrossRef]
15. Iozumi, K.; Hoganson, G.E.; Pennella, R.; Everett, M.A.; Fuller, B.B. Role of tyrosinase as the determinant of pigmentation in cultured human melanocytes. *J. Investig. Dermatol.* **1993**, *100*, 806–811. [CrossRef] [PubMed]
16. Li, G.; Ju, H.K.; Chang, H.W.; Jahng, Y.; Lee, S.H.; Son, J.K. Melanin biosynthesis inhibitors from the bark of *Machilus thunbergii*. *Biol. Pharm. Bull.* **2003**, *26*, 1039–1041. [CrossRef]
17. Pillaiyar, T.; Manickam, M.; Jung, S.H. Inhibitors of melanogenesis: A patent review (2009–2014). *Expert. Opin. Ther. Pat.* **2015**, *25*, 775–788. [CrossRef]

18. Schiaffino, M.V. Signaling pathways in melanosome biogenesis and pathology. *Int. J. Biochem. Cell Biol.* **2010**, *42*, 1094–1104. [CrossRef]
19. Pillaiyar, T.; Manickam, M.; Namasivayam, V. Skin whitening agents: Medicinal chemistry perspective of tyrosinase inhibitors. *J. Enzym. Inhib. Med. Chem.* **2017**, *32*, 403–425. [CrossRef]
20. Huang, H.C.; Chang, T.M. Antioxidative properties and inhibitory effect of *Bifidobacterium adolescentis* on melanogenesis. *World J. Microbiol. Biotechnol.* **2012**, *28*, 2903–2912. [CrossRef]
21. Huang, H.C.; Lee, I.J.; Huang, C.; Chang, T.M. Lactic acid bacteria and lactic acid for skin health and melanogenesis inhibition. *Curr. Pharm. Biotechnol.* **2020**, *21*, 566–577. [CrossRef]
22. Lim, H.Y.; Jeong, D.; Park, S.H.; Shin, K.K.; Hong, Y.H.; Kim, E.; Yu, Y.G.; Kim, T.R.; Kim, H.; Lee, J.; et al. Antiwrinkle and antimelanogenesis effects of tyndallized *Lactobacillus acidophilus* KCCM12625P. *Int. J. Mol. Sci.* **2020**, *21*, 1620. [CrossRef]
23. Chaiprasongsuk, A.; Onkoksoong, T.; Pluemsamran, T.; Limsaengurai, S.; Panich, U. Photoprotection by dietary phenolics against melanogenesis induced by UVA through Nrf2-dependent antioxidant responses. *Redox Biol.* **2016**, *8*, 79–90. [CrossRef]
24. Liu, W.S.; Kuan, Y.D.; Chiu, K.H.; Wang, W.K.; Chang, F.H.; Liu, C.H.; Lee, C.H. The extract of *Rhodobacter sphaeroides* inhibits melanogenesis through the MEK/ERK signaling pathway. *Mar. Drugs* **2013**, *11*, 1899–1908. [CrossRef]
25. Kim, H.; Kim, J.T.; Barua, S.; Yoo, S.Y.; Hong, S.C.; Lee, K.B.; Lee, J. Seeking better topical delivery technologies of moisturizing agents for enhanced skin moisturization. *Expert. Opin. Drug Deliv.* **2018**, *15*, 17–31. [CrossRef]
26. Harding, C.R.; Watkinson, A.; Rawlings, A.V.; Scott, I.R. Dry skin, moisturization and corneodesmolysis. *Int. J. Cosmet. Sci.* **2000**, *22*, 21–52. [CrossRef] [PubMed]
27. Ra, J.; Lee, D.E.; Kim, S.H.; Jeong, J.W.; Ku, H.K.; Kim, T.Y.; Choi, I.D.; Jeung, W.; Sim, J.H.; Ahn, Y.T. Effect of oral administration of *Lactobacillus plantarum* HY7714 on epidermal hydration in ultraviolet B-irradiated hairless mice. *J. Microbiol. Biotechnol.* **2014**, *24*, 1736–1743. [CrossRef] [PubMed]
28. Elias, P.M.; Menon, G.K. Structural and lipid biochemical correlates of the epidermal permeability barrier. *Adv. Lipid Res.* **1991**, *24*, 1–26. [PubMed]
29. Holleran, W.M.; Uchida, Y.; Halkier-Sorensen, L.; Haratake, A.; Hara, M.; Epstein, J.H.; Elias, P.M. Structural and biochemical basis for the UVB-induced alterations in epidermal barrier function. *Photodermatol. Photoimmunol. Photomed.* **1997**, *13*, 117–128. [CrossRef] [PubMed]
30. Im, A.R.; Lee, B.; Kang, D.J.; Chae, S. Skin moisturizing and antiphotodamage effects of tyndallized *Lactobacillus acidophilus* IDCC 3302. *J. Med. Food* **2018**, *21*, 1016–1023. [CrossRef]
31. Papakonstantinou, E.; Roth, M.; Karakiulakis, G. Hyaluronic acid: A key molecule in skin aging. *Derm.-Endocrinol.* **2012**, *4*, 253–258. [CrossRef]
32. Baba, H.; Masuyama, A.; Yoshimura, C.; Aoyama, Y.; Takano, T.; Ohki, K. Oral intake of *Lactobacillus helveticus*-fermented milk whey decreased transepidermal water loss and prevented the onset of sodium dodecylsulfate-induced dermatitis in mice. *Biosci. Biotechnol. Biochem.* **2010**, *74*, 18–23. [CrossRef] [PubMed]
33. Segre, J.A. Epidermal barrier formation and recovery in skin disorders. *J. Clin. Investig.* **2006**, *116*, 1150–1158. [CrossRef] [PubMed]
34. Jung, Y.O.; Jeong, H.; Cho, Y.; Lee, E.O.; Jang, H.W.; Kim, J.; Nam, K.; Lim, K.M. Lysates of a probiotic, *Lactobacillus rhamnosus*, can improve skin barrier function in a reconstructed human epidermis model. *Int. J. Mol. Sci.* **2019**, *20*, 4289. [CrossRef]
35. Draelos, Z.D. New treatments for restoring impaired epidermal barrier permeability: Skin barrier repair creams. *Clin. Dermatol.* **2012**, *30*, 345–348. [CrossRef]
36. Cho, S. The role of functional foods in cutaneous anti-aging. *J. Lifestyle Med.* **2014**, *4*, 8–16. [CrossRef]
37. Trojahn, C.; Dobos, G.; Lichterfeld, A.; Blume-Peytavi, U.; Kottner, J. Characterizing facial skin ageing in humans: Disentangling extrinsic from intrinsic biological phenomena. *Biomed Res. Int.* **2015**, *2015*, 318586. [CrossRef]
38. Mora Huertas, A.C.; Schmelzer, C.E.; Hoehenwarter, W.; Heyroth, F.; Heinz, A. Molecular-level insights into aging processes of skin elastin. *Biochimie* **2016**, *128–129*, 163–173. [CrossRef] [PubMed]
39. Gervason, S.; Napoli, M.; Dreux-Zhiga, A.; Lazzarelli, C.; Garcier, S.; Briand, A.; Albouy, M.; Thepot, A.; Berthon, J.Y.; Filaire, E. Attenuation of negative effects of senescence in human skin using an extract from *Sphingomonas hydrophobicum*: Development of new skin care solution. *Int. J. Cosmet. Sci.* **2019**, *41*, 391–397.
40. Dolan, D.W.; Zupanic, A.; Nelson, G.; Hall, P.; Miwa, S.; Kirkwood, T.B.; Shanley, D.P. Integrated stochastic model of DNA damage repair by non-homologous end joining and p53/p21-mediated early senescence signalling. *PLoS Comput. Biol.* **2015**, *11*, e1004246. [CrossRef]
41. Penner, A.S.; Rock, M.J.; Kielty, C.M.; Shipley, J.M. Microfibril-associated glycoprotein-2 interacts with fibrillin-1 and fibrillin-2 suggesting a role for MAGP-2 in elastic fiber assembly. *J. Biol. Chem.* **2002**, *277*, 35044–35049. [CrossRef]
42. Sheng, W.; Wang, G.; Wang, Y.; Liang, J.; Wen, J.; Zheng, P.S.; Wu, Y.; Lee, V.; Slingerland, J.; Dumont, D.; et al. The roles of versican V1 and V2 isoforms in cell proliferation and apoptosis. *Mol. Biol. Cell* **2005**, *16*, 1330–1340. [CrossRef]
43. Fisher, G.J.; Kang, S.; Varani, J.; Bata-Csorgo, Z.; Wan, Y.; Datta, S.; Voorhees, J.J. Mechanisms of photoaging and chronological skin aging. *Arch. Dermatol.* **2002**, *138*, 1462–1470. [CrossRef] [PubMed]
44. Lee, D.E.; Huh, C.S.; Ra, J.; Choi, I.D.; Jeong, J.W.; Kim, S.H.; Ryu, J.H.; Seo, Y.K.; Koh, J.S.; Lee, J.H.; et al. Clinical evidence of effects of *Lactobacillus plantarum* HY7714 on skin aging: A randomized, double blind, placebo-controlled study. *J. Microbiol. Biotechnol.* **2015**, *25*, 2160–2168. [CrossRef] [PubMed]

45. Im, A.R.; Lee, B.; Kang, D.J.; Chae, S. Protective effects of tyndallized *Lactobacillus acidophilus* IDCC 3302 against UVB-induced photodamage to epidermal keratinocytes cells. *Int. J. Mol. Med.* **2019**, *43*, 2499–2506. [CrossRef] [PubMed]
46. You, G.E.; Jung, B.J.; Kim, H.R.; Kim, H.G.; Kim, T.R.; Chung, D.K. Lactobacillus sakeilipoteichoic acid inhibits MMP-1 induced by UVA in normal dermal fibroblasts of human. *J. Microbiol. Biotechnol.* **2013**, *23*, 1357–1364. [CrossRef]
47. Small, R. Botulinum toxin injection for facial wrinkles. *Am. Fam. Physician* **2014**, *90*, 168–175.
48. Park, S.H.; Lee, K.H.; Han, C.S.; Kim, K.H.; Kim, Y.H. Inhibitory effects of carex humilis extract on elastase activity and matrix metalloproteinase-1 expression. *J. Soc. Cosmet. Sci. Korea* **2010**, *36*, 129–136.
49. Kim, H.M.; Lee, D.E.; Park, S.D.; Kim, Y.T.; Kim, Y.J.; Jeong, J.W.; Jang, S.S.; Ahn, Y.T.; Sim, J.H.; Huh, C.S.; et al. Oral administration of *Lactobacillus plantarum* HY7714 protects hairless mouse against ultraviolet B induced photoaging. *J. Microbiol. Biotechnol.* **2014**, *24*, 1583–1591. [CrossRef] [PubMed]
50. Szántó, M.; Dózsa, A.; Antal, D.; Szabó, K.; Kemény, L.; Bai, P. Targeting the gut-skin axis-Probiotics as new tools for skin disorder management? *Exp. Dermatol.* **2019**, *28*, 1210–1218. [CrossRef]
51. Bowe, W.P.; Filip, J.C.; DiRienzo, J.M.; Volgina, A.; Margolis, D.J. Inhibition of propionibacterium acnes by bacteriocin-like inhibitory substances (BLIS) produced by *Streptococcus salivarius*. *J. Drugs Dermatol.* **2006**, *5*, 868–870.
52. Bowe, W.P.; Logan, A.C. Acne vulgaris, probiotics and the gut-brain-skin axis-back to the future? *Gut Pathog.* **2011**, *3*, 1. [CrossRef]
53. Kim, B.J.; Lee, S.Y.; Kim, H.B.; Lee, E.; Hong, S.J. Environmental changes, microbiota, and allergic diseases. *Allergy Asthma Immunol. Res.* **2014**, *6*, 389–400. [CrossRef]
54. Hidalgo-Cantabrana, C.; Gómez, J.; Delgado, S.; Requena-López, S.; Queiro-Silva, R.; Margolles, A.; Coto, E.; Sánchez, B.; Coto-Segura, P. Gut microbiota dysbiosis in a cohort of patients with psoriasis. *Br. J. Dermatol.* **2019**, *181*, 1287–1295. [CrossRef] [PubMed]
55. Nam, J.H.; Yun, Y.; Kim, H.S.; Kim, H.N.; Jung, H.J.; Chang, Y.; Ryu, S.; Shin, H.; Kim, H.L.; Kim, W.S. Rosacea and its association with enteral microbiota in Korean females. *Exp. Dermatol.* **2018**, *27*, 37–42. [CrossRef]
56. De Pessemier, B.; Grine, L.; Debaere, M.; Maes, A.; Paetzold, B.; Callewaert, C. Gut-skin axis: Current knowledge of the interrelationship between microbial dysbiosis and skin conditions. *Microorganisms* **2021**, *9*, 353. [CrossRef] [PubMed]
57. D'Elios, S.; Trambusti, I.; Verduci, E.; Ferrante, G.; Rosati, S.; Marseglia, G.L.; Drago, L.; Peroni, D.G. Probiotics in the prevention and treatment of atopic dermatitis. *Pediatr. Allergy Immunol.* **2020**, *31*, 43–45. [CrossRef]
58. Habeebuddin, M.; Karnati, R.K.; Shiroorkar, P.N.; Nagaraja, S.; Asdaq, S.M.B.; Khalid Anwer, M.; Fattepur, S. Topical probiotics: More than a skin deep. *Pharmaceutics* **2022**, *14*, 557. [CrossRef] [PubMed]
59. Atabati, H.; Esmaeili, S.A.; Saburi, E.; Akhlaghi, M.; Raoofi, A.; Rezaei, N.; Momtazi-Borojeni, A.A. Probiotics with ameliorating effects on the severity of skin inflammation in psoriasis: Evidence from experimental and clinical studies. *J. Cell. Physiol.* **2020**, *235*, 8925–8937. [CrossRef]
60. Yu, Y.; Dunaway, S.; Champer, J.; Kim, J.; Alikhan, A. Changing our microbiome: Probiotics in dermatology. *Br. J. Dermatol.* **2020**, *182*, 39–46. [CrossRef]
61. França, K. Topical probiotics in dermatological therapy and skincare: A concise review. *Dermatol. Ther.* **2021**, *11*, 71–77. [CrossRef]
62. Ohland, C.L.; Macnaughton, W.K. Probiotic bacteria and intestinal epithelial barrier function. *Am. J. Physiol. Gastrointest. Liver Physiol.* **2010**, *298*, G807–G819. [CrossRef] [PubMed]
63. Eslami, M.; Bahar, A.; Keikha, M.; Karbalaei, M.; Kobyliak, N.; Yousefi, B. Probiotics function and modulation of the immune system in allergic diseases. *Allergol. Immunopathol.* **2020**, *48*, 771–788. [CrossRef]
64. Mantis, N.J.; Rol, N.; Corthésy, B. Secretory IgA's complex roles in immunity and mucosal homeostasis in the gut. *Mucosal. Immunol.* **2011**, *4*, 603–611. [CrossRef] [PubMed]
65. Yokoyama, S.; Hiramoto, K.; Koyama, M.; Ooi, K. Impairment of skin barrier function via cholinergic signal transduction in a dextran sulphate sodium-induced colitis mouse model. *Exp. Dermatol.* **2015**, *24*, 779–784. [CrossRef] [PubMed]
66. Salem, I.; Ramser, A.; Isham, N.; Ghannoum, M.A. The gut microbiome as a major regulator of the gut-skin axis. *Front. Microbiol.* **2018**, *9*, 1459. [CrossRef]
67. Lebeer, S.; Vanderleyden, J.; De Keersmaecker, S.C. Host interactions of probiotic bacterial surface molecules: Comparison with commensals and pathogens. *Nat. Rev. Microbiol.* **2010**, *8*, 171–184. [CrossRef]
68. Ueno, N.; Fujiya, M.; Segawa, S.; Nata, T.; Moriichi, K.; Tanabe, H.; Mizukami, Y.; Kobayashi, N.; Ito, K.; Kohgo, Y. Heat-killed body of lactobacillus brevis SBC8803 ameliorates intestinal injury in a murine model of colitis by enhancing the intestinal barrier function. *Inflamm. Bowel Dis.* **2011**, *17*, 2235–2250. [CrossRef]
69. Smits, H.H.; Engering, A.; van der Kleij, D.; de Jong, E.C.; Schipper, K.; van Capel, T.M.; Zaat, B.A.; Yazdanbakhsh, M.; Wierenga, E.A.; van Kooyk, Y.; et al. Selective probiotic bacteria induce IL-10-producing regulatory T cells in vitro by modulating dendritic cell function through dendritic cell-specific intercellular adhesion molecule 3-grabbing nonintegrin. *J. Allergy Clin. Immunol.* **2005**, *115*, 1260–1267. [CrossRef]
70. Lee, M.J.; Kang, M.J.; Lee, S.Y.; Lee, E.; Kim, K.; Won, S.; Suh, D.I.; Kim, K.W.; Sheen, Y.H.; Ahn, K.; et al. Perturbations of gut microbiome genes in infants with atopic dermatitis according to feeding type. *J. Allergy Clin. Immunol.* **2018**, *141*, 1310–1319. [CrossRef]

71. Kaikiri, H.; Miyamoto, J.; Kawakami, T.; Park, S.B.; Kitamura, N.; Kishino, S.; Yonejima, Y.; Hisa, K.; Watanabe, J.; Ogita, T.; et al. Supplemental feeding of a gut microbial metabolite of linoleic acid, 10-Hydroxy-Cis-12-Octadecenoic acid, alleviates spontaneous atopic dermatitis and modulates intestinal microbiota in NC/Nga mice. *Int. J. Food Sci. Nutr.* **2017**, *68*, 941–951. [CrossRef] [PubMed]
72. Johnson, A.M.F.; DePaolo, R.W. Window-of-opportunity: Neonatal gut microbiota and atopy. *Hepatobiliary Surg. Nutr.* **2017**, *6*, 190–192. [CrossRef] [PubMed]
73. Jin, U.H.; Lee, S.O.; Sridharan, G.; Lee, K.; Davidson, L.A.; Jayaraman, A.; Chapkin, R.S.; Alaniz, R.; Safe, S. Microbiome-derived tryptophan metabolites and their aryl hydrocarbon receptor-dependent agonist and antagonist activities. *Mol. Pharmacol.* **2014**, *85*, 777–788. [CrossRef]
74. Cryan, J.F.; Dinan, T.G. Mind-altering microorganisms: The impact of the gut microbiota on brain and behaviour. *Nat. Rev. Neurosci.* **2012**, *13*, 701–712. [CrossRef]
75. Zipperer, A.; Konnerth, M.C.; Laux, C.; Berscheid, A.; Janek, D.; Weidenmaier, C.; Burian, M.; Schilling, N.A.; Slavetinsky, C.; Marschal, M.; et al. Human commensals producing a novel antibiotic impair pathogen colonization. *Nature* **2016**, *535*, 511–516. [CrossRef]
76. Di Marzio, L.; Cinque, B.; De Simone, C.; Cifone, M.G. Effect of the lactic acid bacterium *Streptococcus thermophilus* on ceramide levels in human keratinocytes in vitro and stratum corneum in vivo. *J. Investig. Dermatol.* **1999**, *113*, 98–106. [CrossRef]
77. Di Marzio, L.; Cinque, B.; Cupelli, F.; De Simone, C.; Cifone, M.G.; Giuliani, M. Increase of skin-ceramide levels in aged subjects following a short-term topical application of bacterial sphingomyelinase from Streptococcus thermophilus. *Int. J. Immunopathol. Pharmacol.* **2008**, *21*, 137–143. [CrossRef] [PubMed]
78. Di Marzio, L.; Centi, C.; Cinque, B.; Masci, S.; Giuliani, M.; Arcieri, A.; Zicari, L.; De Simone, C.; Cifone, M.G. Effect of the lactic acid bacterium *Streptococcus thermophilus* on stratum corneum ceramide levels and signs and symptoms of atopic dermatitis patients. *Exp. Dermatol.* **2003**, *12*, 615–620. [CrossRef]
79. Lee, Y.B.; Byun, E.J.; Kim, H.S. Potential role of the microbiome in acne: A comprehensive review. *J. Clin. Med.* **2019**, *8*, 987. [CrossRef]
80. Ambrożej, D.; Kunkiel, K.; Dumycz, K.; Feleszko, W. The use of probiotics and bacteria-derived preparations in topical treatment of atopic dermatitis—A systematic review. *J. Allergy Clin. Immunol. Pract.* **2021**, *9*, 570–575. [CrossRef]
81. Park, S.B.; Im, M.; Lee, Y.; Lee, J.H.; Lim, J.; Park, Y.H.; Seo, Y.J. Effect of emollients containing vegetable-derived lactobacillus in the treatment of atopic dermatitis symptoms: Split-body clinical trial. *Ann. Dermatol.* **2014**, *26*, 150–155. [CrossRef] [PubMed]
82. Myles, I.A.; Castillo, C.R.; Barbian, K.D.; Kanakabandi, K.; Virtaneva, K.; Fitzmeyer, E.; Paneru, M.; Otaizo-Carrasquero, F.; Myers, T.G.; Markowitz, T.E.; et al. Therapeutic responses to Roseomonas mucosa in atopic dermatitis may involve lipid-mediated TNF-related epithelial repair. *Sci. Transl. Med.* **2020**, *12*, eaaz8631. [CrossRef] [PubMed]
83. Guéniche, A.; Hennino, A.; Goujon, C.; Dahel, K.; Bastien, P.; Martin, R.; Jourdain, R.; Breton, L. Improvement of atopic dermatitis skin symptoms by *Vitreoscilla filiformis* bacterial extract. *Eur. J. Dermatol. EJD* **2006**, *16*, 380–384. [PubMed]
84. Volz, T.; Skabytska, Y.; Guenova, E.; Chen, K.M.; Frick, J.S.; Kirschning, C.J.; Kaesler, S.; Röcken, M.; Biedermann, T. Nonpathogenic bacteria alleviating atopic dermatitis inflammation induce IL-10-producing dendritic cells and regulatory Tr1 cells. *J. Investig. Dermatol.* **2014**, *134*, 96–104. [CrossRef] [PubMed]
85. Reygagne, P.; Bastien, P.; Couavoux, M.P.; Philippe, D.; Renouf, M.; Castiel-Higounenc, I.; Gueniche, A. The positive benefit of *Lactobacillus paracasei* NCC2461 ST11 in healthy volunteers with moderate to severe dandruff. *Benef. Microbes* **2017**, *8*, 671–680. [CrossRef]
86. Knackstedt, R.; Knackstedt, T.; Gatherwright, J. The role of topical probiotics in skin conditions: A systematic review of animal and human studies and implications for future therapies. *Exp. Dermatol.* **2020**, *29*, 15–21. [CrossRef]
87. Knackstedt, R.; Knackstedt, T.; Gatherwright, J. The role of topical probiotics on wound healing: A review of animal and human studies. *Int. Wound J.* **2020**, *17*, 1687–1694. [CrossRef]
88. Fortuna, M.C.; Garelli, V.; Pranteda, G.; Romaniello, F.; Cardone, M.; Carlesimo, M.; Rossi, A. A case of Scalp Rosacea treated with low dose doxycycline and probiotic therapy and literature review on therapeutic options. *Dermatol. Ther.* **2016**, *29*, 249–251. [CrossRef]
89. Levitt, E.L.; Keen, J.T.; Wong, B.J. Augmented reflex cutaneous vasodilatation following short-term dietary nitrate supplementation in humans. *Exp. Physiol.* **2015**, *100*, 708–718. [CrossRef]
90. Mishra, M.; Kumar, A.; Satsangi, G.P.; Bhatnagar, A.K.; Shrivastava, J.N. Inhibitory effects of antibiotics from *Nitrobacter* spp. against Tinea capitis. *Allelopath. J.* **2007**, *19*, 535–542.
91. Shrivastava, J.N.; Shukla, J.P.; Kumar, V. Antibacterial Potential of Nitrobacter species against *Staphylococcus aureus*. *VEGETOS* **2011**, *24*, 26–28.
92. Opländer, C.; Suschek, C.V. New aspects of nitrite homeostasis in human skin. *J. Investig. Dermatol.* **2009**, *129*, 820–822. [CrossRef] [PubMed]
93. Opländer, C.; Suschek, C.V. The role of photolabile dermal nitric oxide derivates in ultraviolet radiation (UVR)-induced cell death. *Int. J. Mol. Sci.* **2012**, *14*, 191–204. [CrossRef] [PubMed]

94. Sun, Z.; Harris, H.M.B.; McCann, A.; Guo, C.; Argimo´n, S.; Zhang, W.; Yang, X.; Jeffery, I.B.; Cooney, J.C.; Kagawa, T.F.; et al. Expanding the biotechnology potential of lactobacilli through comparative genomics of 213 strains and associated genera. *Nat. Commun.* **2015**, *6*, 8322. [CrossRef]
95. Holz, C.; Benning, J.; Schaudt, M.; Heilmann, A.; Schultchen, J.; Goelling, D.; Lang, C. Novel bioactive from *Lactobacillus brevis* DSM17250 to stimulate the growth of *Staphylococcus epidermidis*: A pilot study. *Benef. Microbes* **2017**, *8*, 121–131. [CrossRef]
96. Gueniche, A.; Benyacoub, J.; Philippe, D.; Bastien, P.; Kusy, N.; Breton, L.; Blum, S.; Castiel-Higounenc, I. *Lactobacillus paracasei* CNCM I-2116 (ST11) inhibits substance P-induced skin inflammation and accelerates skin barrier function recovery in vitro. *Eur. J. Dermatol.* **2010**, *20*, 731–737. [PubMed]
97. Patra, V.; Byrne, S.N.; Wolf, P. The skin microbiome: Is it affected by UV-induced immune suppression? *Front. Microbiol.* **2016**, *7*, 1235. [CrossRef] [PubMed]
98. Fabbrocini, G.; Bertona, M.; Picazo, O.; Pareja-Galeano, H.; Emanuele, E. Supplementation with *Lactobacillus rhamnosus* SP1 normalises skin expression of genes implicated in insulin signalling and improves adult acne. *Benef. Microbes* **2016**, *7*, 625–630. [CrossRef]
99. Im, A.R.; Kim, H.S.; Hyun, J.W.; Chae, S. Potential for tyndalized *Lactobacillus acidophilus* as an effective component in moisturizing skin and anti-wrinkle products. *Exp. Ther. Med.* **2016**, *12*, 759–764. [CrossRef]
100. Park, M.J.; Bae, Y.S. Fermented acanthopanax koreanum root extract reduces UVB- and H_2O_2-induced senescence in human skin fibroblast cells. *J. Microbiol. Biotechnol.* **2016**, *26*, 1224–1233. [CrossRef]
101. Satoh, T.; Murata, M.; Iwabuchi, N.; Odamaki, T.; Wakabayashi, H.; Yamauchi, K.; Abe, F.; Xiao, J.Z. Effect of Bifidobacterium breve B-3 on skin photoaging induced by chronic UV irradiation in mice. *Benef. Microbes* **2015**, *6*, 497–504. [CrossRef] [PubMed]
102. Allen, S.J.; Jordan, S.; Storey, M.; Thornton, C.A.; Gravenor, M.B.; Garaiova, I.; Plummer, S.F.; Wang, D.; Morgan, G. Probiotics in the prevention of eczema: A randomised controlled trial. *Arch. Dis. Child.* **2014**, *99*, 1014–1019. [CrossRef]
103. Makarova, K.S.; Grishin, N.V.; Shabalina, S.A.; Wolf, Y.I.; Koonin, E.V. A putative RNA-interference-based immune system in prokaryotes: Computational analysis of the predicted enzymatic machinery, functional analogies with eukaryotic RNAi, and hypothetical mechanisms of action. *Biol. Direct Mar.* **2006**, *16*, 7. [CrossRef]
104. Kano, M.; Masauoka, N.; Kaga, C.; Sugimoto, S.; Iizuka, R.; Manabe, K.; Sone, T.; Oeda, K.; Nonaka, C.; Miazaki, K.; et al. Consecutive intake of fermented milk containing *Bifidobacterium breve* strain yakult and galacto-oligosaccharides benefits skin condition in healthy adult women. *Biosci. Microbiota Food Health* **2013**, *32*, 33–39. [CrossRef] [PubMed]
105. Shin, D.; Lee, Y.; Huang, Y.H.; Lim, H.W.; Jang, K.; Kim, D.D.; Lim, C.J. Probiotic fermentation augments the skin anti-photoaging properties of *Agastache rugosa* through up-regulating antioxidant components in UV-B-irradiated HaCaT keratinocytes. *BMC Complement. Altern. Med.* **2018**, *18*, 196. [CrossRef] [PubMed]
106. Ishii, Y.; Sugimoto, S.; Izawa, N.; Sone, T.; Chiba, K.; Miyazaki, K. Oral administration of *Bifidobacterium breve* attenuates UV-induced barrier perturbation and oxidative stress in hairless mice skin. *Arch. Dermatol. Res.* **2014**, *306*, 467–473. [CrossRef]
107. Kang, Y.M.; Hong, C.H.; Kang, S.H.; Seo, D.S.; Kim, S.O.; Lee, H.Y.; Sim, H.J.; An, H.J. Anti-photoaging effect of plant extract fermented with lactobacillus buchneri on CCD-986sk fibroblasts and HaCaT keratinocytes. *J. Funct. Biomater.* **2020**, *11*, 3. [CrossRef]
108. Chen, H.; Li, Y.; Xie, X.; Chen, M.; Xue, L.; Wang, J.; Ye, Q.; Wu, S.; Yang, R.; Zhao, H.; et al. Exploration of the molecular mechanisms underlying the anti-photoaging effect of limosilactobacillus fermentum XJC60. *Front. Cell. Infect. Microbiol.* **2022**, *12*, 838060. [CrossRef]
109. Yau, Y.F.; El-Nezami, H.; Galano, J.M.; Kundi, Z.M.; Durand, T.; Lee, J.C. Lactobacillus rhamnosus GG and oat beta-glucan regulated fatty acid profiles along the gut-liver-brain axis of mice fed with high fat diet and demonstrated antioxidant and anti-inflammatory potentials. *Mol. Nutr. Food Res.* **2020**, *64*, e2000566. [CrossRef]
110. Mai, C.; Qiu, L.; Zeng, Y.; Tan, X. *Lactobacillus casei* strain shirota enhances the ability of geniposide to activate SIRT1 and decrease inflammation and oxidative stress in septic mice. *Front. Physiol.* **2021**, *12*, 678838. [CrossRef]
111. Ansel, J.C.; Luger, T.A.; Green, I. The effect of in vitro and in vivo UV irradiation on the production of ETAF activity by human and murine keratinocytes. *J. Investig. Dermatol.* **1983**, *81*, 519–523. [CrossRef] [PubMed]
112. Kupper, T.S.; Groves, R.W. The interleukin-1 axis and cutaneous inflammation. *J. Investig. Dermatol.* **1995**, *105*, 62s–66s. [CrossRef] [PubMed]
113. Khmaladze, I.; Butler, É.; Fabre, S.; Gillbro, J.M. Lactobacillus reuteri DSM 17938-A comparative study on the effect of probiotics and lysates on human skin. *Exp. Dermatol.* **2019**, *28*, 822–828. [CrossRef] [PubMed]
114. Keshari, S.; Balasubramaniam, A.; Myagmardoloonjin, B.; Herr, D.R.; Negari, I.P.; Huang, C.M. Butyric acid from probiotic *Staphylococcus epidermidis* in the skin microbiome down-regulates the ultraviolet-induced pro-inflammatory IL-6 cytokine via short-chain fatty acid receptor. *Int. J. Mol. Sci.* **2019**, *20*, 4477. [CrossRef]
115. Hong, K.B.; Jeong, M.; Han, K.S.; Hwan Kim, J.; Park, Y.; Suh, H.J. Photoprotective effects of galacto-oligosaccharide and/or *Bifidobacterium longum* supplementation against skin damage induced by ultraviolet irradiation in hairless mice. *Int. J. Food Sci. Nutr.* **2015**, *66*, 923–930. [CrossRef] [PubMed]
116. Kober, M.M.; Bowe, W.P. The effect of probiotics on immune regulation, acne, and photoaging. *Int. J. Womens Dermatol.* **2015**, *1*, 85–89. [CrossRef]

117. Goodarzi, A.; Mozafarpoor, S.; Bodaghabadi, M.; Mohamadi, M. The potential of probiotics for treating acne vulgaris: A review of literature on acne and microbiota. *Dermatol. Ther.* **2020**, *33*, e13279. [CrossRef]
118. Kim, D.; Lee, K.R.; Kim, N.R.; Park, S.J.; Lee, M.; Kim, O.K. Combination of bifidobacterium longum and galacto-oligosaccharide protects the skin from photoaging. *J. Med. Food* **2021**, *24*, 606–616. [CrossRef]
119. Lavker, R.M.; Zheng, P.; Dong, G. Morphology of aged skin. *Clin. Geriatr. Med.* **1989**, *5*, 53–67. [CrossRef]
120. Shirzad, M.; Hamedi, J.; Motevaseli, E.; Modarressi, M.H. Anti-elastase and anti-collagenase potential of Lactobacilli exopolysaccharides on human fibroblast. *Artif. Cells Nanomed. Biotechnol.* **2018**, *46*, 1051–1061. [CrossRef]
121. Negari, I.P.; Keshari, S.; Huang, C.M. Probiotic activity of Staphylococcus epidermidis induces collagen type i production through FFaR2/p-ERK signaling. *Int. J. Mol. Sci.* **2021**, *22*, 1414. [CrossRef] [PubMed]

MDPI AG
Grosspeteranlage 5
4052 Basel
Switzerland
Tel.: +41 61 683 77 34

Nutrients Editorial Office
E-mail: nutrients@mdpi.com
www.mdpi.com/journal/nutrients

www.ingramcontent.com/pod-product-compliance
Lightning Source LLC
Chambersburg PA
CBHW041017270926
51850CB00024B/1366